Mechanics and Control

Mechanics and Control

Edited by

Ramesh S. Guttalu

University of Southern California
Los Angeles, California

Plenum Press • New York and London

Library of Congress Cataloging-in-Publication Data

Workshop on Control Mechanics (5th : 1992 : University of Southern California)
Mechanics and control / edited by Ramesh S. Guttalu.
p. cm.
Proceedings of the Fifth Workshop on Control Mechanics, in honor of the seventieth birthday of Professor Angelo Miele, held January 21-13, 1992, at the University of Southern California, Los Angles.
Includes bibliographical references and index.
ISBN 0-306-44669-3
1. Automatic control--Congresses. 2. Mechanics, Applied--Congresses. I. Guttalu, R. S. II. Title.
TJ212.2.W67 1992
629.8--dc20 93-48595
CIP

Proceedings of the Fifth Workshop on Control Mechanics, in honor of the Seventieth Birthday of Professor Angelo Miele, held January 21–23, 1992, at the University of Southern California, Los Angeles, California

ISBN 0-306-44669-3

A Division of Plenum Publishing Corporation
233 Spring Street, New York, N.Y. 10013

PREFACE

The Workshop on Control Mechanics has been held at the University of Southern California annually since 1988 under the leadership of late Professor Janislaw M. Skowronski. The primary goal of Professor Skowronski in organizing this series of workshops was to promote the use of advanced mechanics method in control theory with a special emphasis on the control of nonlinear mechanical systems subject to uncertainty. This goal has been achieved through a consistent participation of a large number of researchers in the field of control and mechanics and an intensive exchange of their ideas. Professor Skowronski passed away unexpectedly on March 21, 1992, after the conclusion of the Fifth Workshop. The great success of the Fifth Workshop as well as the entire Control Mechanics Workshops over the years is almost exclusively due to his dedication, enthusiasm, and organizational capabilities. His untimely demise is a great loss to us and to the mechanics and control community.

The proceedings of the Fifth Workshop presented in this volume are dedicated to Professor Angelo Miele, one of the pioneers and a leading contributor in many fields of control theory and its applications. His contribution spans a wide range of topics such as optimization theory, flight mechanics, astrodynamics, ocean engineering, and numerical methods. The presentations in the workshop reflected many of the areas in which Professor Miele has been active.

The papers included in this volume are divided into three major groups of topics. The first group consists of papers dealing with the development of control methods. Corless and Leitmann consider the problem of stabilizing uncertain/nonlinear systems using Lyapunov stability theory. Blaquiére and Sigal-Pauchard treat the geometric approach to entropy minimization problems and its connection to Schrödinger's bridge and a conditional Sanov theorem. Hull and D'Souza formulate second variation conditions for the optimal control problem with normalized final time. Olas and Olas develop a method for finding an optimal Lyapunov function for systems with structured uncertainties. A new tool for robust control of linear, time invariant, uncertain systems is developed by Soldatos, Chung and Auslander. Udwadia and Kalaba formulate the equations of motion for constrained mechanical systems and use them to develop control laws. Lastly, Bell presents his current findings on partially singular control trajectories.

The second group contains papers dealing with applications of control theory to aerospace systems. Greenwood uses the Lyapunov's approach to formulate pursuit-evasion games applied to a missile guidance problem. Ilgen and Speyer employ regular perturbation strategies for aeroassisted plane change mission. A minimum-fuel transfer between elliptic orbits using Green's theorem is studied by Mease and Rao. Sachs and Mehlhorn study the problem of periodic optimal endurance cruise. Finally, Vinh and Medepalli investigate optimal thrust and aerodynamic controls in hypersonic flights.

The subject of the third group of papers is control laws and applications to mechanical systems. Gawronski presents a design procedure for deep space antennas. A method for maneuvering and control of space structures is introduced by Meirovitch. Necsulescu, Skowronski and Shaban-Zanjani treat the problem of low-speed motion control of a Hamiltonian system. Ahmed and Skowronski study boundary stabilization of nonlinear flexible systems. A differential dynamic programming approach to problems with state variable inequality constraint and its application to manipulator control is presented by Ruxton. Sira-Ramírez and Llanes-Santiago investigate dynamical pulse-width-modulation as applied to control of rigid and flexible manipulators. A chaotic controller for a double pendulum is demonstrated by Vincent, Schmitt and Vincent. Zribi, Ahmad and Luo conclude this group with the development of an adaptive control law for redundant multiple robots in cooperative motion.

This volume also includes papers by Stadler on the calculation of the Jacobian needed for kinematic analysis of manipulators, de Figueiredo on a rigorous basis for neural modeling and identification of nonlinear dynamical systems, Zufiria and Guttalu on the global dynamic behavior of Branin's method for determining the zeros of a vector function, and Cheng and Stubberud on a numerical solution for a nonlinear eigenvalue problem.

The organizing committee of the Fifth Workshop would like to thank all the authors for their active participation and cooperation during the preparation of this volume. With great pleasure, we thank the Department of Mechanical Engineering and the University of Southern California, in particular Professor Satwinder S. Sadhal (Acting Chairman) and Mrs. Jacquette Givens (Administrative Officer) for their hospitality and invaluable help in organizing this workshop.

Ramesh S. Guttalu
Los Angeles, July 1993

CONTENTS

CONTROL METHODOLOGY

APPLICATIONS TO AEROSPACE SYSTEMS

CONTROL OF MECHANICAL SYSTEMS

COMPUTATIONAL TECHNIQUES

Professor Angelo Miele

IN HONOR OF ANGELO MIELE

George Leitmann

College of Engineering
University of California
Berkeley, CA 94720

It is always a privilege to be invited to make the opening remarks at a meeting held in honor of a distinguished colleague. When that individual is also a dear friend, then the privilege is further enhanced and becomes a rare pleasure. And so it is this morning when we meet to honor Angelo Miele on the occasion of his upcoming 70th birthday.

Angelo Miele was born in Formia, Italy, in August 1922, the son of Salvatore and Elena (nee Marino) Miele and grandson of Angelo Miele, a former mayor of Formia. His late brother, Franco, a well-known painter, was a professor of art at the Accademia delle Belle Arti in Rome as well as an art critic and author. His other sibling, Tina, is married to a Fiat engineer and lives in Turin.

Angelo Miele was educated at the University of Rome, receiving the degrees of Dr.Ing. in Civil Engineering in 1944 and Dr.Ing. in Aeronautical Engineering in 1946. After completing his formal education, he moved to Argentina where he held an industrial position as Project Engineer in the Military Aircraft Factory at Cardoba from 1947 to 1952. During this period he was also a Lecturer at the School of Military Aviation and the School of Aerospace Engineering of the Argentinian Air Force, as well as at the University of Cardoba. In 1952, Angelo Miele came to the United States to join the faculty of the Polytechnic Institute of Brooklyn as an Assistant Professor. In 1955, he moved to the mid-West to become an Associate Professor of Aeronautical Engineering at Purdue University; after only three years, he was promoted to full Professor and remained at Purdue until 1959. Between 1959 and 1964, he returned to industry as Director of Astrodynamics and Flight Mechanics at the Boeing Scientific Research Laboratory in Seattle; however, he did not sever his academic connections and held an appointment as Visiting Professor at the University of Washington from 1961-64. Having moved as far West as possible in the continental U.S., he returned to the East in 1964 to join the faculty of Rice University in Houston as Professor of Aerospace Sciences and Mathematical Sciences; in 1988 he was honored with an endowed chair as A. J. Foyt Family Professor in Engineering. During his years in the U.S., while holding his various academic positions, he never neglected his role as a practicing engineer; he has consulted, and continues to do so, for various companies, including the Douglas Aircraft Company, the General Motors Corporation, the Boeing Airplane Company, and the United States Aviation Underwriters.

So much for the bare bones of Angelo Miele's professional career. To recount in detail the many accomplishments which have made him a pioneer and an internationally recognized leader in the aerospace field would take up all the allotted pages of the Festschrift which will be the Proceedings of this Workshop on Control Mechanics. So allow me to recount but some of his salient contributions.

In the area of *Flight Mechanics*, he investigated and obtained basic results in the analysis of trajectories of jet-propelled and rocket-powered vehicles, optimum climb trajectories, stratospheric cruise of turbojet aircraft, optimal cruise of hypervelocity gliders, and the general theory of optimum flight paths of rocket-powered aircraft, missiles and satellite carriers. In recent years, he made fundamental contributions to aircraft control for takeoff and landing under windshear conditions, to windshear identification and detection, and to trajectory optimization for the National Aerospace Plane.

In the field of *Astrodynamics*, he studied lunar trajectories, derived the so-called "image trajectory" theorem in earth-moon space, developed optimal trajectories and guidance schemes for aeroassisted orbital transfer (AOT) and for the aeroassisted flight experiment (AFE). His AOT and AFE research is of considerable interest in the design of Mars exploration/return vehicles and lunar return vehicles.

His concern for *Applied Aerodynamics* led him to study optimum aerodynamic shapes in various flow regimes, and in particular to the derivation of optimum longitudinal and transversal contours for bodies at hypersonic speeds.

In the area of *Optimization Theory* he investigated the so-called "singular" control problems and utilized Green's theorem to arrive at sufficient conditions for such controls. He also contributed to the theory of extremizing products of powers of functionals, to minmax problems, and to identification problems.

Always concerned with the applicability of mathematical tools, Angelo Miele became interested in computational aspects of optimization theory, that is, in *Numerical Methods*. He developed mathematical programming algorithms, such as the memory-gradient algorithm for extremizing unconstrained functions, and the sequential gradient-restoration and modified quasilinearization algorithms for constrained function minimization. In view of the boundary-value nature of necessary conditions for optimal control, he deduced various techniques for numerical solutions of two-point and multipoint boundary-value problems for ordinary differential equations; his method of particular solutions has the same degree of generality as the methods of complementary functions, adjoint variables, and Riccati transformation. For optimal control problems, he developed various sequential gradient-restoration techniques, a modified quasilinearization algorithm, and even numerical algorithms for problems with nondifferential constraints. In particular, the family of sequential gradient-restoration algorithms for optimal control problems has become widely known and used in the USA (NASA-JSC) and abroad (Japan).

Finally, to emphasize the wide range of Angelo Miele's interests, let me end this cursory account of his contributions by mentioning his work in *Ocean Engineering*. This research dealt with the wave parameter identification problem for ocean test structure data. A byproduct of this research was the determination of the optimal location of sensors for an offshore structure.

Of course, Angelo Miele's list of publications is long and distinguished. He has published scores of articles in archival journals, a large number of these in co-authorship with his graduate students and post-doctoral fellows (some 70 of them). In addition to a number of edited volumes, one must note his seminal book "Flight Mechanics, Volume I: Theory of Flight Paths", first published by Addison-Wesley in 1962 and republished by Pergamon Press in 1963 (Russian edition published by Nauka, Moscow, in 1965).

Angelo Miele has also employed his considerable talents for the good of the profession as Editor-in-Chief of the Journal of Optimization Theory and Applications, Associate Editor of other leading journals, Editor of the Plenum series "Mathematical Concepts and Methods in Science and Engineering", and membership on numerous editorial boards and technical committees.

For his many outstanding contributions to research, teaching and professional service, his peers have bestowed on him a plethora of honors and prizes. He is a member of the International Academy of Astronautics, a Corresponding Member of the Academy of Science of Turin, a Fellow of the American Institute of Aeronautics and Astronautics (AIAA) and of the American Astronautical Society (AAS), and a Knight Commander of the Order of Merit of the Republic of Italy. He is the recipient of the Pendray Aerospace Literature Award and the Mechanics and Control of Flight Award of the AIAA, the Brouwer Award of the AAS, the Levy Medal of the Franklin Institute, and the Schuck Award of the American Automatic Control Council.

Angelo Miele's success in life is in no small measure due to the loving support of his parents and of his spouse Gymme Elizabeth (nee Odom) of Houston, whom he married in 1970 (after a five year courtship, as befits a prudent Italian). His warm relations with his graduate students are exemplary. Angelo has always looked at them as an extension of his family.

I am happy to know that Angelo stems from a clan of long-lived people; ten members of his immediate family survived well into their eighties, and two of his aunts still live in Rome at the ages of 92 and 96, respectively. In wishing Angelo "Happy Birthday" (albeit, well ahead of his 70th birthday in August), let me express our hope, indeed our conviction, that he will continue his family's tradition of long and productive lives.

A TRIBUTE TO ANGELO MIELE

Allen R. Stubberud

Department of Electrical Engineering
University of California, Irvine
Irvine, CA 92717

When Professor Skowronski asked me to say a few words about Professor Angelo Miele after dinner tonight, I was both pleased and honored. And perhaps, a little surprised as well.

I was pleased because Professor Miele strongly deserves having this Workshop on Control Mechanics dedicated to him. His many contributions in flight mechanics, aerodynamics, astrodynamics, and optimization make him one of the most visible researchers in this field.

As pleased as I am that Professor Miele is being honored here, I must say that I am personally most highly honored to be asked to speak on his behalf. I can still recall very clearly during my days as a graduate student reading papers written by Angelo Miele. His early contribution to the field of Optimal Control were extremely valuable to me, especially, after I joined the faculty at UCLA and taught courses in optimal control. One of my prime references for theses courses was the book *Optimization Techniques*, edited by George Leitmann and published in 1962. Professor Miele contributed two chapters to this book: "Extremization of Linear Integrals by Green's Theorem"; "The Calculus of Variations in Applied Aerodynamics and Flight Mechanics". The second chapter, in particular, had a very strong effect on my approach to the teaching of optimal control.

In fact, very recently, after Professor Skowronski asked me to speak here tonight, I pulled out my copy of that book and looked through that chapter and was quite surprised how my approach to teaching optimal control still follows this work.

I met Professor Miele personally while still a junior faculty member and we have crossed paths occasionally at technical meetings. I must say that I have always been a little awed by being acquainted with one of the true stars of this age of engineering. He was, of course, always very cordial and truly a gentleman. However, my real insight to Angelo came in 1982. I was on my way from Irvine to Texas A&M to attend an ASEE meeting in June of that year. My youngest son was at that time about to enter his senior year in high school and was looking for an appropriate engineering school to attend after graduation from high school. Appropriate meant among other things a school that his older brother was not attending, which dropped UCLA from consideration, or a school where his father was on the faculty, which dropped out UC Irvine. Rice University had been strongly recommended to him, so I took him along to talk to some of the people in Admission and Counseling at Rice. While there, I was talking to an administrator who asked who I might know on the Rice faculty and I said among others, I knew Professor Miele. He asked if I would like to talk with him and I said if he was available and not too busy, I would drop by and say hello. He called Professor Miele and was told that he would be delighted to see us. So my son and I went to his office to see him. We were greeted, both of us, as long lost friends and Professor Miele proceeded to give us, but my son primarily, a brief description of the wonders of Rice University. He also told us that he was leaving almost immediately for the airport to catch a plane. Since I had told my son what a famous man Professor Miele was, he was most impressed with the courtesy and kindness he had shown to a high school student, especially when he was so pressed for time.

My most recent "encounter" with Angelo occurred when one of my recent graduate students, in casting about for a problem to be used as an application for a system identification problem asked me what might be an important class of problems in this area. I told him about some work, which I had seen while I was with the U.S. Air Force, in windshear. He thought that sounded quite interesting, so I told him to go to the library and look for any recent publications which might be relevant. Sure enough, when he came back he reported that the name of Angelo Miele was well established in the literature as a leader in flight control during a windshear. Thus after 30 years, I was still following Professor Miele. As a follower, I am extremely honored to pay my tribute to him tonight.

Now let me discuss some of Professor Miele's accomplishments. Professor Leitmann discussed some of his technical achievements earlier today, thus I will not speak of those.

Professor Miele was born in Italy in August 1922 where he attended the University of Rome receiving the degrees of Dr.Ing. in Civil Engineering in 1944 and Dr.Ing. in Aeronautical Engineering in 1946. By the way I am always pleased to see a successful Leo, since I am also a Leo.

For the past four decades, Dr. Miele has been engaged in engineering, teaching, research and consulting in Italy, South America, and the United States. As an academic in the U.S. he has been associated with Polytechnic Institute of Brooklyn, Purdue University, and the University of Washington and is presently at Rice University.

At Rice University he is a Professor of Aerospace Sciences and Mathematical Sciences and since 1988 he has held a chaired position as the A. J. Foyt Family Professor in Engineering. He also is the director of the Aero-Astronautics Group.

In addition to his many technical and research activities, Professor Miele has contributed extensively in the area of professional activities. As a consultant he has provided his services to: the Guided Missiles Division of Douglas Aircraft Company; the Allison Division of the General Motors Corporation; the U.S. Aviation Underwriters; the Boeing Commercial Airplane Company. Also in his earlier professional life he took a hiatus from academia and spent some five years as Director of Astrodynamics and Flight Mechanics at the Boeing Scientific Research Laboratories.

This close association with the non-academic engineering world is strongly reflected in his research where he believes in tackling real problems as opposed to purely academic problems.

Other than these activities, Professor Miele is very active in professional societies, activities from which all the rest of us benefit. He has served on several national committees: the Flight Mechanics panel of the Institute of Aerospace Sciences; the Astrodynamics Committee of the American Rocket Society; the Astrodynamics Committee of the American Astronautical Society; the Astrodynamics Committee of the American Institute of Aeronautics and Astronautics.

He has also served on a number of international committees including: the Mathematics of Control Committee of the International Federation of Automatic Control; the Committee on Future Space System Evolution of the IAA; the Committee on Space Plans and Policies of IAA; the Mars Exploration Subcommittee of IAA.

Professor Leitmann earlier today described Professor Miele's editorial activities which, we all know, are numerous and time consuming. I would, however, just like to mention again that he has been editor-in-chief of the excellent Journal of Optimization Theory and Applications. The huge success of this highly regarded journal is certainly due to his outstanding leadership.

Finally, let me point out that in addition to the many honors and awards which Professor Leitmann described to you earlier today, Professor Miele has been recognized by the following special honors: in 1982, the Journal of the Astronautical Sciences dedicated a special issue to him dealing with numerical methods of optimization; in 1983, the journal Optimal Control Applications and Methods dedicated a special issue to him dealing with Optimal Control; in 1992, the Technion-Israel Institute of Technology awarded the Honorary Doctorate of Science to Angelo Miele.

We are, of course, dedicating the three days of this special workshop to him. Those of us in academe, who live by publishing, should be most thankful to Angelo Miele since because of him we are getting additional publication opportunities which we might not otherwise receive.

In all sincerity, Angelo, it is a pleasure to know you and to have had an opportunity to follow your work for the past 30 years.

CONTROL METHODOLOGY

ROBUST EXPONENTIAL CONVERGENCE WITH BOUNDED CONTROLLERS

Martin Corless[1] and G. Leitmann[2]

[1]School of Aeronautics and Astronautics
Purdue University
West Lafayette, Indiana 47907

[2]College of Engineering
University of California at Berkeley
Berkeley, CA 94720

ABSTRACT

We consider the problem of stabilizing an uncertain/nonlinear system when the norm of the control input is bounded by a prespecified constant. We treat continuous-time dynamical systems whose nominal part is linear and whose uncertain/nonlinear part is norm-bounded by a known affine function of the norm of the system state and the norm of the control input. Given a prespecified rate of convergence and a ball containing the origin of the state space, we present controllers which guarantee that, for all allowable uncertainties and nonlinearities, there is a region of attraction from which all solutions converge to the given ball with the prespecified convergence rate.

INTRODUCTION

In recent years, much effort has been devoted to the problem of obtaining stabilizing controllers for uncertain systems. In much of this research, the uncertainties are modelled deterministically, rather than stochastically, and they are characterized by certain structural conditions and known bounds; see, e. g., [2, 3, 11] and the references therein. Such uncertainties could be due to uncertain disturbance inputs, uncertain parameters, or nonlinear elements which are difficult to characterize.

In most of the literature, no consideration is given to control constraints; exceptions are [4, 5, 6, 8, 9, 10, 12]. Here we consider the problem of stabilizing an uncertain system when the norm of the control input is bounded by a prespecified constant. We treat continuous-time dynamical systems whose nominal part is linear and whose uncertain/uncertain part satisfies a matching condition and is norm-bounded by an affine function of the norm of the state and the norm of the control input.

Given a prespecified rate of convergence and a ball containing the origin of the state space, we present controllers which guarantee that, for all allowable uncertainties and

Mechanics and Control, Edited by R.S. Guttalu
Plenum Press, New York, 1994

nonlinearities, there is a region of attraction from which all solutions converge to the given ball with the prespecified convergence rate.

The results are illustrated by application to an inverted pendulum which is subject to a bounded control torque.

PROBLEM STATEMENT

We consider uncertain/nonlinear systems described by

$$\dot{x}(t) = Ax(t) + Bu(t) + \Delta F(t, x(t), u(t)) \tag{1}$$

where $t \in \mathbb{R}$ is "time", $x(t) \in \mathbb{R}^n$ is the state and $u(t) \in \mathbb{R}^m$ is the control input. The constant matrices A, B are known; they define the *nominal system*

$$\dot{x}(t) = Ax(t) + Bu(t). \tag{2}$$

All the uncertainty, nonlinearity and time-dependence in the system is represented by the term ΔF which is assumed to be a continuous function.

The control is subject to the hard constraint

$$||u(t)|| \leq \bar{\rho} \tag{3}$$

where $\bar{\rho} > 0$ is prescribed. We shall consider the control to be generated by a memoryless state feedback controller p, i.e.,

$$u(t) = p(x(t)). \tag{4}$$

The resulting *closed loop system* is described by

$$\dot{x}(t) = f(t, x(t)) \tag{5}$$

with

$$f(t, x) := Ax + Bp(x) + \Delta F(t, x, p(x)). \tag{6}$$

For any $r \geq 0$, the *ball of radius* r is defined by

$$\mathcal{B}(r) := \{x \in \mathbb{R}^n : ||x|| \leq r\}.$$

Suppose that $\alpha > 0$ and $\mathcal{R} \subset \mathbb{R}^n$ contains a neighborhood of 0.

Definition 1 System (5) is *uniformly exponentially convergent* to $\mathcal{B}(r)$ with *rate* α and *region of attraction* $\mathcal{R}$ iff it has the following properties.

(i) *Existence of solutions.* For each $t_0 \in \mathbb{R}$, $x_0 \in \mathcal{R}$, there exists a solution $x(\cdot) : [t_0, t_1) \to \mathbb{R}^n$, $t_0 < t_1$, of (5) with $x(t_0) = x_0$.

(ii) *Indefinite extension of solutions.* Each solution $x(\cdot) : [t_0, t_1) \to \mathbb{R}^n$ of (5), with $x(t_0) \in \mathcal{R}$, has an extension $\bar{x}(\cdot) : [t_0, \infty) \to \mathbb{R}^n$, i.e., $\bar{x}(t) = x(t)$ for all $t \in [t_0, t_1)$ and $\bar{x}(\cdot)$ is a solution of (5).

(iii) *Uniform exponential convergence of solutions.* There exists a real scalar $\beta \geq 0$ such that every solution $x(\cdot) : [t_0, \infty) \to \mathbb{R}^n$ of (5) with $x(t_0) \in \mathcal{R}$ satisfies

$$||x(t)|| \leq r + \beta ||x(t_0)|| \exp[-\alpha(t - t_0)]$$

for all $t \geq t_0$.

Remark 1 If (5) satisfies the requirements of the above definition with $r = 0$ then it is said to be *uniformly exponentially stable* with region of attraction $\mathcal{R}$. The scalar r can be regarded as a measure of the distance of the system behavior from that of exponential stability. If r is "small", then for practical purposes the system can be considered exponentially stable.

Definition 2 System (5) is *globally uniformly exponentially convergent to* $\mathcal{B}(r)$ iff it is exponentially convergent with $\mathbb{R}^n$ as a region of attraction.

The problem we wish to consider is as follows.

Problem Statement Consider a system described by (1), subject to control constraint (3), and let $\alpha > 0$ and $r \geq 0$ be specified real numbers. Find a memoryless state feedback controller p which renders the closed loop system (5) exponentially convergent to $\mathcal{B}(r)$ with rate α. In addition, it is desired that the region of attraction $\mathcal{R}$ be "as large as possible."

A Lyapunov Sufficient Condition for Exponential Convergence

The following theorem provides a sufficient condition which is useful in guaranteeing that a given system be exponentially convergent.

Theorem 1 *Consider a system described by (5) with $f(\cdot)$ continuous and suppose there exists a C^1 function $V(\cdot) : \mathbb{R}^n \to \mathbb{R}$ with the following properties.*

(i) There are scalars $\omega_1, \omega_2 > 0$ such that for all $t \in \mathbb{R}$ and $x \in \mathbb{R}^n$

$$\omega_1 \|x\|^2 \leq V(x) \leq \omega_2 \|x\|^2 \ .$$

(ii) There are scalars $\overline{V}, \underline{V}$, with $\overline{V} > \underline{V} \geq 0$ such that for all $t \in \mathbb{R}$,

$$\underline{V} < V(x) < \overline{V} \quad \Rightarrow \quad DV(x)f(t,x) \leq -2\alpha[V(x) - \underline{V}] \ .$$

Then, letting

$$r := [\underline{V}/\omega_1]^{\frac{1}{2}},$$

system (5) is uniformly exponentially convergent to $\mathcal{B}(r)$ with rate α and region of attraction

$$\mathcal{R} = \{x \in \mathbb{R}^n : V(x) \leq \overline{V}\}.$$

Proof The proof is very similar to that of Theorem 1 in [1].

We will refer to a function V which satisfies the hypotheses of the above theorem as a *Lyapunov function.*

Remark 2 If the hypotheses of the above theorem hold with $\overline{V} = \infty$, then system (5) is *globally* uniformly exponentially convergent to $\mathcal{B}(r)$.

ASSUMPTIONS

To guarantee that the nominal system (2) can be stabilized with an arbitrary rate α, the following assumption is necessary.

Assumption 1 The nominal system (2) is controllable.

The next assumption, which is sometimes referred to as *a matching condition*, is common in the literature on control of uncertain systems. To achieve stabilization of an uncertain system with an arbitrary convergence rate, this condition is necessary in some cases; see [13].

Assumption 2 There is a function $e(\cdot)$ such that

$$\Delta F = Be. \tag{7}$$

Assumption 3 There exist non-negative real scalars k_0, k_1, k_2 with

$$k_2 < 1$$

such that

$$||e(t,x,u)|| \leq k_0 + k_1||x|| + k_2||u|| \tag{8}$$

for all $t \in \mathbb{R}, x \in \mathbb{R}^n, u \in \mathbb{R}^m$.

Assumption 4 The control bound $\bar{\rho}$ satisfies

$$\bar{\rho} > k_0/(1-k_2)\ . \tag{9}$$

If this last assumption is not satisfied, the uncertain term can completely cancel the effect of the controller. To see this, suppose $\bar{\rho} \leq k_0/(1-k_2)$ and consider

$$e(t,x,u) = -(1-k_2)p(x) - k_2 u\ .$$

The closed loop system reduces to

$$\dot{x}(t) = Ax(t)$$

and

$$\begin{aligned} \|e(t,x,u)\| &\leq (1-k_2)\,\|p(x)\| + k_2\,\|u\| && (10)\\ &\leq (1-k_2)\bar{\rho} + k_2\,\|u\| && (11)\\ &\leq k_0 + k_2\,\|u\|\,, && (12) \end{aligned}$$

i.e., the uncertain term satisfies Assumption 3.

UNCONSTRAINED CONTROLLERS AND GLOBAL CONVERGENCE

In this section, we ignore control constraint (3) and present controllers which yield exponential convergence which is global. First, we need the following result.

Lemma 1 *Suppose (A,B) is a controllable pair. Then, for each $\sigma, \alpha > 0$ and $k_1 \geq 0$, there exists a matrix X with*

$$X^T = X > 0 \tag{13}$$

such that the following matrix inequality holds:

$$X(A+\alpha I) + (A+\alpha I)^T X - \sigma XBB^T X + {k_1}^2 I \leq 0. \tag{14}$$

The proof of the above result is established in the next section where it is shown how to construct such matrices. In what follows, the matrix

$$P := \frac{1}{2}(1+\sigma)X, \tag{15}$$

where X satisfies conditions (13) and (14), will be referred to as a *Lyapunov matrix.*

Neglecting the control constraint, the proposed controllers consist of a linear part and a bounded nonlinear part; namely,

$$p(x) = p^{\epsilon}(x) := -Kx - \rho s(\epsilon^{-1}B^T Px), \tag{16}$$

where P is a Lyapunov matrix, $\epsilon > 0$ is arbitrary,

$$K := (1-k_2)^{-1}B^T P, \quad \rho := (1-k_2)^{-1}k_0, \tag{17}$$

and the function $s(\cdot)$ is given by

$$s(y) := (1 + ||y||)^{-1}y \ . \tag{18}$$

Theorem 2 *Consider an uncertain system described by (1), satisfying Assumptions 1 - 4, and subject to control given by (4),(16). Then the resulting closed loop system (5) is globally uniformly exponentially convergent to* $\mathcal{B}(r_\epsilon)$ *with rate* α *where*

$$r_\epsilon := [\epsilon k_0/\alpha\lambda_{min}(P)]^{\frac{1}{2}} \ . \tag{19}$$

Proof The proof consists of demonstrating that the function $V(\cdot)$, given by $V(x) = x^T Px$, is a Lyapunov function for the closed loop system. First, note that for all $x \in \mathbb{R}^n$,

$$\lambda_{min}(P) \, \|x\|^2 \leq V(x) \leq \lambda_{max}(P) \, \|x\|^2 .$$

Utilizing (15),(14),(6),(16), one obtains that (the detailed calculations are given in the Appendix of [4]), for all $t \in \mathbb{R}, x \in \mathbb{R}^n$,

$$DV(x)f(t,x) = 2x^T Pf(t,x) \leq -2\alpha V(x) + 2\epsilon k_0 \ .$$

Applying Theorem 1 with $\underline{V} = \epsilon k_0/\alpha$ and $\overline{V} = \infty$ (see Remark 2) yields the desired result.

Exponential Stabilization

It should be clear from (19) that the radius r_ϵ of the ball of convergence can be made arbitrary small, but non-zero, by choosing $\epsilon > 0$ sufficiently small. Convergence to zero , i.e. exponential stability, requires $k_0 = 0$. If $k_0 = 0$, the proposed controllers are linear and they are given by

$$p(x) = -Kx.$$

In this case, exponential stability is achieved and we have the following corollary.

Corollary 1 *Suppose the hypotheses of Theorem 2 are satisfied and* $k_0 = 0$. *Then the closed loop system (5) is globally uniformly exponentially stable with rate* α.

LYAPUNOV MATRICES

The General Case: $k_1 > 0$

Since (A, B) is controllable, it follows that $(A + \alpha I, B)$ is controllable and the following Riccati equation has a unique positive definite symmetric solution for X:

$$X(A + \alpha I) + (A + \alpha I)^T X - \sigma X B B^T X + {k_1}^2 I = 0; \tag{20}$$

hence, X satisfies (14).

Special Case: $k_1 = 0$

With $k_1 = 0$, (20) may have no positive definite solution. However, if one chooses any matrix $Q \in \mathbb{R}^{n \times n}$, with $Q^T = Q \geq 0$ and (A, Q) observable, then the following Riccati equation has a unique positive definite symmetric solution for X:

$$X(A + \alpha I) + (A + \alpha I)^T X - \sigma X B B^T X + Q = 0; \tag{21}$$

clearly, X satisfies (14).

Special Case: $k_1 = 0$ and $A + \alpha I$ Antistable

If α is chosen sufficiently large, specifically

$$\alpha > -\min\{Re(\lambda) : \lambda \in \sigma(A)\},$$

where $\sigma(A)$ denotes the set of eigenvalues of A, then $Re(\lambda) > 0$ for all $\lambda \in \sigma(A + \alpha I)$; hence, $A + \alpha I$ is antistable, i.e. $-A - \alpha I$ is exponentially stable. Since (A, B) is controllable, the pair $(-A^T - \alpha I,\ B^T)$ is observable; this last fact and the stability of $-A^T - \alpha I$ guarantee that the following Lyapunov equation has a unique solution for S:

$$(A + \alpha I)S + S(A + \alpha I)^T - \sigma B B^T = 0; \tag{22}$$

also, this solution is symmetric and positive definite. Letting

$$X = S^{-1} \tag{23}$$

and pre- and post-multiplying (22) by X yields equality in (14). Since it is linear, a Lyapunov equation is easier to solve than a Riccati equation.

CONSTRAINED CONTROLLERS AND MAIN RESULT

Taking control constraint (3) into account, the proposed controllers are simply saturating versions of those presented in the previous section. They are given by

$$p(x) = \bar{p}^{\,\epsilon}(x) := -\tilde{\rho}\ \mathrm{sat}(\tilde{\rho}^{-1} K x) - \rho s(\epsilon^{-1} B^T P x), \tag{24}$$

where K, and P are as defined for the unconstrained controllers, the *saturation function* $\mathrm{sat}(\cdot)$ is given by

$$\mathrm{sat}(y) := \begin{cases} y & \text{if } \ ||y|| \leq 1 \\ ||y||^{-1} y & \text{if } \ ||y|| > 1 \end{cases}, \tag{25}$$

ϵ is any positive real scalar which satisfies

$$\epsilon < \epsilon^* := \alpha(1 - k_2)^2 \tilde{\rho}^2 / \lambda_{max}(BB^T P) k_0 \tag{26}$$

and

$$\tilde{\rho} := \bar{\rho} - \rho. \tag{27}$$

with ρ given by (17).

Before introducing the main result, consider any $k \geq 0$ and define the *Lyapunov ellipsoid*

$$\mathcal{E}(k) := \{x \in \mathbb{R}^n : x^T P x \leq k\}. \tag{28}$$

Theorem 3 *Consider an uncertain system described by (1), satisfying Assumptions 1 - 4, and subject to bounded control given by (4),(24). Then the resulting closed loop system (5) is uniformly exponentially convergent to $\mathcal{B}(r_\epsilon)$ with rate α and region of attraction $\mathcal{R} = \mathcal{E}(k_{\tilde{\rho}})$, where*

$$k_{\tilde{\rho}} := (1 - k_2)^2 \tilde{\rho}^2 / \lambda_{max}(BB^T P).$$

Proof. First note that, whenever

$$\left\| B^T P x \right\| \leq (1 - k_2)\tilde{\rho}, \tag{29}$$

one has $\|\tilde{\rho}^{-1} K x\| \leq 1$ and hence, recalling (16),(24),

$$\bar{p}^{\,\epsilon}(x) = p^\epsilon(x). \tag{30}$$

Since

$$\left\| B^T P x \right\|^2 = x^T P B B^T P x \leq \lambda_{max}(BB^T P) x^T P x,$$

it follows that, whenever

$$x^T P x \leq k_{\tilde{\rho}},$$

(29) holds and hence (30) holds. Proceeding now as in the proof of Theorem 2, one shows that the hypotheses of Theorem 1 hold with $V(x) = x^T P x$, $\underline{V} = \epsilon k_0/\alpha$, and $\overline{V} = k_{\tilde{\rho}}$. The constraint (26) on ϵ guarantees that $\underline{V} < \overline{V}$.

Corollary 2 *Suppose the hypotheses of Theorem 3 are satisfied. Then the closed loop system (5) is uniformly exponentially convergent to $\mathcal{B}(r_\epsilon)$ with rate α and region of attraction $\mathcal{R} = \mathcal{B}(r_{\tilde{\rho}})$, where*

$$r_{\tilde{\rho}} := (1 - k_2)\tilde{\rho} / [\lambda_{max}(P) \lambda_{max}(BB^T P)]^{\frac{1}{2}}.$$

Proof. Since $x^T P x \leq \lambda_{max}(P) \|x\|^2$, it follows that $\mathcal{B}(r_{\tilde{\rho}}) \subset \mathcal{E}(k_{\tilde{\rho}})$. The result now follows from Theorem 3.

EXAMPLE: STABILIZATION OF AN INVERTED PENDULUM WITH A BOUNDED CONTROLLER

Consider a pendulum $\mathcal{P}$ (or one-link manipulator) which is subject to a bounded control torque T; see Figure 1. Its motion is described by

$$I\ddot{\theta} - mgl \sin\theta = T \tag{31}$$

where I is the moment of inertia of $\mathcal{P}$ about its axis of rotation, m is the mass of $\mathcal{P}$, l is the distance between the center of mass of $\mathcal{P}$ and support point O, and g is the gravitational acceleration constant of the planet (or moon) in which $\mathcal{P}$ resides. We consider the situation in which

$$|T(t)| \leq \overline{T}, \tag{32}$$

where the control torque bound $\overline{T}$ is specified and satisfies

$$\overline{T} > mgl. \tag{33}$$

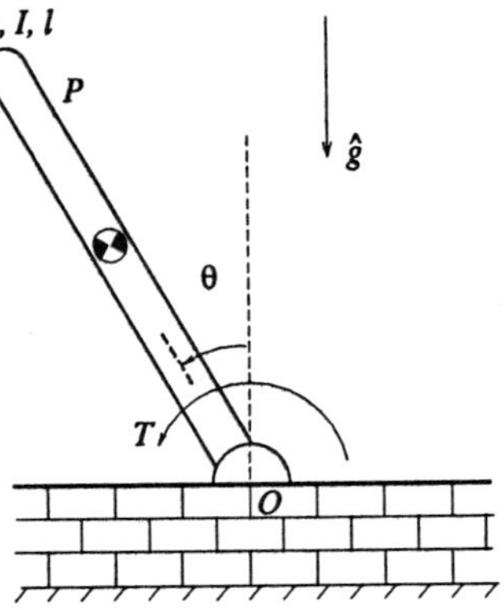

Figure 1. One-link manipulator

If we let

$$x_1 := \theta, \; x_2 := \dot{\theta}, \; u := T/I,$$

then

$$\begin{aligned} \dot{x}_1 &= x_2 \\ \dot{x}_2 &= k_0 \sin x_1 + u, \end{aligned} \tag{34}$$

where

$$k_0 := mgl/I$$

and

$$|u(t)| \leq \bar{\rho} := \overline{T}/I.$$

Clearly, description (34) is in the form of (1) with

$$A = \begin{bmatrix} 0 & 1 \\ 0 & 0 \end{bmatrix}, \quad B = \begin{bmatrix} 0 \\ 1 \end{bmatrix}, \quad \Delta F(t, x, u) = \begin{bmatrix} 0 \\ k_0 \sin x_1 \end{bmatrix}.$$

Also, Assumptions 1-4 are assured with

$$e(t, x, u) = k_0 \sin x_1$$

and

$$k_1 = k_2 = 0.$$

Since $A + \alpha I$ is antistable for any $\alpha > 0$, a Lyapunov matrix P for this system can be obtained from (15) and (23), where S solves the Lyapunov equation (22). This yields

$$P = 2\kappa\alpha \begin{bmatrix} 2\alpha^2 & \alpha \\ \alpha & 1 \end{bmatrix}, \quad \kappa := 1 + \sigma^{-1}. \tag{35}$$

Utilizing the results of the previous section, a controller given by (24) yields exponential convergence to $\mathcal{B}(r_\epsilon)$ with rate α and region of attraction $\mathcal{B}(r_{\tilde{\rho}})$, where

$$r_\epsilon = (\epsilon k_0)^{\frac{1}{2}} / \alpha\kappa^{\frac{1}{2}} [1 + 2\alpha^2 - (1 + 4\alpha^4)^{\frac{1}{2}}]^{\frac{1}{2}}$$

and

$$r_{\tilde{\rho}} = \tilde{\rho} / \sqrt{2}\kappa\alpha [1 + 2\alpha^2 + (1 + 4\alpha^4)^{\frac{1}{2}}]^{\frac{1}{2}}. \tag{36}$$

It should be clear from (36) that, as the desired convergence rate α increases the radius $r_{\tilde{\rho}}$ of the region of attraction decreases. When $\alpha \gg 1/\sqrt{2}$, $r_{\tilde{\rho}} \cong \tilde{\rho}/2\sqrt{2}\kappa\alpha^2$.

Note also that as σ increases κ decreases; hence $r_{\tilde{\rho}}$ increases. As σ approaches ∞, κ approaches 1.

ACKNOWLEDGEMENT

The first author was supported by the *US National Science Foundation* under Grant MSS-90-57079.

REFERENCES

1. M. Corless, "Guaranteed Rates of Exponential Convergence for Uncertain Systems," *Journal of Optimization Theory and Applications*, Vol. 64, pp. 481-494, 1990.

2. M. Corless and G. Leitmann, "Controller Design for Uncertain Systems via Lyapunov Functions," Proceedings of the American Control Conference, Atlanta, Georgia, 1988.

3. M. Corless and G. Leitmann, "Deterministic Control of Uncertain Systems: A Lyapunov Theory Approach," in *Deterministic Control of Uncertain Systems*, (Edited by A. Zinober), Peter Peregrinus Ltd., London, pp. 220-251, 1990.

4. M. Corless and G. Leitmann, "Bounded Controllers for Robust Exponential Convergence," *Journal of Optimization Theory and Applications*, Vol. 76, pp. 1-12, 1993.

5. R. M. Dolphus and W. E. Schmitendorf, "Stability Analysis for a Class of Linear Controllers Under Control Constraints," Proceedings of the 30th IEEE Conference on Decision and Control, Brighton, England, 1991.

6. M. Hached, S. M. Madani-Esfahani and S. H. Zak, "On the Stability and Estimation of Ultimate Boundedness of Nonlinear/Uncertain Dynamic Systems with Bounded Controllers," Proceedings of the American Control Conference, San Diego, CA, 1990.

7. S. M. Madani-Esfahani, H. Hached and S. Zak, "Estimation of Sliding Mode Domains of Uncertain Variable Structure Systems with Bounded Controllers," to appear.

8. S. M. Madani-Esfahani, S. Hui and S. H. Zak, "On the Estimation of Sliding Domains and Stability Regions of Variable Structure Control Systems with Bounded Controllers," Proceedings of the 26th Allerton Conference on Communication, Control, and Computing, Monticello, IL, 1988.

9. S. M. Madani-Esfahani and S. H. Zak, "Variable Structure Control of Dynamical Systems with Bounded Controllers," Proceedings of the American Control Conference, 1987.

10. P.-O. Gutman and P. Hagander, "A New Design of Constrained Controllers for Linear Systems," IEEE Transactions on Automatic Control, Vol. AC-30, pp. 22-33, 1985.

11. G. Leitmann, "Deterministic Control of Uncertain Systems via a Constructive Use of Lyapunov Stability Theory," in *System Modelling and Optimization*, (Edited by H.-J. Sebastian and K. Tammer), Lecture Notes in Control and Information Sciences, 143, Springer Verlag, Berlin, pp. 38 - 55, 1990.

12. A. G. Soldatos and M. Corless, "Stabilizing Uncertain Systems with Bounded Control," Dynamics and Control, Vol. 1, pp. 227-238, 1991.

13. S. M. Swei and M. Corless, "On the Necessity of the Matching Condition in Robust Stabilization," Proceedings of the 30th IEEE Conference on Decision and Control, Brighton, England, 1991.

ENTROPY MINIMIZATION PROBLEMS WITH LINEAR CONSTRAINTS, SCHRÖDINGER BRIDGE AND A CONDITIONAL SANOV THEOREM

Austin Blaquière[1] and M. Sigal-Pauchard[2]

[1]Université de Paris
22 rue de la Prairie
92410 Ville d'Avray (France)

[2]Université du Maine, Faculté des Sciences,
Département de Mathématique et d'Informatique
Avenue Olivier Messiaen BP 535 72017 Le Mans Cedex

Dedicated to Angelo Miele

ABSTRACT

This paper accounts for Csiszar's geometric approach to entropy minimization problems, connection with Schrödinger's bridge and a conditional Sanov theorem.

INTRODUCTION

In his remarkable papers published in 1931/1932, Schrödinger has solved the following problem : knowing the position of a Brownian particle in a Euclidean space M at times a and b, a < b, what is the probability for this particle to have passed through some prescribed domain of the space at some intermediate time? A generalization of this problem by prescribing probability distributions P_a and P_b at the initial and terminal time, respectively, has led to the concept of *Schrödinger bridge*, and it has been shown to be a problem of convex optimization.

The concept of Schrödinger bridge has been approached from different points of view. The main ones are: *the theory of reciprocal processes*, *information theory and statistics* with the central concept of *entropy* (or *negentropy*), a modern approach placing Schrödinger's bridge in the framework of *large deviations problems* for empirical distributions on path space in the light of Sanov's theorem, a *variational approach* comprising a class of stochastic optimal control problems.

Mechanics and Control, Edited by R.S. Guttalu
Plenum Press, New York, 1994

The most concise formulation of the convex optimization problem mentioned above is probably the one of Csiszar : In Csiszar's geometric approach, there is given a convex set Ξ of probability measures on some measurable space (Ω, F), a reference measure R, and one wishes to find a probability measure $P^* \in \Xi$, whenever it exists, for which the relative entropy

$$H(P;R) := \int (\log dP/dR)\, dP \quad \text{if } P << R,$$

$$H(P;R) = +\infty \text{ otherwise}, \qquad P \in \Xi,$$

is minimum; that is, such that $H(P^*;R) = \min_{P \in \Xi} H(P;R)$.

Schrödinger's problem fits nicely this framework if one considers R as the common distribution of *independent, identically distributed* (i.i.d.) random trajectories, and one defines the set Ξ through given probability constraints: initial (or/and) terminal conditions, or whatsoever. The events whose probabilities satisfy the constraints can be regarded as *rare events* and handled through the *theory of large deviations.* Another way to look at them is to consider them as the outcome of a proper control in which case the concerned areas are the ones of *controllability* and (stochastic) *optimal control.* This is the way we have approached the problem in earlier publications. These two points of view are related by a change of probability measure through Girsanov's formula.

In Schrödinger's views the reference measure R is the Wiener measure (associated with the Brownian transition probability density). Generalizing: the reference measure R is associated with a Markov process $(\xi_t,\ a \le t \le b)$ (possibly a diffusion process) with "creation" and "killing" introduced as usual through a multiplicative functional of Dynkin's type $\exp\left(\int_{[a,b]} c(r,\xi_r)dr\right)$, where c is a potential function. In the earlier publications we have supposed c bounded. As shown in a first part of this paper results concerning the case where the potential function c is unbounded and possibly singular are easily obtained through Csiszar's geometric approach, with the help of Dynkin's theory of multiplicative functionals. They proceed from our Theorem 5 which links together Csiszar's main existence theorem and Schrödinger's bridge. In a second part we also derive from Csiszar's geometric approach a conditional Sanov theorem.

RELATIVE ENTROPY

Let $M_1(\Omega)$ be the space of probability measures on a given measurable space (Ω,F). We shall consider subsets of $M_1(\Omega)$ endowed with the topology of the variation distance defined by

$$\| P - Q \| = \int \left| \frac{dP}{dR} - \frac{dQ}{dR} \right| dR\,, \quad P \in M_1(\Omega),\ Q \in M_1(\Omega),$$

where R is any probability in $M_1(\Omega)$ with $P << R$, $Q << R$.

Definition 1. The *relative entropy of probability measures* P and Q on a measurable space, also called *Kullback-Leibler information number* H(P;Q), *I-divergence*, or *information gain*, is defined as

$$H(P;Q) = \int \log \frac{dP}{dQ}\, dP = \int \frac{dP}{dQ} \log \frac{dP}{dQ}\, dQ \qquad \text{if } P << Q, \tag{1}$$

$$= +\infty \qquad \text{otherwise.}$$

In computing H(P;Q), the conventions about infinity are $\log 0 = -\infty$, $\log (a/0) = +\infty$, $0.(\infty) = 0$. It follows from Jensen's inequality that H(P;Q) is nonnegative and vanishes only for P = Q. Let us note that the relative entropy is not a metric, and cannot be used in general to define a topology.

One can easily prove the following well-known:

Proposition 1. Let $\{P_n\}$ be a sequence of elements of $M_1(\Omega)$, and $R \in M_1(\Omega)$, with $P_n << R$, n = 1, 2, ... If $\int | (dP_m)/dR) - (dP_n/dR) | dR \to 0$, whenever m, n $\to \infty$, then there exists a $Q \in M_1(\Omega)$, $Q << R$, such that $\int | (dP_n/dR) - (dQ/dR) | dR \to 0$, as $n \to \infty$.

Proposition 2. The variation distance $\| P - Q \|$ can be estimated by

$$\| P - Q \| \le [2\, H(P;Q)]^{1/2} \tag{2}$$

Definition 2. (Csiszar [4]) If E is a convex subset of $M_1(\Omega)$ containing at least one element P with finite entropy relative to some $R \in M_1(\Omega)$, a $Q \in E$ satisfying

$$H(Q;R) = \min_{P \in E} H(P;R)$$

is called the I-projection of R on E.

From now on we will call the I-projection *Csiszar's projection.* If Csiszar's projection Q of R on E exists, the convexity of E together with the strict convexity of H(P;R) in P, for P << R, guarantees its uniqueness.

GEOMETRIC RESULTS

Theorems 1 and 2 below are proved in Csiszar [4] :

Theorem 1. If the subset E of $M_1(\Omega)$ is convex and variation-closed, and if E contains at least one element P with finite entropy relative to some $R \in M_1(\Omega)$, then R has one and only one Csiszar's projection on E.

Theorem 2. An element Q of the convex subset E of $M_1(\Omega)$, with finite entropy relative to $R \in M_1(\Omega)$, is the Csiszar's projection of R on E iff for every $P \in E$, with $H(P;R) < \infty$

$$\int \log \frac{dQ}{dR}\, dP \ge H(Q;R)\,, \tag{3}$$

or, equivalently, iff

$$H(P;R) \geq H(P;Q) + H(Q;R) \qquad \text{for every } P \in E. \tag{4}$$

Now let us state:

Theorem 3. Let E and E_k, k = 1, 2, ... , be subsets of $M_1(\Omega)$. Assume that E contains at least one element **P** with finite entropy relative to some R $\in M_1(\Omega)$, that the E_k's, k = 1, 2, .. and E are convex and variation closed, and that

$$E_k \supset E_{k+1},\ k = 1, 2, \ldots, \tag{5}$$

$$\cap_{k=1}^{\infty} E_k = E. \tag{6}$$

Then

(i) There is a unique Csiszar's projection of R on each set E_k, k = 1, 2, ... and on E, namely Q_k, k = 1, 2, ... and Q*, respectively.

(ii) $H(Q_k;R) \to H(Q_\infty;R)$, and $H(Q_\infty;Q_k) \to 0$ as $k \to \infty$.

(iii) Q_k converges in variation to Q_∞, $Q_\infty = Q^*$, as $k \to \infty$.

Proof. (i) follows from Theorem 1. (5) implies

$$H(Q_k;R) \leq H(Q_{k+1}) \leq \ldots \leq H(Q^*;R) \leq H(\mathbf{P};R) < \infty. \tag{7}$$

From (5) and Theorem 2, one deduces

$$H(Q_m;Q_n) + H(Q_n;R) \leq H(Q_m;R) \qquad \text{for } m \geq n. \tag{8}$$

From (7) it follows that $H((Q_n;R)$ converges nondecreasingly to a limit c as $n \uparrow \infty$, that is $H(Q_n;R) \uparrow c \leq H(Q^*;R)$ which, together with (8), implies that

$$H(Q_m;Q_n) \to 0 \text{ as } m, n \to \infty \text{ with } m \geq n. \tag{9}$$

From (9) and Proposition 2, it follows that

$$\| Q_m - Q_n \| \to 0 \text{ as } m, n \to \infty \text{ with } m \geq n. \tag{10}$$

From (10) and Proposition 1, it follows that Q_k converges in variation to a limit Q_∞ as $k \to \infty$. Let K $\in$ { 1, 2, ... }. By (5) we have $Q_n \in E_K$ for n > K. E_K is variation closed and Q_n converges in variation to Q_∞ as $n \to \infty$. Therefore $Q_\infty \in E_K$. Since this holds for any K $\in$ { 1, 2, ... },

$$Q_\infty \in \cap_{k=1}^{\infty} E_k = E.$$

(5) and (6) with $Q_\infty \in E$ imply $H(Q_k;R) \leq H(Q_\infty;R)$ for all k = 1, 2, ...Therefore,

$$H(Q_\infty;R) \geq \lim H(Q_k;R). \tag{11}$$

Since Q_k converges in variation to Q_∞, by Fatou's lemma there is a subsequence $\{(dQ_{k_r}/dR)\}$ of $\{(dQ_k/dR)\}$ converging to the limit (dQ_∞/dR) as $k_r \to \infty$, R- a. s. , and, again by Fatou's lemma

$$\lim\ H(Q_{kr};R) \geq \int \lim \left(\frac{dQ_{kr}}{dR} \log \frac{dQ_{kr}}{dR} \right) dR = H(Q_\infty;R) \ . \tag{12}$$

Comparing (11) and (12) gives

$$\lim\ H(Q_k;R) = H(Q_\infty;R) \quad \text{as } k \to \infty \ . \tag{13}$$

On the other hand, by Theorem 2,

$$H(P;Q_k) + H(Q_k;R) \leq H(P;R) \quad \text{for every } P \in E_k\ ,\ k = 1, 2, \ldots . \tag{14}$$

Therefore by substituting Q_∞ for P, we get

$$H(\ Q_\infty;Q_k) + H(Q_k;R) \leq H(\ Q_\infty;R)\ , \qquad k = 1, 2, \ldots . \tag{15}$$

At last, by letting $k \to \infty$ and using (13), we obtain $\lim H(\ Q_\infty;Q_k) = 0$ as $k \to \infty$. From (13) and (7) it follows that $H(Q_\infty;R) \leq H(Q^*;R)$. Since $Q_\infty \in E$, we must have $H(Q_\infty;R) = H(Q^*;R)$. At last, the uniqueness of Csiszar's projection entails $Q^* = Q_\infty$, which ends the proof of Theorem 3.

MINIMIZING RELATIVE ENTROPY UNDER LINEAR CONSTRAINTS

As in Föllmer [12], let there be given a sequence $\{f_i : i = 1, 2, \ldots\}$ of bounded measurable real valued functions, a sequence $\{c_i : i = 1, 2, \ldots\}$ of real numbers, and let E be the subset of $M_1(\Omega)$ defined by

$$E := \{P \in M_1(\Omega) : \int f_i\, dP = c_i\ , i = 1, 2, \ldots\}. \tag{16}$$

Proposition 3. E defined by (16) is convex and variation closed.

Proof. Convexity of E follows readily from $\int f_i\, (\alpha\, dP_1 + (1 - \alpha)\, dP_2) = \alpha\, c_i + (1 - \alpha)\, c_i = c_i$, $0 \leq \alpha \leq 1$, $i = 1, 2, \ldots$ whenever P_1, $P_2 \in E$. Further, consider a sequence $\{P_n\}$ of elements of E, and a $Q \in M_1(\Omega)$, with

$$\int |\, (dP_n/dR)\ - (dQ/dR)\, |\, dR\ \to 0\ , \qquad \text{as } n \to \infty \tag{17}$$

for some $R \in M_1(\Omega)$. Since the f_i's, $i = 1, 2, \ldots$, are bounded functions, (17) implies

$$|\, c_i - \int f_i\, dQ\, | = |\int f_i\, dP_n - \int f_i\, dQ\, | \leq \int |\, f_i\, |\ |\, (dP_n/dR)\ - (dQ/dR)\, |\, dR\ \to 0, \text{ as } n \to \infty\, .$$

Therefore $\int f_i\, dQ = c_i$, $i = 1,2, \ldots$, which proves that $Q \in E$, and completes the proof.

In addition we shall suppose that E contains at least one element **P** with finite entropy relative to some $R \in M_1(\Omega)$, and we shall let

$$E_k := \{P \in M_1(\Omega) : \int f_i\, dP = c_i, i = 1, 2, \ldots k\},\ \ 1 \leq k < \infty\ , \tag{18}$$

Note that $E_k \downarrow E$; that is

$$E_k \supset E_{k+1}, \ k = 1, 2, \dots, \tag{19}$$

and

$$\cap_{k=1}^{\infty} E_k = E. \tag{20}$$

By the same argument as above, each of the E_k's, k = 1, 2, ..., is convex and variation closed, and, by (20), contains **P**. We see that E and the E_k's, k = 1, 2, ... satisfy the assumptions of Theorem 3.

Therefore we have:

Theorem 4. (Special case of Theorem 3.1 of Csiszar [4]) R has Csiszar's projection Q* on E defined by (16), and the R-density of Q* is of form

$$\frac{dQ^*}{dR}(\omega) = c \exp g(\omega) \qquad \text{if } \omega \notin N \tag{21}$$

$$= 0 \qquad \text{if } \omega \in N \tag{22}$$

where P(N) = 0 for every P ∈ E such that H(P;R) < ∞ , 0 < c < ∞, and g belongs to the closure in $L^1(Q^*)$ of the subspace spanned by the f_i 's. On the other hand, if a Q ∈ E with H(Q;R) < ∞ has R-density of the form (21),(22), with g belonging to the subspace spanned by the f_i 's, then Q is the Csiszar's projection of R on E.

If the subspace spanned by the f_i 's is closed in $L^1(Q)$, $Q \in M_1(\Omega)$, which is the case in particular if the set $\{f_i\}$ is finite, then Theorem 4 provides a necessary and sufficient condition for Q to be the Csiszar's projection of R on E; that is we have:

Corollary 1. If the subspace spanned by the f_i 's, i = 1, 2, ... is closed in $L^1(Q)$, Q ∈ E with H(Q;R) < ∞, then Q is the Csiszar's projection of R on E iff it has R-density of the form (dQ/dR) = c exp g, where 0 < c < ∞ and g is a finite linear combination of the f_i 's, except possibly for a subset N of $M_1(\Omega)$ where (dQ/dR) vanishes and P(N) = 0 for every P ∈ E such that H(P;R) < ∞.

LINEAR CONSTRAINTS DEFINED BY GIVEN MARGINALS

Suppose $(\Omega,F) := (X_1,\Sigma_1)\otimes(X_2,\Sigma_2)$, and E consists of the probability measures P on (Ω,F) with given marginals P_a and P_b. Assume

(A0) $X_1 = X_2 = M$ *where* M *is a separable metric space and* $\Sigma_1 = \Sigma_2 = \Sigma$, *its σ-field of Borel sets.*

In view of the countability properties of M, one can prove

Proposition 4. Suppose (A0) holds. The fact that P on $(X_1,\Sigma_1)\otimes(X_2,\Sigma_2)$ has marginals P_a and P_b can be expressed by constraints of the form $\int f_i \, dP = c_i$, i = 1, 2, ... , where the f_i

are bounded measurable real valued functions only depending on one of the two arguments $x \in X_1$, $y \in X_2$.

For later use we shall need the following:

Lemma 1. (Donsker, Varadhan [6]) Let (X_1,Σ_1, α) and (X_2,Σ_2, β) be two probability measure spaces, and let (Z, Γ, μ) be the product space with product measure. Let $\{h_n\}$ be a sequence of functions in Z with $h_n(x,y) = f_n(x)/g_n(y)$, $(x,y) \in Z$, where, for each n, f_n and g_n are bounded strictly positive functions on X and Y, respectively. If $\lim h_n(x,y) = h(x,y)$, as $n \to \infty$, exists μ- a. s., and $0 \leq h(x,y) < \infty$ μ- a. s. , then $h(x,y) = f(x)/g(y)$ where $0 \leq f(x) < \infty$, α- a. s. and $0 < g(y) \leq \infty$, β- a. s.

MINIMUM ENTROPY, SCHRÖDINGER SYSTEM

Theorem 5. Suppose M is a σ-compact metric space, and that P_a, P_b are probability measures on its σ-field Σ of Borel sets. Let E denote the set of all probability measures on $\Sigma\otimes\Sigma$ with marginals P_a, P_b, and let q be a measurable function on $M\otimes M$, bounded away from zero below $(P_a\otimes P_b)$-a.s., and $(P_a\otimes P_b)$-integrable. Then there is a unique pair Q^*, π of measures on $\Sigma\otimes\Sigma$ for which

(a) Q^* is a probability measure and π is a finite product measure;

(b) $Q^* \in E$;

(c) $(dQ^*/dp) = q$;

(d) $H(Q^*;R^*) \leq H(P;R^*)$ for all $P \in E$, where $dR^* := q\, d(P_a\otimes P_b)/\int q\, d(P_a\otimes P_b)$.

(e) $d\pi = [1/\int q\, d(P_a\otimes P_b)]\, \phi\, \gamma\, d(P_a\otimes P_b)$, where $\log \phi \in L^1(P_a)$, $\log \gamma \in L^1(P_b)$.

(f) If P_a and P_b are absolutely continuous with respect to σ-finite measures λ_a and λ_b, respectively, and $\log (dP_a/d\lambda_a) \in L^1(P_a)$, $\log (dP_b/d\lambda_b) \in L^1(P_b)$, then $dQ^* = \varphi_a\, q\, \varphi_b\, d\lambda_a\, d\lambda_b$, where $\log \varphi_a \in L^1(P_a)$, $\log \varphi_b \in L^1(P_b)$.

For proving Theorem 5, we need:

Lemma 2. If R is a probability measure on $\Sigma\otimes\Sigma$ for which there exists some $\mathbf{P} \in E$ with $H(\mathbf{P} ; R) < \infty$, then there is a unique probability measure $\mathbf{Q} \in E$ such that $H(\mathbf{Q} ; R) \leq H(P ; R)$ for all $P \in E$. Further, if $\mathbf{P}$ and R are measure-theoretically equivalent *to a same probability measure product* - say $\alpha\otimes\beta$ - on $\Sigma\otimes\Sigma$, then $d\mathbf{Q}/dR$ is such that $(d\mathbf{Q}/dR)(x,y) = \phi(x)\gamma(y)$, $(x,y) \in M\otimes M$, with $0 \leq \phi(x) < \infty$ α - a.s., $0 \leq \gamma(y) < \infty$ β - a.s.

Proof of Lemma 2. E being convex and variation closed, the first statement in Lemma 2 is an application of Theorem 1. The second statement is based on the fact that, M being a σ-compact and metric space, its σ-field of Borel sets Σ is separable, that is, it is generated by a countable class of sets. It then follows that

$$E = \{P \in \Pi : \int f_i\, dP = \int f_i\, dP_a ,\ \int g_i\, dP = \int g_i\, dP_b, i = 1, 2, ...\} \tag{23}$$

where Π denotes the set of all probability measures on $\Sigma\otimes\Sigma$, and the f_i 's and g_i 's are bounded measurable real valued functions, only depending on one of the two arguments x, y $\in$ M, respectively. This remark enables us to use an argument of Föllmer [12]; that is, we consider a sequence of subsets of Π defined by finite sets of linear constraints, namely $\{E_n, n = 1, 2, ...\}$ with $E_n := \{P \in \Pi : \int f_i \, dP = \int f_i \, dP_a , \int g_i \, dP = \int g_i \, dP_b, i = 1,...n\}$.

According to Föllmer's argument, $E_n \downarrow E$ as $n \to \infty$ and, for each n, there is a Csiszar's projection Q_n of R on E_n which converges in variation to $\mathbf{Q}$ as $n \to \infty$. According to Corollary 1, the Q_n 's have densities with respect to R of the form $dQ_n/dR = \phi_n \gamma_n$, where ϕ_n and γ_n are bounded strictly positive functions of x and y, respectively, except possibly for a subset $\mathcal{N}_n$ of $M\otimes M$ where dQ_n/dR vanishes and $P_n(\mathcal{N}_n) = 0$ for every $P_n \in E_n$ with $H(P_n;R) < \infty$. Thus, for every such P_n we have

$$0 < (dQ_n/dR)(x,y) = \phi_n(x) \, \gamma_n(y), \quad P_n \text{ - a.s.} \tag{24}$$

Further $\int | (d\mathbf{Q}/dR) - (dQ_n/dR) | \, dR \to 0$ as $n \to \infty$, so that, by Fatou's lemma there is a subsequence $\{dQ_{nk}/dR\}$ for which

$$\lim_{nk} | (d\mathbf{Q}/dR) - (dQ_n/dR) | = 0, \quad R\text{ - a.s. with } 0 \le d\mathbf{Q}/dR < \infty \quad R\text{ - a.s.} \tag{25}$$

Since $P \in E \Rightarrow P \in E_n$, n = 1, 2, ..., by (24) and (25) the assumptions of the Lemma 2.5 of Donsker and Varadhan [7] are satisfied if R is absolutely continuous with respect to some $\mathbf{P} \in E$ with $H(\mathbf{P} ; R) < \infty$, *and is measure-theoretically equivalent to a probability measure product* - say $\alpha\otimes\beta$. If it is the case, then, in view of (24) and (25), an application of that lemma shows that $(d\mathbf{Q}/dR)(x,y) = \phi(x)\gamma(y)$, $(x,y)\in M\otimes M$, with $0\le \phi(x)< \infty$ α - a.s., $0 \le \gamma(y) < \infty$ β - a.s.

This proves the second part of Lemma 2.

<u>Proof of Theorem 5</u>. Let $\mathbf{P} := P_a\otimes P_b$. Then $\mathbf{P} \in E$, and from the assumptions of Theorem 5 and the definition of R* it follows that

$$0 \le H(\mathbf{P} ; R^*) \le | \log \int q \, d(P_a\otimes P_b) | + \int | \log q | \, d(P_a\otimes P_b) < \infty.$$

From the first part of Lemma 2, there is a unique probability measure $Q^* \in E$ satisfying $H(Q^*;R^*) \le H(P;R^*)$ for all $P \in E$. From relation (2.6) and Lemma 2.1 of Csiszar [4], we have

$$H(P;R^*) - H(P;Q^*) = \int \log \frac{dQ^*}{dR^*} \, dP, \text{ and } \int \log \frac{dQ^*}{dR^*} \, dP \ge H(Q^*;R^*) \ge 0$$
$$\text{for each } P \in E \text{ such that } H(P;R^*) < \infty.$$

It follows that

$$0 \le \int \log \frac{dQ^*}{dR^*} \, dP < \infty \quad \text{for each } P \in E \text{ such that } H(P;R^*) < \infty.$$

In particular

$$0 \le \int \log \frac{dQ^*}{dR^*} \, d(P_a \otimes P_b) < \infty. \tag{26}$$

Let $m := q / \int q \, d(P_a \otimes P_b)$, then $dR^* = m \, d(P_a \otimes P_b)$. Since $m > 0$ $(P_a \otimes P_b)$-a.s., the probability measure R^* is measure-theoretically equivalent to $P_a \otimes P_b$. Accordingly, we deduce from Lemma 2 that $(dQ^*/dR^*)(x,y) = \phi(x)\gamma(y)$, $(x,y) \in M \otimes M$, with $0 \le \phi(x) < \infty$ P_a-a.s., $0 \le \gamma(y) < \infty$ P_b-a.s. but, since (26) implies that $(P_a \otimes P_b)\{(x,y): (dQ^*/dR^*)(x,y) = 0\} = 0$, we have in fact

$$\frac{dQ^*}{dR^*}(x,y) = \phi(x)\gamma(y), \ (x,y) \in M \otimes M, \ 0 < \phi(x) < \infty \ (P_a\text{-a.s.}), \ 0 < \gamma(y) < \infty \ (P_b\text{-a.s.}) \tag{27}$$

or equivalently

$$\log \frac{dQ^*}{dR^*}(x,y) = \log \phi(x) + \log \gamma(y), \tag{28}$$

$$\text{with } |\log \phi(x)| < \infty \ (P_a\text{ - a.s.}), \ |\log \gamma(y)| < \infty \ (P_b\text{ - a.s.}).$$

Then, by substituting (28) in (26), one easily finds that

$$\log \phi \in L^1(P_a), \quad \log \gamma \in L^1(P_b). \tag{29}$$

From (27) and the definition of R^* we get $dQ^*/d\pi = q$, where π is the product measure defined by

$$d\pi = [1/\int q \, d(P_a \otimes P_b)] \, \phi \, \gamma \, d(P_a \otimes P_b) := d\pi_0 \, d\pi_1. \tag{30}$$

We have also

$$dQ^* = q \, d\pi = [q/\int q \, d(P_a \otimes P_b)] \, \phi \, \gamma \, d(P_a \otimes P_b). \tag{31}$$

The fact that π is finite is a consequence of $d\pi = dQ^*/q$ with $(1/q)$ bounded above $(P_a \otimes P_b)$- a.s.

Under the assumptions of (f) of Theorem 5 relation (31) becomes

$$dQ^* = \varphi_a \, q \, \varphi_b \, d\lambda_a \, d\lambda_b, \tag{32}$$

where $\log \varphi_a = \text{const.} + \log \phi + \log (dP_a/d\lambda_a) \in L^1(P_a)$, $\log \varphi_b = \text{const.} + \log \gamma + \log (dP_b/d\lambda_b) \in L^1(P_b)$.

This ends the proof of Theorem 5.

Corollary 2. Under the assumptions of Theorem 5, and $(dP_a/d\lambda_a) := \Phi_a$, $(dP_b/d\lambda_b) := \Phi_b$, with $\log \Phi_a \in L^1(P_a)$ and $\log \Phi_b \in L^1(P_b)$, there is a unique (up to multiplicative constants) nonnegative solution (φ_a, φ_b) for the Schrödinger system :

$$\Phi_a(x) = \varphi_a(x) \int q(x,y) \, \varphi_b(y) \, dy$$

$$\Phi_b(y) = \varphi_b(y) \int q(x,y) \, \varphi_a(x) \, dx \tag{33}$$

Theorem 6. Let q satisfy the assumptions of Theorem 5, and suppose that P_a and P_b satisfy the assumptions of part (f) of that theorem. Then Q* given by (32) satisfies $H(Q^*;R) \leq H(P;R)$ for all $P \in E$ and all R given by

$$dR(x,y) = f(x)g(y)q(x,y)d\lambda_a(x)d\lambda_a(y)\ ,\ (x,y) \in M\otimes M, \qquad (34)$$
$$\text{with } \log f \in L^1(P_a) \text{ and } \log g \in L^1(P_b).$$

<u>Proof of Theorem 6</u>. In view of (32) and (34), we have

$$(dQ^*/dR)(x,y) = \varphi_a(x)\varphi_b(y)/f(x)g(y),\ \ (x,y) \in M\otimes M,\ \ (P_a\otimes P_b)\text{- a.s.} \qquad (35)$$

Since $\log \varphi_a$ and $\log f$ are in $L^1(P_a)$, and $\log \varphi_b$ and $\log g$ are in $L^1(P_b)$, Corollary 3.1 of Csiszar [4] applies telling that $H(Q^*;R) \leq H(P;R)$ for all $P \in E$.

APPLICATION TO A PROCESS WITH CREATION AND KILLING

Now suppose that, in addition to be σ-compact and metric, the space M is complete. Since a σ-compact and metric space is separable, M is now a Polish space. Suppose $\{\xi_t\ , a \leq t < \infty\}$ is a (M,Σ)-valued continuous Markov process in the sense of Fleming and Rishel [11] with underlying complete probability space $(\Omega, F, \mathbf{P})$, Markov transition probability function $P(s,x;t,B) := \mathbf{P}(\xi_t \in B \mid \xi_s = x) := \mathbf{P}_{sx}(\xi_t \in B)$, $a \leq s < t < \infty$, $x \in M$, $B \in \Sigma$, and initial distribution measure P_a. We assume that the transition probability function P(.) is given by a positive density relative to some σ-finite measure λ on Σ; that is, there is a strictly positive function $p(s,x;t,y)$ defined for $a \leq s < t < \infty$ and $(x,y) \in M\otimes M$, $\Sigma\otimes\Sigma$-measurable in (x,y) for each s and t, and for which

$$P(s,x;t,B) = \int_B p(s,x;t,y)\ \lambda(dy), \quad a \leq t < \infty,\ \ x \in M, B \in \Sigma.$$

Let $\mathbf{B}$ be a nonempty open relatively compact subset of M . Then, for prescribed terminal time $b > a$, the part $\eta := \{\eta_t\ , a \leq t \leq b\}$ of the process $\xi := \{\xi_t\ , a \leq t \leq b\}$ on the set $\mathbf{B}$ has transition density $\mathbf{p}(s,x;t,y)$ defined for $a \leq s < t \leq b$ and $(x,y) \in \mathbf{B}\otimes\mathbf{B}$ by

$$\mathbf{p}(s,x;t,y) = \mathbf{E}_{sx}\ [1_{t<\tau_s} \mid \eta_t = y]\ p(s,x;t,y)$$

where τ_s is defined by $\tau_s := \inf\{t>s : \xi_t \in M-\mathbf{B}\}$ if it exists, $\tau_s := \infty$ otherwise.

Note that $\int_B \mathbf{p}(s,x;t,y)\ \lambda(dy) = \mathbf{P}_{sx}(\ \xi_t\ \in B,\ \tau_s > t\ )$, $a \leq s < t \leq b$, $x \in \mathbf{B}$, $B \in \Sigma$.

Let c be a measurable function on $[a, b]\otimes M$ with values in $[-\infty, +\infty]$. Let $D := \{(s,x) \in [a, b]\otimes M :\ |c(s,x)| < \infty\}$. Assume

(A1) P_a *and* P_b *have densities,* Φ_a *and* Φ_b *respectively, with respect to* λ, *and* Φ_a, Φ_b *are continuous functions with compact supports* K_a *and* K_b, *respectively; further* $\mathbf{B} \supset K_a \cup K_b$.

(A2) *The map* $(x,y) \to \mathbf{p}(a,x;b,y)$ *is continuous on* $\mathbf{B}\otimes\mathbf{B}$, *and* $\mathbf{p}(a,x;b,y)$ *is strictly positive on* $K_a \otimes K_b$.

(A3) c *is finite and continuous on* $N := [a, b]\otimes(\text{cl }\mathbf{B})$, *where* cl $\mathbf{B}$ *denotes the closure of* $\mathbf{B}$.

(A4) $\int [\int \exp (\int_{[a,b]} c(r, \xi_r)\, dr)\, 1_{b<T}\, dP_{ax}]\, \Phi_a(x)\, \lambda(dx) < \infty$, *where* T, *defined by* $T := \inf \{t>a : |c(t,\xi_t)| = \infty\}$ *if it exists*, $T := \infty$ *otherwise, is supposed to be a measurable function.*

Theorem 7. Let (A1)-(A4) hold. Then the function q given by

$$q(x,y) := \mathbf{E}_{ax} [\exp (\int_{[a,b]} c(r, \xi_r)\, dr)\, 1_{b<T} \mid \xi_b = y]\, p(a,x;b,y), \quad (x,y) \in M \otimes M,$$

satisfies the assumptions of Theorem 5.

The proof of Theorem 7 will be given in another publication.

Here the following question arises : does q derive from the transition density function of a Markov process? One can prove at least that it is given by a quasi-transition density function (in the case of homogeneous processes, see e.g. Dynkin [9], p. 283), under additional assumptions.

Proposition 5. For M a σ-compact metric complete space, under assumptions (B1)-(B4) below, the function $q(s,x;t,y)$, $a \le s < t \le b$, $(x,y) \in M \otimes M$, given by

$$q(s,x;t,y) := \mathbf{E}_{sx} [\exp (\int_{[s,t]} c(r, \xi_r)\, dr)\, 1_{t<T_s} \mid \xi_t = y]\, p(s,x;t,y), \tag{36}$$

where T_s is defined by $T_s := \inf \{t>s : |c(t,\xi_t)| = \infty\}$ if it exists, $T_s := \infty$ otherwise, is a quasi-transition density function.

Let c' be a measurable function on $[a, \infty) \otimes M$ with values in $[-\infty, +\infty]$. Let

$$D' := \{(s,x) \in [a, \infty) \otimes M : | c'(s,x) | < \infty\}.$$

For $a \le s$ let $\zeta_s := \inf \{t>s : |c'(t,\xi_t)| = \infty\}$ if it exists, $\zeta_s := \infty$ otherwise. Let $\mathcal{N}_t^s := \sigma(\xi_u, a \le s \le u \le t < \infty)$. The aforementioned assumptions are

(B1) $$\{\omega : \zeta_s(\omega) > t\} \in \mathcal{N}_t^s, \quad a \le s < t < \infty, \tag{37}$$

which condition is satisfied if the boundary of D' *in* $[a, \infty) \otimes M$ *is regular enough,*

(B2) c' *is continuous on* D'.

Then, according to the proof of Theorem 3.7 of Dynkin [8],

$$\beta_t^s := \exp (\int_{[s,t]} c'(r, \xi_r)\, dr), \quad t < \zeta_s(\omega),$$

is a (possibly non-contracting) multiplicative functional.

(B3) $\mathbf{E}_{sx}[\exp (\int_{[s,t]} c'(r, \xi_r)\, dr)\, 1_{t<\zeta_s}] < \infty$, for *all* $a \le s < t < \infty$, $x \in M$, (i.e., relation 9.2.B. of Dynkin [9], p. 283, in the case of homogeneous processes).

(B4) $c'(s,x) = c(s,x)$ *for all* $(s,x) \in [a, b] \otimes M$.

The proof of Proposition 5 will be given in another publication. It is based on the properties of Dynkin's multiplicative functionals $\int_{[s,t]} c(r, \xi_r)\, dr$ $(t < T_s(\omega))$ and $1_{t<T_s}$ which transform the Markov process ξ into a Markov process with creation and killing.

LARGE DEVIATIONS, CONDITIONAL SANOV PROPERTY

Let (Ω,F) be an arbitrary measurable space. In this paragraph we denote by $M_1(\Omega)$ the set of probability measures on (Ω,F) endowed with the Csiszar's τ_0-topology :

Definition 3. (Csiszar [5]) The *τ_0-topology* on $M_1(\Omega)$ is defined by the basic neighborhoods of an element $P \in M_1(\Omega)$:

$$U_0(P, Z, \varepsilon) := \{Q \in M_1(\Omega) : | P(\Omega_i) - Q(\Omega_i) | < \varepsilon , Q(\Omega_i) = 0 \text{ if } P(\Omega_i) = 0, i = 1, \ldots , k\} ,$$

where Z ranges over all finite measurable partitions $(\Omega_1 \ldots , \Omega_k)$ of (Ω,F), and ε ranges over the positive numbers.

Approximating Sets

As before, let there be given a sequence $\{f_1, f_2, \ldots\}$ of bounded measurable real valued functions, a sequence $\{c_1, c_2, \ldots\}$ of real numbers, and let E be the subset of $M_1(\Omega)$ defined by $E := \{P \in M_1(\Omega) : \int f_i \, dP = c_i, i = 1, 2, \ldots\}$. We fix a probability measure $R \in M_1(\Omega)$, and we assume :

(C1) *The set* E *contains at least one element* **P** *with finite entropy relative to* R.

In order to define subsets which approximate the set E, we consider a family of subsets $A(\varepsilon, k)$ of $M_1(\Omega)$, for $\varepsilon > 0$ and $k = 1, 2, \ldots$:

$$A(\varepsilon, k) := \{ P \in M_1(\Omega) : \ |\int f_i \, dP - c_i| \leq (\varepsilon/2^k), \quad i = 1, \ldots k\}.$$

One can easily prove the following:

Proposition 6. The set $A(\varepsilon, k)$ is convex and variation closed; and

Proposition 7. For given $\varepsilon > 0$, $A(\varepsilon, k) \downarrow E$ as $k \uparrow \infty$.

from which by Theorem 3 we have:

Proposition 8. For given $\varepsilon > 0$
(i) There is a unique Csiszar's projection of R on each set $A(\varepsilon, k)$, $k = 1, 2, \ldots$ and on E, namely $Q_{\varepsilon k}$, $k = 1, 2, \ldots$ and Q^*, respectively.
(ii) $Q_{\varepsilon k}$, $k = 1, 2, \ldots$, converges to Q^* in entropy; that is, $H(Q^*;Q_{\varepsilon k}) \to 0$ as $k \to \infty$.
(iii) $Q_{\varepsilon k}$, $k = 1, 2, \ldots$, converges in variation to Q^*, as $k \to \infty$.
(iv) $H(Q_{\varepsilon k};R) \to H(Q^*;R)$ as $k \to \infty$.

In the following we shall also use Csiszar's notation $H(\Gamma;R) := \inf_{P \in \Gamma} H(P;R)$, for a subset Γ of $M_1(\Omega)$.

Definition 4. (Csiszar [5]) A subset **E** of $M_1(\Omega)$ is *completely convex* if for every probability space $(\Theta,\mathbf{A},\mu)$ and Markov probability kernel η on $\Theta \otimes F$ such that $\eta(\lambda, .) \in \mathbf{E}$ for each $\lambda \in \Theta$, the probability measure $\mu\eta$ defined by $\mu\eta(B) = \int \eta(.,B) \, d\mu$ $(B \in F)$ also belongs to **E**.

Proposition 9. The set $A(\varepsilon, k)$ is completely convex.

Proof. Let $(\Theta,\mathcal{A},\mu)$ be an arbitrary probability space and η be a Markov probability kernel on $\Theta\otimes F$ such that $\eta(\lambda, .) \in A(\varepsilon, k)$ for each $\lambda \in \Theta$. Then $| \int_\Omega f_i(\omega)\, \eta(\lambda, d\omega) - c_i | \leq (\varepsilon/2^k)$, $i = 1, \ldots k$, $\lambda \in \Theta$.

This implies

$$\int_\Theta [\, | \int_\Omega f_i(\omega)\, \eta(\lambda, d\omega) - c_i |\,]\, \mu(d\lambda) \leq (\varepsilon/2^k), \quad i = 1, \ldots k,$$

which in turn implies

$$| \int_\Theta [\int_\Omega f_i(\omega)\, \eta(\lambda, d\omega) - c_i]\, \mu(d\lambda) | = | \int_\Theta [\int_\Omega f_i(\omega)\, \eta(\lambda, d\omega)]\, \mu(d\lambda) - c_i] \; | \leq (\varepsilon/2^k), \quad i = 1, \ldots k.$$

Since $\eta(\lambda, B)$, for $B \in F$, is, as a function of λ, measurable $\mathcal{A}$, and the f_i 's are measurable F, by extended Fubini's theorem (in the probability case) we have

$$\int_\Theta [\int_\Omega f_i(\omega)\, \eta(\lambda, d\omega)]\, \mu(d\lambda) = \int_\Omega f_i(\omega) [\int_\Theta \eta(\lambda, d\omega)\, \mu(d\lambda)] = \int_\Omega f_i(\omega)\, \mu\eta(d\omega).$$

Therefore

$$| \int_\Omega f_i(\omega)\, \mu\eta(d\omega) - c_i | \leq (\varepsilon/2^k), \qquad i = 1, \ldots k,$$

so that $\mu\eta \in A(\varepsilon, k)$; that is, the set $A(\varepsilon, k)$ is completely convex.

From now on we shall let

(C2) $f_i := 1_{A_i}$, $i = 1, 2, \ldots$ *where* $\{A_1, A_2, \ldots\}$ *is a sequence of measurable sets.*

Let us associate with each $P \in M_1(\Omega)$ the family of subsets $U(P,\varepsilon,k)$, of $M_1(\Omega)$, for $\varepsilon > 0$ and $k = 1, 2, \ldots$:

$$U(P,\varepsilon,k) := \{Q \in M_1(\Omega) : \; | \int f_i\, dP - \int f_i\, dQ) | < \varepsilon, \quad i = 1, \ldots k\} = \{Q \in M_1(\Omega) : \; | P(A_i) - Q(A_i) | < \varepsilon, \quad i = 1, \ldots k\}.$$

Let us denote by $A^o(\varepsilon, k)$ the subset of $A(\varepsilon, k)$ consisting of all the probability measures P for which $A(\varepsilon, k) \supset U(P,\varepsilon',k')$ for some $\varepsilon'>0$ and some positive integer k'.

Proposition 10. For each $\varepsilon > 0$ and $k = 1, 2, \ldots$ $U(P,\varepsilon,k)$ is a neighborhood of P in the τ_o-topology.

Proof. Let Δ be the (measurable and finite) partition of Ω generated by $\{A_1, \ldots A_k\}$, say $\Delta = \{\Omega_1, \ldots \Omega_{k'}\}$, and consider the τ_o-neighborhood $U_o(P, \Delta, \varepsilon')$ of P :

$$U_o(P, \Delta, \varepsilon') := \{Q \in M_1(\Omega) \; : | P(\Omega_i) - Q(\Omega_i) | < \varepsilon', Q(\Omega_i) = 0 \text{ if } P(\Omega_i) = 0, \quad i = 1, \ldots k'\}.$$

Each A_i, $i = 1, \ldots k$, is the union of elements of Δ, say:

$$A_i = \cup_{j=1}^{m} \Omega_{ij}, \;\; \Omega_{ij} \in \Delta ,$$

and we have

$$| P(A_i) - Q(A_i) | = | \Sigma_{j=1}^{m} P(\Omega_{ij}) - \Sigma_{j=1}^{m} Q(\Omega_{ij}) | \leq \Sigma_{j=1}^{m} | P(\Omega_{ij}) - Q(\Omega_{ij}) | \leq k'\varepsilon', \quad i = 1, \dots k.$$

Therefore, provided that $\varepsilon' < (\varepsilon/k')$,

$$U(P,\varepsilon,k) \supset U_o(P, \Delta, \varepsilon'),$$

which proves Proposition 10.

Proposition 10 has the following corollary :

Corollary 3. Let $A^\circ(\varepsilon, k)$ denote the interior of the set $A(\varepsilon, k)$ with respect to the τ_o-topology. Then $A^\circ(\varepsilon, k) \supset A^o(e, k)$.

Further we have:

Proposition 11. $A^\circ(\varepsilon_2, k) \supset A^o(\varepsilon_2, k) \supset A(\varepsilon_1, k)$ if $0 < \varepsilon_1 < \varepsilon_2$.

<u>Proof</u>. Let $P \in A(\varepsilon_1, k)$. Then

$$| P(A_i) - c_i | \leq (\varepsilon_1/2^k) , i = 1, \dots k.$$

For $Q \in U(P,\varepsilon,k)$, $\varepsilon > 0$, we have

$$| P(A_i) - Q(A_i) | < \varepsilon, \quad i = 1, \dots k.$$

Further

$$| Q(A_i) - c_i | \leq | Q(A_i) - P(A_i) | + | P(A_i) - c_i | < \varepsilon + (\varepsilon_1/2^k) , i = 1, \dots k.$$

We can choose ε such that $\varepsilon + (\varepsilon_1/2^k) < (\varepsilon_2/2^k)$. Then

$$Q \in U(P,\varepsilon,k) \Rightarrow Q \in A(\varepsilon_2, k), \text{ i.e. } A(\varepsilon_2, k) \supset U(P,\varepsilon,k).$$

Therefore, if $P \in A(\varepsilon_1, k)$ there is an $\varepsilon > 0$ and a neighborhood $U(P,\varepsilon,k)$ of P contained in $A(\varepsilon_2, k)$; that is $A^o(\varepsilon_2, k) \supset A(\varepsilon_1, k)$. This inclusion together with Corollary 3 proves Proposition 11.

Proposition 12. Let (C1) hold. then $\lim H(A^\circ(\varepsilon, k);R) = \lim H(A^o(\varepsilon, k);R) = H(Q^*;R)$ for $k \to \infty$, for any $\varepsilon > 0$.

<u>Proof</u>. It follows from Proposition 11 that

$$A(\varepsilon_2, k) \supset A^\circ(\varepsilon_2, k) \supset A^o(\varepsilon_2, k) \supset A(\varepsilon_1, k) \quad \text{if} \quad 0 < \varepsilon_1 < \varepsilon_2,$$

and therefore

$$H(A(\varepsilon_2, k);R) \leq H(A^\circ(\varepsilon_2, k);R) \leq H(A^o(\varepsilon_2, k);R) \leq H(A(\varepsilon_1, k);R) \text{ if } 0 < \varepsilon_1 < \varepsilon_2 . \quad (38)$$

Letting $k \to \infty$ and using (iv) of Proposition 8, according to which, for any $\varepsilon > 0$

$$\lim H(A(\varepsilon, k);R) = H(Q^*;R) \quad \text{for } k \to \infty ,$$

Proposition 12 is proved.

Conditional Sanov Property

Let $\mathcal{A}$ denote the subset of $M_1(\Omega)$ consisting of all probability measures P such that

$$P(B) = \Sigma_{i=1}^{k} \alpha_i 1_B(\omega_i), \ \omega_i \in \Omega, \ \alpha_i > 0, i = 1, 2, ...k; \quad \Sigma_{i=1}^{k} \alpha_i = 1$$

for every $B \in F$.

The *empirical distribution of a sample* $\omega = (\omega_1, \omega_2, ... \omega_n) \in \Omega^n$ is the probability measure $L_n(\omega, .) \in \mathcal{A}$ defined by

$$L_n(\omega, B) := \frac{1}{n} \Sigma_{i=1}^{n} 1_B(\omega_i), \quad B \in F.$$

Let $X_1, ... X_n$ be n independent identically distributed (i.i.d.) Ω-valued random variables with common distribution R . Following Csiszar [5] we regard the nth Cartesian power of (Ω, F, R) as the sample space of $X^n = (X_1, ... X_n)$ and by probabilities of events determined in terms of X^n we mean R^n-measures of subsets of Ω^n where R^n is the nth Cartesian power of R. In particular, for any subset Γ of $M_1(\Omega)$ such that $\{\omega : L_n(\omega, .) \in \Gamma\} \in F^n$ we define the probability that the empirical distribution L_n of $(X_1, ... X_n)$ belongs to Γ by $\mathbf{P}\{L_n \in \Gamma\} := R^n\{\omega : L_n(\omega, .) \in \Gamma\}$.

For any subset $A \in F^n$ of Ω^n with $R^n(A) > 0$, we designate by $R^n[\ .\ |\ A\ ]$ the conditional probability measure relative to A defined by

$$R^n[\ B\ |\ A\ ] := R^n(B \cap A)/ R^n(A) \quad \text{for all } B \in F^n.$$

From the definitions of $A(\varepsilon, k)$ and $L_n(\omega, .)$ it readily follows that $\{\omega : L_n(\omega, .) \in A(\varepsilon, k)\} \in F^n$. Further, under (C1), *for sufficiently large* n, $R^n\{\omega : L_n(\omega, .) \in A(\varepsilon, k)\} > 0$ as a consequence of the following:

Proposition 13. Let (C1) hold. $\mathbf{P}\{\ L_n \in A^\circ(\varepsilon, k)\ \} \geq (n+1)^{-k'} \exp [\ - n\ H(A^\circ(\varepsilon, k);R)\ ]$ for some positive integer k' and for sufficiently large n, and

$\liminf_{n\to\infty} (1/n) \log \mathbf{P}\{\ L_n \in A^\circ(\varepsilon, k)\ \} \geq - H(A^\circ(\varepsilon, k);R)$.

Proposition 13 is a special case of Lemma 4.1 of Csiszar [5].

For prescribed $\varepsilon > 0$, and for n such that $\mathbf{P}\{L_n \in A^\circ(e, k)\} > 0$, let us denote

$$\mathbf{P}^{(n,\varepsilon,k)}[\ .\ ] := R^n[\ .\ |\ L_n \in A(\varepsilon, k)\ ] .$$

Now we are ready to apply the basic theorem 1 of Csiszar [5]. Since according to

Proposition 9, the set $A(\varepsilon, k)$ is completely convex, and $H(A(\varepsilon, k);R) := H(Q_{\varepsilon k};R) < \infty$, in the case we consider that theorem reads:

Theorem 8. Let (C1) hold. Let $\varepsilon > 0$ and $k \in \{1, 2, ...\}$. Then, for sufficiently large n,

$$(1/n) \log \mathbf{P}\{L_n \in A(\varepsilon, k)\} \leq - H(A(\varepsilon, k);R), \text{ and} \tag{39}$$

$$(1/n) \log \mathbf{P}\{L_n \in A(\varepsilon, k)\} \leq - H(A(\varepsilon, k);R) - (1/n)\, H(\mathbf{P}^{(n,\varepsilon,k)};Q^n{}_{\varepsilon k}), \tag{40}$$

where $Q^n{}_{\varepsilon k}$ denotes the n-product of $Q_{\varepsilon k}$.

Theorem 9. Let (C1) hold. For $k \in \{1, 2, ...\}$ and $\varepsilon > 0$ except possibly for countably many ε,

$$\lim_{n\to\infty}(1/n) \log \mathbf{P}\{L_n \in A(\varepsilon, k)\} = - H(A(\varepsilon, k);R); \tag{41}$$

in other words, the set $A(\varepsilon,k)$ has the *Sanov property* with respect to R.

Proof. From (39) and Proposition 13, taking account of

$$\{\omega : L_n(\omega, .) \in A(\varepsilon, k)\} \supset \{\omega : L_n(\omega, .) \in A^\circ(\varepsilon, k)\},$$

from which

$$\mathbf{P}\{L_n \in A^\circ(\varepsilon, k)\} \leq \mathbf{P}\{L_n \in A(\varepsilon, k)\} \tag{42}$$

follows, we deduce

$$- H(A^\circ(\varepsilon, k);R) - (k'/n) \log (1+n) \leq (1/n) \log \mathbf{P}\{L_n \in A^\circ(\varepsilon, k)\} \leq \\ (1/n) \log \mathbf{P}\{L_n \in A(\varepsilon, k)\} \leq - H(A(\varepsilon, k);R), \tag{43}$$

for $\varepsilon > 0$, $k \in \{1, 2, ...\}$, for some positive integer k' and for sufficiently large n.

By letting $n\to \infty$ in (43) we obtain

$$- H(A^\circ(\varepsilon, k);R) \leq \liminf_{n\to\infty} (1/n) \log \mathbf{P}\{L_n(\omega, .) \in A(\varepsilon, k)\} \leq \\ \limsup_{n\to\infty} (1/n) \log \mathbf{P}\{L_n(\omega, .) \in A(\varepsilon, k)\} \leq - H(A(\varepsilon, k);R). \tag{44}$$

Now, it follows from (38) that, for given k, $H(A(\varepsilon,k);R)$ is a nonincreasing function of ε. Therefore, following a remark of Dawson, Gorostiza and Wakolbinger [6], this function is continuous except possibly for countably many ε. If ε_1 is a continuity point, by letting $\varepsilon_2 \to \varepsilon_1$ in (38) we arrive at

$$H(A^\circ(\varepsilon_1, k);R) = H(A(\varepsilon_1, k);R). \tag{45}$$

Therefore, if ε in relation (44) is a continuity point of $H(A(\varepsilon, k);R)$, for given k, (44) reduces to (41), which ends the proof of Theorem 9.

Theorem 10. Let (C1) hold. For $k \in \{1, 2, ...\}$ and $\varepsilon > 0$ except possibly for countably many ε,

$$\lim_{n\to\infty}(1/n)\, H(\mathbf{P}^{(n,\varepsilon,k)};Q^n{}_{\varepsilon k}) = 0; \tag{46}$$

in other words, $X_1, \dots X_n$ are *asymptotically quasi-independent under the condition* $L_n \in A(\varepsilon, k)$ *with limiting distribution* $Q_{\varepsilon k}$ (cf. Csiszar [5], p. 772).

Proof. From (iv) of Proposition 8 and Proposition 12, and by letting $k \to \infty$ in (44), we obtain

$$\lim_{k\to\infty}\liminf_{n\to\infty}(1/n) \log \mathbf{P}\{L_n(\omega, .) \in A(\varepsilon, k)\} =$$
$$= \lim_{k\to\infty}\limsup_{n\to\infty}(1/n) \log \mathbf{P}\{L_n(\omega, .) \in A(\varepsilon, k)\} = - H(Q^*;R). \tag{47}$$

Then, (40), (42) and (43) result in

$$0 \le (1/n)\, H(\mathbf{P}^{(n,\varepsilon,k)};Q^n{}_{\varepsilon k}) \le H(A^\circ(\varepsilon, k);R) - H(A(\varepsilon, k);R) + (k'/n) \log (1+n),$$

from which we deduce, by letting $n \to \infty$

$$0 \le \liminf_{n\to\infty}(1/n)\, H(\mathbf{P}^{(n,\varepsilon,k)};Q^n{}_{\varepsilon k}) \le (1/n) \limsup_{n\to\infty}(H(\mathbf{P}^{(n,\varepsilon,k)};Q^n{}_{\varepsilon k}) \le$$
$$H(A^\circ(\varepsilon, k);R) - H(A(\varepsilon, k);R). \tag{48}$$

Again, if ε is a continuity point of $H(A(\varepsilon, k);R)$, for given k, (48) reduces to (46).

Example. Let $(\Omega, F) := (C, \mathfrak{B})$, where C denotes the set of all continuous functions from [a, b] into M endowed with the usual sup norm, and $\mathfrak{B}$ is the σ-field of Borel subsets of C. We denote by X_t the coordinate function on C; that is, $X_t(x(.)) = x(t)$ for $x(.) \in C$. Let

$$E := \{P \in M_1(\Omega) : \ P[X_a \in B_i] = P_a(B_i), \quad P[X_b \in B_i] = P_b(B_i), \quad i = 1, 2, \dots\},$$

where the B_i 's are Borel sets of M. Here we have

$$U(P,\varepsilon,k) :=$$
$$\{Q \in M_1(\Omega) : \ | P[X_a \in B_i] - Q[X_a \in B_i] | < \varepsilon, \ | P[X_b \in B_i] - Q[X_b \in B_i] | < \varepsilon,$$
$$i = 1, \dots k\}. \tag{5.13}$$

$$A(\varepsilon, k) :=$$
$$\{P \in M_1(\Omega) : \ | P[X_a \in B_i] - P_a(B_i) | \le (\varepsilon/2^k), | P[X_b \in B_i] - P_b(B_i) | \le (\varepsilon/2^k),$$
$$i = 1, \dots k\}. \tag{49}$$

By the same argument as in Dawson, Gorostiza and Wakolbinger [6] one can prove that, under (C1), the Csiszar's projection $Q_{\varepsilon k}$ of R on $A(\varepsilon, k)$ coïncides with the Schrödinger's bridge $Q^*{}_{\varepsilon k}$ specified by R and the marginals $Q_{a\varepsilon k} := Q_{\varepsilon k}[X_a \in .]$ and $Q_{b\varepsilon k} := Q_{\varepsilon k}[X_b \in .]$ of $Q_{\varepsilon k}$. In fact, it suffices to show that $Q^*{}_{\varepsilon k} \in A(\varepsilon, k)$.

The remarks above and theorems 9 and 10 imply readily:

Theorem 11. Consider n i.i.d. trajectories $X_1, \dots X_n$, i.e. elements of $(C, \mathcal{B})$, with common distribution R . Then, under (C1)

(i) $X_1, \dots X_n$ are, for $n \to \infty$, asymptotically quasi-independent under the condition $L_n \in A(\varepsilon, k)$, i.e.

$$\left| \frac{1}{n} \Sigma_{i=1}^{n} \delta_{Xia}(Bj) - P_a(Bj) \right| \le (\varepsilon/2^k), \quad \left| \frac{1}{n} \Sigma_{i=1}^{n} \delta_{Xib}(Bj) - P_b(Bj) \right| \le (\varepsilon/2^k), \quad j = 1,\dots k,$$

$$\delta_{Xia}(Bj) := 1_{Bj}\,(Xia), \quad \delta_{Xib}(Bj) := 1_{Bj}\,(Xib),$$

and their limiting distribution is a Schrödinger bridge specified by R and $Q_{a\varepsilon k}$ and $Q_{b\varepsilon k}$, satisfying

$$\left| Q_{a\varepsilon k}\,(Bj) - P_a(Bj) \right| \le (\varepsilon/2^k), \left| Q_{b\varepsilon k}\,(Bj) - P_b(Bj) \right| \le (\varepsilon/2^k), \quad j = 1, \dots k;$$

(ii) the set $A(\varepsilon, k)$ defined by (49) has the Sanov property with respect to R; for k = 1, 2, ..., and for $\varepsilon > 0$, except possibly for countably many ε.

Theorem 11 is similar to Theorem 2 of Dawson, Gorostiza and Wakolbinger [6], except that they use the topology of the Prohorov distance.

REFERENCES

1. R. Aebi and M. Nagasawa, "Large Deviations and the Propagation of Chaos for Schrödinger Process (Preprint).

2. A. Blaquière, "Controllability of a Fokker-Planck Equation, the Schrödinger System, and a Related Stochastic Optimal Control (Revised Version)", in *Dynamics and Control* , vol.2, no. 3, pp. 235-253, 1992.

3. A. Blaquière and M. Sigal-Pauchard, "Stochastic Control Approach to the Control of a Forward Parabolic Equation, Reciprocal Process and Minimum Entropy", in *Analysis and Optimization of Systems : State and Frequency Domain Approaches for Infinite-Dimensional Systems* (Ed. R.F. Curtain, Honorary Eds. A. Bensoussan, J.L. Lions) Lecture Notes in Control and Information Sciences 185, Springer-Verlag: Berlin Heidelberg, pp. 476-488, 1993; Proc.10th Int. Conf. on *Analysis and Optimization of Systems*, INRIA-Sophia-Antipolis, France, 1992.

4. I. Csiszar, "*I*-Divergence Geometry of Probability Distributions and Minimization Problems" *The Annals of Probability*, vol. 3, no. 1, pp. 146-158, 1975.

5. I. Csiszar, "Sanov Property, Generalized *I*-Projection and a Conditional Limit Theorem". *The Annals of Probability*, vol. 12, no. 3, pp. 768-793, 1984.

6. D. Dawson, L. Gorostiza, and A. Wakolbinger, "Shrödinger processes and large deviations", *J. Math. Phys.*, vol 31, no. 10, pp. 2385-2388, 1990.

7. M.D. Donsker and S.R. Varadhan, "Asymptotic Evaluation of Certain Markov Process Expectations for Large Time -III", *Communications on pure and applied Mathematics*, vol. XXIX, pp. 389-461, 1976.

8. E.B. Dynkin, *Theorie des Processus Markoviens*, Dunod: Paris, 1963.

9. E.B. Dynkin, *Markov Processes*, Volume I, Springer-Verlag: Berlin, Heidelberg, 1965.

10. E.B. Dynkin, *Markov Processes*, Volume II, Springer-Verlag: Berlin, Heidelberg, 1965.

11. W. H. Fleming and R. Rishel, *Deterministic and Stochastic Optimal Control*, Springer-Verlag: Berlin, 1975.

12. H. Föllmer, "Random fields and diffusion processes", in *Ecole d'été de Saint Flour XV-XVII (1985-1987)*, Lecture Notes in Mathematics 1362, Springer-Verlag: Berlin, 1988.

13. B. Jamison, "Reciprocal Processes", *Z. Wahrscheinlichkeitstheorie ver. Gebiete* vol. 30, 65, 1974.

14. R. S. Liptser and A. N. Shiryayev, *Statistics of Random Processes I, General Theory*, Springer-Verlag: Berlin Heidelberg, 1977.

15. M. Pavon and A. Wakolbinger, "On Free Energy, Stochastic Control, and Schrödinger Processes", in Proceedings Workshop on Modeling and Control of Uncertain Systems, Birkhäuser, Boston, May 1991.

16. E. Schrödinger, *Sitzungsbericht der Preubischen Akademie, Phys. Math.* Classe, 144, 1931.

17. E. Schrödinger, "Une analogie entre la mécanique ondulatoire et quelques problèmes de probabilités en physique classique", *Annales de l'Institut Henri Poincaré*, 11, 300, 1932.

18. D. W. Stroock and S. R. S. Varadhan, Multidimensional Diffusion Processes, Springer-Verlag, Berlin Heidelberg, 1979

19. A. Wakolbinger, "A Simplified Variational Characterisation of Schrödinger Processes", *J. Math. Phys.*, 30, 2943, 1989.

20. A. Wakolbinger, "Schrödinger Bridges from 1931 to 1991", in: E. Cabana et al. (eds) *Proc. of the* 4th *Latin American Congress in Probability and Mathematical Statistics*, Mexico City 1990, Contribuciones en probabilidad y estadistica matematica **3**, pp. 61-79, 1992.

SECOND VARIATION CONDITIONS FOR THE OPTIMAL CONTROL PROBLEM WITH NORMALIZED FINAL TIME

David G. Hull[1] and Christopher N. D'Souza[2]

[1]Department of Aerospace Engineering and Engineering Mechanics
The University of Texas
Austin, TX 78712
[2]Wright Labs
Eglin AFB, FL 32542

ABSTRACT

Second variation conditions for the free final time optimal control problem formulated with a normalized final time are derived. This formulation leads to a fixed final time problem with a parameter. The formalism of the neighboring extremal path obtained by the sweep method is used to write the second variation as a perfect square and derive conditions for a minimum. The sufficient conditions are the strengthened Legendre-Clebsch condition, the conjugate point condition, and a condition associated with the parameter.

INTRODUCTION

There are at least three ways to formulate the free final time optimal control problem. The first approach is to use the time as the running variable. The second approach is to normalize the time relative to the final time thereby converting the problem to a fixed final time problem with a parameter. The third approach is to convert the parameter to a state which leads to a fixed final time problem with a free initial state. While the second formulation is considered here, it is a goal of the author to develop consistent second variation conditions for all three approaches and show that they are equivalent.

In recent times, there have been a number of studies of the second variation conditions for the free final time problem with t as the running variable using the sweep method. Wood's first derivation of second variation conditions (Ref. 4) used the artifice of creating additional final conditions so that a particular matrix has full rank. In Ref. 3, Wood rearranges the final conditions so that it is always possible to solve for the final time perturbation. Finally, in Ref. 2, Hull solves for the end point multiplier perturbation and final time perturbation simultaneously. In all cases, the

Mechanics and Control, Edited by R.S. Guttalu
Plenum Press, New York, 1994

sufficient conditions for a minimum are the strengthen Legendre-Clebsch condition and the conjugate point condition. The approaches differ in the way the conjugate point is calculated.

The most general derivation of second variation conditions for the case of normalized time is to use the approach of Ref. 2. At this writing, however, it has not been possible to do so. Instead an approach consistent with that of the fixed final time problem (Ref. 1) is used.

In Section 2, the normalized time optimal control problem is defined, and the first variation conditions are derived in Section 3. The neighboring extremal is discussed in Section 4, and it is used in Section 5 to make the second variation a perfect square. Sufficient conditions for a minimum are obtained in Section 6, and applications are discussed in Section 7. Finally, Section 8 contains the conclusions.

OPTIMAL CONTROL PROBLEM WITH NORMALIZED FINAL TIME

The optimal control problem being considered here is the free final time problem with the change in integration variable from t to $\tau = t/t_f$. Hence, the optimization problem is to find the control history $u(\tau)$ and the parameter t_f which minimize the performance index

$$J = \phi(x_f, t_f) + \int_0^1 L(\tau, x, u, t_f)d\tau \tag{1}$$

subject to the differential constraints

$$x' = f(\tau, x, u, t_f) \tag{2}$$

the initial conditions

$$\tau_0 = 0, \quad x_0 = x_{0_s} \tag{3}$$

and the final conditions

$$\tau_f = 1, \quad \psi(x_f, t_f) = 0 \tag{4}$$

where $(\ )' = d(\ )/d\tau$ and x_{0_s} denotes a specified quantity. Also, the dimensions of x, u, and ψ are $n \times 1$, $m \times 1$, and $p \times 1$, respectively. Note that this is now a fixed final time problem.

A standard approach to the derivation of second variation conditions is to consider a comparison path from a perturbed initial point τ_0, δx_0 to the final constraint manifold $\tau_f = 1$, $\delta\psi = 0$. Hence, both the first and second variations must be developed for $\delta x_0 \neq 0$.

FIRST VARIATION AND FIRST VARIATION CONDITIONS

Adjoining the constraints by Lagrange multipliers ν and $\lambda(\tau)$ leads to the augmented performance index

$$\bar{J} = G(x_f, t_f, \nu) + \int_{\tau_0}^{\tau_f} [H(\tau, x, u, t_f, \lambda) - \lambda^T x']d\tau \tag{5}$$

where

$$G = \phi + \nu^T\psi, \qquad H = L + \lambda^T f\ . \tag{6}$$

Then, taking the variation of Eq. (5), integrating by parts, and combining terms leads to

$$\begin{aligned} \delta\bar{J} \;=\; & (G_{x_f} - \lambda_f^T)\delta x_f + \left[G_{t_f} + \int_{\tau_0}^{\tau_f} H_{t_f} d\tau\right]\delta t_f + \psi^T\delta\nu \\ & + \lambda_0^T\delta x_0 + \int_{\tau_0}^{\tau_f} [(H_x + \lambda'^T)\delta x + H_u\delta u + (f - x')^T\delta\lambda]d\tau\ . \end{aligned} \tag{7}$$

By standard reasoning, the first variation (7) leads to the following equations for an extremal path ($\delta x_{0_s} = 0$)

$$\begin{aligned} x' &= f(\tau, x, u, t_f) & (8)\\ \lambda' &= -H_x^T(\tau, x, u, t_f, \lambda) & (9)\\ 0 &= H_u^T(\tau, x, u, t_f, \lambda) & (10)\\ \mu' &= -H_{t_f}(\tau, x, u, t_f, \lambda) & (11) \end{aligned}$$

and the boundary conditions

$$\tau_0 = 0, \qquad x_0 = x_{0_s}, \qquad \mu_0 = 0 \tag{12}$$

$$\tau_f = 1, \quad \psi(x_f, t_f) = 0, \quad \lambda_f = G_{x_f}^T(x_f, t_f, \nu), \quad \mu_f = G_{t_f}^T(x_f, t_f, \nu) \tag{13}$$

The actual condition for t_f is the coefficient of δt_f in Eq. (7) equal to zero. It has been converted into a differential equation and boundary conditions by defining $\mu(\tau)$ as $\mu' = -H_{t_f}$.

Note that if the problem does not explicitly involve τ, the first integral $H = \text{Const}$ holds.

NEIGHBORING EXTREMAL PATH

In the derivation of the second variation conditions, use is made of the equations for a neighboring extremal path satisfying the following boundary conditions:

$$\tau_0 = 0, \quad \delta x_0 = \delta x_{0_s}, \quad \tau_f = 1, \quad \delta\psi = 0\ . \tag{14}$$

The equations for a neighboring extremal path are obtained by solving the accessory minimum problem or taking the variation of the first variation conditions (8) through (13). The perturbed extremal conditions are given by the equations

$$\begin{aligned} \delta x' &= f_x \delta x + f_u \delta u + f_{t_f} \delta t_f & (15)\\ \delta\lambda' &= -H_{xx}\delta x - H_{xu}\delta u - f_x^T \delta\lambda - H_{xt_f}\delta t_f & (16)\\ 0 &= H_{ux}\delta x + H_{uu}\delta u + f_u^T \delta\lambda + H_{ut_f}\delta t_f & (17)\\ \delta\mu' &= -H_{t_f x}\delta x - H_{t_f u}\delta u - f_{t_f}^T \delta\lambda - H_{t_f t_f}\delta t_f\ , & (18) \end{aligned}$$

the initial conditions at $\tau_0 = 0$

$$\begin{aligned} \delta x_0 &= \delta x_{0_s} & (19)\\ \delta\mu_0 &= 0\ , & (20) \end{aligned}$$

and the final conditions at $\tau_f = 1$

$$\begin{aligned} \delta\psi &= \psi_{x_f}\delta x_f + \psi_{t_f}\delta t_f = 0 & (21)\\ \delta\lambda_f &= G_{x_f x_f}\delta x_f + \psi_{x_f}\delta\nu + G_{x_f t_f}\delta t_f & (22)\\ \delta\mu_f &= G_{t_f x_f}\delta x_f + \psi_{t_f}^T \delta\nu + G_{t_f t_f}\delta t_f\ . & (23) \end{aligned}$$

To solve these equations, the strengthened Legendre-Clebsch condition $H_{uu} > 0$ is assumed to hold. Then, Eq. (17) can be solved for δu as

$$\delta u = -H_{uu}^{-1}(H_{ux}\delta x + f_u^T \delta\lambda + H_{ut_f}\delta t_f) \tag{24}$$

and eliminated from Eqs. (15), (16), and (18). This process leads to the differential equations

$$
\begin{aligned}
\delta x' &= A\delta x - B\delta\lambda + D\delta t_f && (25)\\
\delta\lambda' &= -C\delta x - A^T\delta\lambda - E\delta t_f && (26)\\
\delta\mu' &= -E^T\delta x - D^T\delta\lambda - F\delta t_f && (27)
\end{aligned}
$$

where

$$
\begin{aligned}
A &= f_x - f_u H_{uu}^{-1} H_{ux} && (28)\\
B &= f_u H_{uu}^{-1} f_u^T && (29)\\
C &= H_{xx} - H_{xu} H_{uu}^{-1} H_{ux} && (30)\\
D &= f_{t_f} - f_u H_{uu}^{-1} H_{ut_f} && (31)\\
E &= H_{xt_f} - H_{xu} H_{uu}^{-1} H_{ut_f} && (32)\\
F &= H_{t_f t_f} - H_{t_f u} H_{uu}^{-1} H_{ut_f} && (33)
\end{aligned}
$$

Eqs. (25) through (27) and the boundary conditions (21) through (23) constitute a linear, two-point, boundary-value problem which can formally be solved by the sweep method. Here, a solution of the form of the boundary conditions, that is,

$$
\begin{aligned}
\delta\lambda &= S\delta x + R\delta\nu + m\delta t_f && (34)\\
\delta\psi &= R^T\delta x + Q\delta\nu + n\delta t_f = 0 && (35)\\
\delta\mu &= m^T\delta x + n^T\delta\nu + \alpha\delta t_f && (36)
\end{aligned}
$$

is assumed. The differential equations for the functions S, R, Q, m, n, and α are obtained by differentiating Eqs. (34) through (36) and combining them with Eqs. (25) through (27). The final conditions are obtained from Eqs. (21) through (23). This process leads to the following equations:

$$
\begin{aligned}
S' &= SBS - A^TS - SA - C && S_f = G_{x_f x_f} && (37)\\
R' &= (SB - A^T)R && R_f = \psi_{x_f}^T && (38)\\
Q' &= R^TBR && Q_f = 0 && (39)\\
m' &= (SB - A^T)m - SD - E && m_f = G_{x_f t_f} && (40)\\
n' &= R^TBm - R^TD && n_f = \psi_{t_f} && (41)\\
\alpha' &= m^TBm - D^Tm - m^TD - F && \alpha_f = G_{t_f t_f} && (42)
\end{aligned}
$$

which can be solved upon problem definition.

If the approach of the fixed final time problem is followed, Eq. (35) is solved at t_0 for $\delta\nu$ as

$$\delta\nu = -Q_0^{-1}R_0^T\delta x_0 - Q_0^{-1}n_0\delta t_f\ . \qquad (43)$$

Then, substitution into Eq. (36) evaluated at t_0 where $\delta\mu_0 = 0$ leads to

$$\delta t_f = -\frac{(m_0 - R_0Q_0^{-1}n_0)^T}{\alpha_0 - n_0^TQ_0^{-1}n_0}\delta x_0\ . \qquad (44)$$

Finally, $\delta\lambda_0$ is obtained from Eq. (34) as

$$\delta\lambda_0 = S_0\delta x_0 + R_0\delta\nu + m_0\delta t_f\ . \qquad (45)$$

Given δx_0, δt_f of the neighboring extremal path is obtained from Eq. (44); $\delta\nu$ is obtained from Eq. (43); and $\delta\lambda_0$, from Eq. (45). Then, given δx_0 and $\delta\lambda_0$, $\delta x(\tau)$ and

$\delta\lambda(\tau)$ follow from the integration of Eqs. (25) and (26). The control perturbation δu follows from Eq. (24).

Another form of the control perturbation is obtained by combining Eqs. (24) and (34), that is,

$$\delta u = -H_{uu}^{-1}[(H_{ux} + f_u^T S)\delta x + f_u^T R\delta\nu + (f_u^T m + H_{ut_f})\delta t_f] \,. \tag{46}$$

SECOND VARIATION

The second variation is obtained by taking the variation of the transpose of the first variation and is given by

$$\begin{aligned} \delta^2 \bar{J} &= [\delta x_f^T \delta t_f] \begin{bmatrix} G_{x_f x_f} & G_{x_f t_f} \\ G_{t_f x_f} & G_{t_f t_f} \end{bmatrix} \begin{bmatrix} \delta x_f \\ \delta t_f \end{bmatrix} \\ &+ \int_{\tau_0}^{\tau_f} [\delta x^T \delta u^T \delta t_f] \begin{bmatrix} H_{xx} & H_{xu} & H_{xt_f} \\ H_{ux} & H_{uu} & H_{ut_f} \\ H_{t_f x} & H_{t_f u} & H_{t_f t_f} \end{bmatrix} \begin{bmatrix} \delta x \\ \delta u \\ \delta t_f \end{bmatrix} d\tau \,. \end{aligned} \tag{47}$$

By considering a particular variation, it is possible to show that the Legendre-Clebsch necessary condition $H_{uu} \geq 0$ must hold. To derive additional second variation conditions for a minimum, it is assumed that $H_{uu} > 0$, and the second variation is written as a perfect square using the forms of the neighboring extremal control perturbation.

The approach followed here is to examine the term $\delta P^T H_{uu} \delta P$ where δP has the form of Eq. (46), that is

$$\delta P = H_{uu}^{-1}[(H_{ux} + f_u^T S)\delta x + f_u^T R\delta\nu + (H_{ut_f} + f_u^T m)\delta t_f] + \delta u \,. \tag{48}$$

For the time being, $\delta\nu$ is arbitrary, and δx, δu and δt_f must satisfy only the following equations:

$$\begin{aligned} &\delta x' = f_x \delta x + f_u \delta u + f_{t_f} \delta t_f \\ &\tau_0 = \tau_{0_s}, \qquad \delta x_0 = \delta x_{0_s} \\ &\delta\psi = \psi_{x_f} \delta_{x_f} + \psi_{t_f} \delta t_f = 0 \end{aligned} \tag{49}$$

For δP to represent the neighboring extremal control perturbation, $\delta\nu$ and δt_f must satisfy Eqs. (43) and (44).

With some manipulation, it can now be shown that

$$\begin{aligned} \delta P^T H_{uu} \delta P &= [\delta x^T \delta u^T \delta t_f] \begin{bmatrix} H_{xx} & H_{xu} & H_{xt_f} \\ H_{ux} & H_{uu} & H_{ut_f} \\ H_{t_f x} & H_{t_f u} & H_{t_f t_f} \end{bmatrix} \begin{bmatrix} \delta x \\ \delta u \\ \delta t_f \end{bmatrix} \\ &+ \frac{d}{d\tau}[\delta x^T S \delta x + \delta x^T m \delta t_f + \delta t_f m^T \delta x + \delta t_f \alpha \delta t_f \\ &+ \delta x^T R \delta\nu + \delta t_f n^T \delta\nu + \delta\nu^T R^T \delta x \\ &+ \delta\nu^T n \delta t_f + \delta\nu^T Q \delta\nu] \end{aligned} \tag{50}$$

so that the second variation (47) becomes

$$\begin{aligned} \delta^2 \bar{J} &= \delta x_0^T S_0 \delta x_0 + \delta x_0^T m_0 \delta t_f + \delta t_f m_0^T \delta x_0 \\ &+ \delta t_f \alpha_0 \delta t_f + \delta x_0^T R_0 \delta\nu + \delta t_f n_0^T \delta\nu \\ &+ \delta\nu^T R_0^T \delta x_0 + \delta\nu^T n_0 \delta t_f + \delta\nu^T Q_0 \delta\nu \\ &+ \int_{\tau_0}^{\tau_f} \delta P^T H_{uu} \delta P d\tau \end{aligned} \tag{51}$$

where all of the final point terms have cancelled.

At this point, $\delta\nu$ is chosen to have the value (43) of the neighboring extremal so that second variation can be rewritten as

$$\begin{aligned}\delta^2 \bar{J} &= \delta x_0^T \bar{S}_0 \delta x_0 + \delta x_0^T \bar{m}_0 \delta t_f + \delta t_f \bar{m}_0^T \delta x_0 \\ &\quad + \delta t_f \bar{\alpha}_0 \delta t_f + \int_{\tau_0}^{\tau_f} \delta P^T H_{uu} \delta P d\tau\end{aligned} \tag{52}$$

where

$$\bar{S} = S - RQ^{-1}R^T, \qquad \bar{m} = m - RQ^{-1}n, \qquad \bar{\alpha} = \alpha - n^T Q^{-1} n \tag{53}$$

and where

$$\begin{aligned}\delta P &= H_{uu}^{-1}[(H_{ux} + f_u^T S)\delta x - f_u^T R Q_0^{-1} R_0^T \delta x_0 \\ &\quad + (H_{ut_f} + f_u^T m - f_u^T R Q_0^{-1} n_0)\delta t_f] + \delta u\end{aligned} \tag{54}$$

Eq. (50) seems to require that all the sweep variables be differentiable over (τ_0, τ_f). However, if δt_f is set equal to zero and $\delta\nu = -QR\delta x$, the differential terms reduce to

$$\frac{d}{d\tau}[\delta x^T (S - RQ^{-1}R^T)\delta x] \tag{55}$$

and only requires that $\bar{S}$ be differentiable.

SECOND VARIATION CONDITIONS

Consider the class of admissible comparison paths for which $\delta t_f = 0$. Here, the second variation reduces to that for the fixed final time problem for which

$$\bar{S} \text{ finite over } (\tau_0, \tau_f) \tag{56}$$

is a necessary condition for a minimum (Ref. 1), that is, $\delta^2 \bar{J} \geq 0$. If the conjugate point τ_{cp} is the time when $\bar{S}$ becomes infinite, another way to state the condition (56) is that there should be no conjugate point, or if a conjugate point exists, τ_{cp} should be $\leq \tau_0$.

Now, assume that

$$\bar{S} \text{ finite over } [\tau_0, \tau_f)\ , \tag{57}$$

and consider the case where $\delta t_f \neq 0$. For $\delta x_0 = 0$, the second variation (52) reduces to

$$\delta^2 \bar{J} = \delta t_f \bar{\alpha}_0 \delta t_f + \int_{\tau_0}^{\tau_f} \delta P^T H_{uu} \delta P d\tau \tag{58}$$

where

$$\begin{aligned}\delta P &= H_{uu}^{-1}[(H_{ux} + f_u^T S)\delta x + \\ &\quad (H_{ut_f} + f_u^T m - f_u^T R Q_0^{-1} n_0)\delta t_f] + \delta u\end{aligned} \tag{59}$$

An additional necessary condition can be obtained if it can be shown that admissible comparison paths exist for which $\delta P = 0$, that is, for which

$$\begin{aligned}\delta u &= -H_{uu}^{-1}[(H_{ux} + f_u^T S)\delta x + \\ &\quad (H_{ut_f} + f_u^T m - f_u^T R Q_0^{-1} n_0)\delta t_f]\ .\end{aligned} \tag{60}$$

Substitution into the differential equation (49) and use of Eqs. (28) through (33) leads to

$$\delta x' = -(BS - A)\delta x - BR\delta\nu - (Bm - D)\delta t_f\ . \tag{61}$$

Then, multiplication by R^T, use of the differential equations (38), (39) and (41), and integration yields

$$R^T \delta x + Q\delta\nu + n\delta t_f = \text{Const.} \tag{62}$$

Evaluated at the final point, it is seen that Const $= 0$. The result is satisfied at the initial point by the definition (43) of $\delta\nu$. Hence, admissible comparison paths exist for which $\delta P = 0$.

For these paths, the second variation reduces to

$$\delta^2 \bar{J} = \delta t_f \bar{\alpha}_0 \delta t_f \tag{63}$$

and requires that

$$\bar{\alpha}_0 \geq 0 \tag{64}$$

for a minimum ($\delta^2 J' \geq 0$).

For arbitrary admissible comparison paths ($\delta P \neq 0$), Eq. (58) shows that

$$H_{uu} > 0 \tag{65}$$

$$\bar{S} \text{ finite over } [\tau_0, \tau_f) \tag{66}$$

$$\bar{\alpha} > 0 \tag{67}$$

are sufficient conditions for a minimum.

REMARKS

The second variation conditions developed here have been applied to the problem of minimizing the distance from the origin to the parabola $x_f^2 + t_f - 1 = 0$ considered in Ref. 2 and solved with t as the variable of integration. Two extremal solutions have been shown to exist: a maximum characterized by $u = 0$ and a minimum characterized by $u = \sqrt{2}$. For both solutions, $H_{uu} > 0$, but the $u = 0$ solution has a conjugate point at $t = 1/2$, making it not a minimum. Condition (56) cannot be applied to the $u = 0$ solution because $Q = 0$ and Q^{-1} does not exist. Hence, sufficient conditions must be developed for the particular case. On the other hand, conditions (65) through (67) are satisfied for the $u = \sqrt{2}$ solution.

Similarly, the results developed here cannot be applied to Zermelo's problem (Ref. 2). Q^{-1} does not exist for the case where the final abscissa is given ($Q = 0$) or the case where both the final abscissa and ordinate are given ($\det Q = 0$). As a consequence, continued effort should be made to develop the second variation conditions for normalized time without having to invert Q.

DISCUSSION AND CONCLUSIONS

Second variation conditions have been developed for the optimal control problem with normalized final time. This optimal control problem is a fixed final time problem with a parameter. The approach followed is to use neighboring extremal results obtained by a sweep method to write the second variation in a perfect square form from necessary condition and sufficient conditions are readily obtained. Sufficient conditions for a minimum are that the strengthened Legendre-Clebsch condition, the conjugate point condition, and the parameter condition be satisfied.

The procedure which has been used to develop the second variation conditions involves a matrix inverse that does not exist for all problems. Additional effort should be made to obtain second variation conditions which do not involve this inverse.

REFERENCES

1. Bryson, A. E. Jr., and Ho, Y. C., *Applied Optimal Control*, Hemisphere Publishing Corporation, New York, 1975.

2. Hull, D. G., "Sufficient Conditions for a Minimum of the Free-Final-Time Optimal Control Problem," *Journal of Optimization Theory and Applications*, Vol. 68, No. 2, February, 1991.

3. Wood, L. J., "Sufficient Conditions for a Local Minimum of the Bolza Problem with Multiple Terminal Point Constraints," AAS/AIAA Astrodynamics Specialist Conference, Durango, Colorado, August, 1991.

4. Wood, L. J., and Bryson, A. E. Jr., "Second-Order Optimality Conditions for Variable End Time Terminal Control Problems," *AIAA Journal*, Vol. 11, No. 9, September, 1973, pp. 1241–1246.

IDENTIFICATION OF THE OPTIMAL LYAPUNOV FUNCTION FOR SYSTEMS WITH STRUCTURED UNCERTAINTIES

Andrew Olas and Ela Olas

Olas & Associates
1135 NW Circle
Corvallis, OR 97330

ABSTRACT

The problem of determination of the optimal Lyapunov function is modified and formulated as the problem of an identification of such a function. The objects considered are systems which are nominally linear but subject to nonlinear, time-varying structured uncertainties.

In the paper the identification problem of the optimal Lyapunov function from the class of quadratic forms is solved. Under some weak conditions the simple and effective identification procedure is formulated. The procedure requires only such numerical routines as inverting positive-definite symmetric matrices and determining the eigenvalues and eigenvectors of symmetric matrices.

It is expected that optimal Lyapunov functions may be utilized in a design of an intelligent robust linear quadratic controller. The examples demonstrate the effectiveness of the method. As shown when considering a 24-dimensional system, the method is effective for large scale systems.

INTRODUCTION

Uncertain systems are recently a subject of a significant scientific interest. The interest is prompted by increasing industry need for properly designed robust controllers for multivariable linear systems. Lyapunov's direct method is widely used in robust stability problems due to the fact that it can easily accommodate nonlinear and time-varying perturbations. The objects discussed at the present paper are systems which are nominally linear and uncertainties are structured ones. The author attempts to formalize the process of identification - an important step in the design procedure of linear state variable feedback controller for above defined systems with structured uncertainties.

Presently there are two Lyapunov-based methods of designing such controllers. The first, used by Siljak in [1] and Singh and Coelho in [2], is to design a controller for a nominal system, to obtain a closed loop equation and then to test the uncertainties bounds for robust stability of the latter. There is an extensive literature on the determination of uncertainties bounds. The first to discuss the procedures for determining the bounds were Siljak in [3] and Patel et al. in [4]. Yedavalli and Liang in [5] using transformations obtained improvements in bounds estimation. Results of [4] and [5] were based on the selection of some quadratic form as the Lyapunov function. As it is known this approach

Mechanics and Control, Edited by R.S. Guttalu
Plenum Press, New York, 1994

has its limitations; Becker and Grimm in [6] proved that applying the small gain theorem provides the stability bound which cannot be improved by state transformations used in [5].

Radziszewski in [7] examining two-dimensional structurally-perturbed system discussed quadratic forms as the class of Lyapunov function candidates and determined the best Lyapunov function in this class. He found that robustness bounds obtained using this function were worse than those obtained by other methods, applicable for two-dimensional system.

Siljak in [1] shows that the bound estimation strongly depends on selection of the system state space. Further, he suggests use of vector Lyapunov functions to reduce conservatism of robustness estimation and provides an example of application. In Olas' paper [8] piecewise quadratic Lyapunov functions are proposed to improve the bound estimation.

The concept of optimal Lyapunov function was introduced in Olas' paper [9], and global optimality was proved in Olas and Ahmadkhanlou paper [10]. Under some weak assumptions the last paper gives the complete solution to the problem of determination of the largest bounds attainable by using the Lyapunov approach with a quadratic form selected as a Lyapunov function candidate. The procedure developed may give better bounds estimates than these obtained by small gain theorem, see example by Ahmadkhanlou in [14].

The drawback of the first method of controller design is that the controller is designed by trial and error, without any accounting for uncertainties. When the closed loop system obtained is unstable, or when it is impossible to prove its robust stability, it is necessary to repeat the controller design and test for robustness again.

The second direct method of controller design is well described and referenced by Chen and Chen in [11]. The controller is designed by determining the gain matrix K such that some Lyapunov function ensures robust stability of the system. Contrary to the previous method, the controller obtained guarantees robust stability. The drawback of this method is that the Lyapunov function determined at the procedure is not an optimal one. Thus the controller designed may be a conservative one. For some applications, numerical results show that the design is still conservative, see Olas and Ahmadkhanlou [10].

The rationale for a present paper is as follows. For a given system it is possible to introduce the concept of the performance measure of Lyapunov function, i.e. the number λ such that the inequality $\dot{V} \leq \lambda V$ is fulfilled. If $\lambda < 0$ the Lyapunov function time constant $\tau = -1/\lambda$ defines the attenuation of the function. It is possible to define the optimal Lyapunov function as the one having the best performance measure. It may be expected that if such a function is used for a controller design, then the resulting controller gain matrix will be the one with the smallest in some sense gains. Moreover, the controller designed may be tested for its conservatism - by determining the Lyapunov function time constant.

The paper deals with the first part of the above assignment - the problem of designing the optimal Lyapunov function for the nominally linear systems with structured uncertainty. While in the previous paper [13] the optimal Lyapunov function was designed on the basis of the fully-described uncertain system (i.e. with given nominal and uncertainty matrices as well as the uncertainties bounds) here we consider the case when the knowledge about the system is limited to the following two facts:

- the dimension of the system is known,
- it is known that the nominal system is LTI and that it is subject to structured uncertainties.

The paper shows that even in the case of so severely limited information the optimal Lyapunov function may be designed. Examples demonstrate the procedure of identification of the optimal Lyapunov function design.

OPTIMAL LYAPUNOV FUNCTION, PROBLEM STATEMENT

Consider a nominally linear system with structured perturbation

$$\dot{x} = A_N x + \sum_{j=1}^{q} p_j A_j x \tag{1}$$

with $p_j = p_j(x,t)$, where p_j is assumed to fulfill conditions on existence and uniqueness of solutions to (1). Let $\mathbf{p} \triangleq [p_1,...,p_q]^T$. Let the bounds $\mathbf{p}^- = [p_1^-,...,p_q^-]^T$, $\mathbf{p}^+ = [p_1^+,...,p_q^+]^T$ imposed on p(x,t) be given, thus $\mathbf{p}^- \leq \mathbf{p}(x,t) \leq \mathbf{p}^+$. Consider the class of quadratic forms $V(x) = x^T S x$, with constant, symmetric, positive definite matrices S as the class of Lyapunov function candidates. Determine the derivative of V along solutions of (1) and introduce a function

$$\Lambda(x,t) = \frac{\dot{V}(t,x)}{V(x)}, \qquad \|x\| \neq 0$$

Define the performance measure λ of the function V as an upper bound of the function $\Lambda(t,x)$

$$\lambda = \sup_{x \in R^n \setminus 0} \sup_{t \in [0,\infty)} \Lambda(x,t) \tag{2}$$

We say that a function V is better than V^* if the performance measure λ of the function V is smaller than the performance measure λ^* of V^*.

Since each quadratic form may be associated with its symmetric matrix we may introduce a space of quadratic Lyapunov functions and define the distance d between two functions $V_1 = x^T S_1 x$ and $V_2 = x^T S_2 x$ by a norm of a difference between matrices S_1 and S_2

$$d = \|S_1 - S_2\|$$

We call the function V optimal if all the neighboring functions at less than some distance d from V are not better than V.

Finally, we call the function V globally optimal if the entire space of functions does not contain the better one.

We define the problem of designing the Lyapunov function for (1) as the assignment to determine the globally optimal function V(x).

Remark. The most common procedure to determine the Lyapunov function is to solve Lyapunov equation. Examples show that such functions are usually bad choices, and optimization often allows to prove the system stability.

We denote

$$\sum_{j=1}^{q} p_j A_j = A_U(p)$$

Let Π denote an uncertainty parallelepiped (box) in R^q

$$\Pi = \{ p \in R^q : p^- \leq p \leq p^+ \}$$

with $p^{(1)}, p^{(2)}, ...$ being the 2^q vertices of Π. Consider the function

$$V(x) = x^T S x$$

where S is a symmetric, positive-definite matrix, as a Lyapunov function candidate.

The derivative $\dot{V}(t,x)$ of this function along the solutions of (1) is a quadratic form of x and linearly depends on parameters $p_i(t,x)$. Thus when determining the performance measure of the function V the 2^q quadratic forms

$$QF_j \triangleq x^T \left(A_N^T S + S A_N + A_U^T(p^{(j)}) S + S A_U(p^{(j)}) \right) x, \;\; j = 1,...,2^q \tag{3}$$

generated by 2^q vertices $p^{(1)}, p^{(2)}, ...$ of the parallelepiped Π are considered. In particular, if the forms QF_j are non-positive for all x then the system is Lyapunov stable for all uncertainties remaining inside assigned bounds.

SOME DEFINITIONS AND LEMMA

A convenient way of analyzing properties of the function $\Lambda(t,x)$ for $x \neq 0$ is to consider it on the unit sphere Φ_1 in R^n. The relation

$$\max_{x \in \Phi_1} \Lambda(x,t) = \sup_{x \in R^n \setminus 0} \Lambda(x,t)$$

shows that the domain of the argument x may be restricted to Φ_1. To find maximum of the function $\Lambda(x,t)$ we consider the functions QF_j/V, $j = 1,...,2^q$. The following relation

$$\max_{x \in \Phi_1} \Lambda(x,t) = \max_{j=1,...2^q} \max_{x \in \Phi_1} \frac{QF_j(x)}{V(x)}$$

holds.

Denoting the matrix of the form QF_j by R_j we are looking for the maximum of the expression $x^T R_j x / x^T S x$ on the unit sphere. Substituting $y = S^{.5}x$ we get $y^T S^{-.5} R_j S^{-.5} y / y^T y$. The maximum value of this form is equal to the largest eigenvalue Λ_{1j} of the matrix $S^{-.5} R_j S^{-.5}$. The corresponding eigenvector $x = \zeta_j$ fulfills the relation

$$\zeta_j (R_j - \Lambda_{1j} S) \zeta_j = 0$$

Denote by Λ_1 the maximal value of Λ_{1j}, $j = 1,...,2^q$. In general case more than one Λ_{1j} may attain the value Λ_1. Further, we accept the following restriction on the forms QFj.

Assumption A. There is k eigenvalues which attain the value Λ_1, such that the corresponding eigenvectors ζ_j are mutually different.

The Assumption A may be removed at the cost of more complex calculations. For the sake of simplicity we assume that it is the first k forms which have eigenvalues attaining the value Λ_1.

The following Lemma discusses the extremal properties of continuous functions defined on the compact set.

Lemma. Let Φ_1 be a compact subset of R^n, let f be a continuous function defined on Φ_1. Let maximum value M of f be attained at the subset Ψ of Φ_1. Let g be a continuous function and $g(x) < 0$ for $x \in \Psi$. Then for any sufficiently small ϵ we have

$$f(x) + \epsilon\, g(x) < M \quad \text{for } x \in \Phi_1$$

Proof. Since $g(x) < 0$ for $x \in \Psi$ then there exists a neighborhood $N(\Psi)$ of Ψ such that $g(x) < 0$ at this neighborhood. Then for any $x \in N(\Psi)$ and any ϵ we have

$$f(x) + \epsilon\, g(x) < 0 \quad \text{for } x \in N(\Psi)$$

Denote by $M_1 < f(\zeta)$ the upper bound of f(x) on $\Phi_1 \setminus N(\Psi)$ and by $M_2 > 0$ the upper bound of g(x) on $\Phi_1 \setminus N(\Psi)$ (for $M_2 < 0$ the problem is trivial). Select $\epsilon = 0.5(M - M_1)/M_2)$. The proof ends by estimating the upper bound on $\Phi_1 \setminus N(\Psi)$

$$f(x) + \epsilon\, g(x) < M_1 + 0.5\left(\frac{M - M_1}{M_2}\right) M_2 = \frac{1}{2} M_1 + \frac{1}{2} M < M$$

CONSTRUCTION OF OPTIMAL LYAPUNOV FUNCTION

Let $V = x^T S x$ be given, and 2^q quadratic forms QF_j be determined, as described previously. Consider arbitrary form $\Delta V = x^T \Delta S x$, and let ΔQF_j, $j = 1,...,2^q$ denote the 2^q forms resulting from entering the vertices $p^{(j)}$ into the derivative of ΔV along solutions of (1). We form the function $V^* = V + \epsilon\, \Delta V$. For small ϵ we have

$$\Lambda^*(t,x) = \frac{\dot{V}(t,x) + \epsilon\Delta\dot{V}(t,x)}{V(x) + \epsilon\Delta V(x)} = \frac{\dot{V}(t,x)}{V(x)} + \epsilon\frac{\Delta V(x)}{V(x)}\left(\frac{\Delta\dot{V}(t,x)}{\Delta V(x)} - \frac{\dot{V}(t,x)}{V(x)}\right)$$

and we estimate

$$\lambda^* = \sup_{t\in[0,\infty),x\in\Phi_1} \Lambda^*(t,x) \le \max_{j=1,\dots,2^q,\, x\in\Phi_1}\left[\frac{QF_j(x)}{V(x)} + \epsilon\left(\frac{\Delta QF_j(x)}{V(x)} - \frac{QF_j(x)}{V(x)}\frac{\Delta V(x)}{V(x)}\right)\right]$$

The first term of the right hand side of the expression on Λ^* is a continuous function having k isolated maximal points ζ_i. By virtue of the Lemma we state that if

$$\frac{\Delta QF_j(\zeta_j)}{V(\zeta_j)} - \frac{QF_j(\zeta_j)}{V(\zeta_j)}\frac{\Delta V(\zeta_j)}{V(\zeta_j)} < 0 \quad j = 1,\dots,k \tag{4}$$

then

$$\lambda^* < \lambda \tag{5}$$

Invoking the previous estimation of the second term of the left hand of (4) we rewrite the condition (4)

$$\Delta QF_j(\zeta_j) - \Lambda_1\,\Delta V(\zeta_j) < 0 \qquad j=1,\dots,k \tag{6}$$

which implies (5). The left side of the above expression is a linear form on the elements of the matrix ΔS. Applying the substitution

$$\mathbf{u} = \begin{bmatrix} u_1 \\ u_2 \\ u_3 \\ \cdot \\ u_{n(n+1)/2} \end{bmatrix} \triangleq \begin{bmatrix} \Delta s_{11} \\ \Delta s_{12} \\ \Delta s_{22} \\ \cdot \\ \Delta s_{nn} \end{bmatrix} \tag{7}$$

and introducing a matrix B of coefficients multiplying the components of the vector u we obtain the system

$$\mathbf{B\,u} < 0$$

In conclusion of this Section we quote the Theorem formulated in [13]:

Theorem 1. Let be a system (1), let V be Lyapunov quadratic function and assumption A be satisfied. If there exists a solution to the system of inequalities

$$\mathbf{B\,u} < 0 \tag{8}$$

then for sufficiently small ϵ the Lyapunov function $V + \epsilon\,\Delta V$ is better than the function V.

OPTIMIZATION

A simple technique of optimization of the Lyapunov function may be based on the proved Theorem 1. Following are the principal stages of the algorithm.

1. Either solve Lyapunov equation or put S = I to define the zero-approximation Lyapunov function matrix.

2. Form the 2^q quadratic forms QF_i. For each form determine the largest eigenvalue Λ_{1i}. Determine Λ_1.
3. Select these k forms QF_i which have eigenvalues Λ_{1i} differing from Λ_1 by less than assigned $\delta < 0$. For each such Λ_{1i} determine the eigenvector ζ_i. Form the matrix B.
4. Solve the inequality (8), determine the corresponding ΔS. If (8) has no solution go to step 6.
5. Create a new Lyapunov function with the matrix $S + \varepsilon\Delta S$, where ε is a small positive constant. Go to step 2.
6. The function is optimal with regard to assumed algorithm. End of the algorithm.

EXAMPLES OF CONSTRUCTION OF OPTIMAL FUNCTION

Example 1. Two-Dimensional System. Radziszewski in [7] considered the robust stability problem for the system

$$\dot{x} = A_N x + p_1(x,t)A_1 x, \qquad x \in R^2 \tag{9}$$

where $p_1(x,t)$ is a scalar function, and

$$A_N = \begin{bmatrix} 0 & 1 \\ -1 & -1 \end{bmatrix}, \quad A_1 = \begin{bmatrix} 0 & 0 \\ -1 & 0 \end{bmatrix}$$

Using the Lyapunov quadratic function

$$V(x) = x_1^2 + x_1 x_2 + x_2^2$$

Radziszewski found that the system is stable if $p_1(x,t)$ remains inside of the bounds $p_1^- = -\sqrt{3}/2$, $p_1^+ = \sqrt{3}/2$.

To carry out the presented procedure we analyze the forms QF_j, $j = 1,2$ which are associated with the vertices $p^{(1)} = p_1^-$, $p^{(2)} = p_1^+$. The forms QF_j have, correspondingly, roots $\zeta_1 = [0.939, 0.344]^T$, $\zeta_2 = [0.5907, -0.8069]^T$. Introducing the function $\Delta V = x^T \Delta S\, x$ we obtain its derivative along solutions of (9)

$$\Delta\dot{V} = \Delta s_{11}(2x_1x_2) + \Delta s_{12}\left(-2x_1^2 - 2p_1(x,t)x_1^2 - 2x_1x_2 + 2x_2^2\right) + \Delta s_{22}\left(-2x_1x_2 - 2p_1(x,t)x_1x_2 - 2x_2^2\right)$$

The forms $\Delta QF_j(x)$, $j = 1,2$ are obtained by entering the vertex $p^{(j)}$ into the above expression. Finally, the components $(\mathbf{Bu})_1$, $(\mathbf{Bu})_2$ of the vector $\mathbf{Bu}$ being the left-hand side of (8) are of the form

$$(\mathbf{Bu})_1 = .646\,\Delta s_{11} - .64562\,\Delta s_{12} - .32322\,\Delta s_{22}$$

$$(\mathbf{Bu})_2 = -.953272\,\Delta s_{11} + .9532355\,\Delta s_{12} + .47665\,\Delta s_{22}$$

After normalizing the matrix **B** takes the form

$$\mathbf{B} = \begin{bmatrix} 1.000 & -1.000 & -.5000 \\ -1.000 & 1.000 & .5000 \end{bmatrix} \tag{10}$$

It is easy to check that the vector **Bu**, where **B** is given by (10) and u is an arbitrary u cannot fulfill the condition

$$\mathbf{Bu} < 0$$

and the Lyapunov function can not be improved.

Example 2. Two-Dimensional System. The same system (9) is considered. We determine the optimal function for the bounds

$$p_1^- = -.999\,, \quad p_1^+ = .999 \tag{11}$$

At the first step, taking S = I the performance measure is equal to .413506. Subsequent iterations bring it to the value .172180 and any further significant improvement is impossible. The corresponding matrix S is

$$S = \begin{bmatrix} .9438 & .3921 \\ .3921 & 1.0432 \end{bmatrix}$$

As it is known, [8], it is possible to construct a Lyapunov function, which implies stability of the system (9) with bounds (11). Such a function does not belong, however, to the class of quadratic forms (it is a piecewise quadratic). The example shows that in some cases quadratic forms are not an efficient tool in stability investigation.

The next example deals with the design of a robust controller to control the dynamics of a helicopter in a vertical plane. In the paper [12] by Narendra and Tripathi the dynamics is described in the state variables x_1 - horizontal velocity, x_2 - vertical velocity, x_3 - pitch rate, x_4 - pitch angle by the following system

$$\dot{x} = A\,x + B\,u$$

$$A = \begin{bmatrix} -.0366 & .0271 & .0188 & -.4555 \\ .0482 & -1.0100 & .0024 & -4.0208 \\ .1002 & .2855 & -.7070 & 1.3299 \\ 0 & 0 & 1 & 0 \end{bmatrix} \tag{12}$$

$$\Delta A = \begin{bmatrix} 0 & 0 & 0 & 0 \\ 0 & 0 & 0 & 0 \\ 0 & p_1 & 0 & p_2 \\ 0 & 0 & 0 & 0 \end{bmatrix} \qquad B = \begin{bmatrix} .4422 & .1761 \\ 3.04475 & -7.5922 \\ -5.5200 & 4.9900 \\ 0 & 0 \end{bmatrix} \qquad \Delta B = \begin{bmatrix} 0 & 0 \\ p_3 & 0 \\ 0 & 0 \\ 0 & 0 \end{bmatrix}$$

where the control inputs are given by u_1 - collective and u_2 - longitudinal cyclic. For the assigned range of air speed a significant change is observed in the values of some elements of matrices A and B. Three parameters appearing in matrices A and B are considered as varying and their bounds are specified as follows (all matrices are shown after symmetrizing the bounds)

$$|p_1| \le 0.2192 \qquad |p_2| \le 1.2031 \qquad |p_3| \le 2.0673 \tag{13}$$

Example 3. Robust Control of Helicopter, Four-Dimensional System. Chen and Chen in [11] reconsidered the robust controller design of the Narendra and Tripathi helicopter model. The task was to obtain a nonconservative controller, i.e., the controller guaranteeing stability for parameters bounded by (13), with the gain matrix containing gains as small as possible. They used Davison-Fletcher-Powell optimization technique to determine the gain matrix, obtaining

$$K = \begin{bmatrix} -.1640 & .2699 & .4511 & .4308 \\ .0364 & .1692 & -.1066 & -.4519 \end{bmatrix} \tag{14}$$

For this gain matrix the closed loop system (with u = Kx) has a form

$$\dot{\mathbf{x}} = \begin{bmatrix} -.1027 & .1762 & .1995 & -.3446 \\ -.7275 - .1640p_3 & -1.428 + .2699p_3 & 2.1852 + .4511p_3 & .7218 + .4308p_3 \\ 1.1689 & -.4446 + p_1 & -3.6757 & -3.0841 + p_2 \\ 0 & 0 & 1 & 0 \end{bmatrix} \mathbf{x} \tag{15}$$

We attempt to optimize a Lyapunov function for the system (15) with bounds (13) and to determine Lyapunov function time constant. The Lyapunov function designed in [10] gives the performance measure -.11635 and time constant τ = 6.1162. Optimization procedure results in the matrix S

$$S = \begin{bmatrix} 3.553 & .223 & -.386 & -2.317 \\ .223 & .54 & .311 & -.425 \\ -.386 & .311 & 2.222 & 1.452 \\ -2.317 & -.425 & 1.452 & 5.911 \end{bmatrix}$$

and the performance measure λ = -0.4333, which gives the Lyapunov function time constant τ = 2.3079. The performance measure unexceedable for time-invariant uncertainties is equal to -0.45.

Example 4. A 24-Dimensional System. The matrices A_n and A_1 of the system with a single uncertainty p_1 were arbitrarily selected. The optimization process begun with the 24 × 24 matrix S of the quadratic Lyapunov function assumed as the unit matrix. The corresponding performance measure was obtained as +4.1382 and was better than the performance measure 16.8087 obtained when S was found by solving the equation $A_N^T S + SA_N = - I$. At the 20th step of the algorithm the performance measure -10.81 was obtained.

IDENTIFICATION, PROBLEM STATEMENT, PRELIMINARIES

We say that the structured-uncertainty-system (1) is un-identified if the only information provided is the dimension of the system and the fact that uncertainties are structured. The dimension q of the uncertainty vector as well as nominal and uncertainties matrices A_N and A_i are unknown. We define the problem of the identification of the optimal Lyapunov function as the problem of determining the optimal Lyapunov function for the un-identified uncertain system (1) by means of organized and limited monitoring of system solutions $x(t,x_o)$.

The solutions $x(t,x_o)$ of the system are assumed to be monitored by means of measuring the points x_i of trajectories at discrete time instants t_i. Since the system is uncertain the monitoring takes place while some uncertainty realization occurs, the realization being a well-defined but unknown function fulfilling the initial assumptions and remaining inside the assigned bounds. Thus estimations of right-hand side of the system, as well as any other estimations, are done with this realization of the uncertainty. The same estimation repeated at the same point x_i, but at the different instant t_i^* when different realization takes place may result in the different value of right-hand side of the system (1). We account for this fact by making all estimations dependent on the additional argument $p_r(x,t)$.

Explicit estimate of the derivative. At each point x_i, such that the next point x_{i+1} belongs to the same trajectory, the derivative of Lyapunov function along the solution is approximated by

$$\dot{V}(x_i,p_r(x_i,t_i)) = \frac{V(x_{i+1}) - V(x_i)}{t_{i+1} - t_i} \qquad i=1,\dots,s \tag{16}$$

Correspondingly the function $\Lambda(x_i,t_i)$ is obtained as

$$\Lambda(x_i,p_r(x_i,t_i)) = \frac{\dot{V}(x_i,p_r(x_i,t_i))}{V(x_i)} \tag{17}$$

The value of the right hand side of the system is then approximated by the expression

$$f(x_i,p_r(x_i,t_i)) = \frac{x_{i+1} - x_i}{t_{i+1} - t_i} \tag{18}$$

Further we adopt the notation x_i for the points at which the above calculations are possible and denote the set of s such points by Z.

Implicit estimate of the derivative. The alternate way of estimating the derivative utilizes the approximation of the right-hand side of a system. It is used when calculating the derivative of the function ΔV along the solutions of the system. For $\Delta V = x^T \Delta S x$ we have

$$\Delta\dot{V}(x_i,p_r(x_i,t_i)) = x_i^T \, \Delta S \, f(x_i,p_r(x_i,t_i)) + f^T(x_i,p_r(x_i,t_i)) \, \Delta S \, x_i \tag{19}$$

This formula makes possible the analytical expression of the derivative as the function solely of the elements of the matrix S. We arrive to the

Conclusion 1. For any given point of the trajectory of the system the expressions (16)-(19) make possible to determine:

(i) the values of Lyapunov function V and its derivative along the solution at this point,

(ii) the function $\Delta V = x^T \Delta S x$ and its derivative along the solution at the point of the trajectory of the system with some uncertainty realization as the function, solely, of the elements of the matrix ΔS.

IDENTIFICATION OF OPTIMAL LYAPUNOV FUNCTION

We introduce a concept of an active vertex. For a given system (1) and function V(x) let Π be, as previously, the uncertainty parallelepiped of (1) and let Λ_1 be, as previously, the maximum value of the function $\Lambda(x,t)$ on the unit circle.

Definition 1. The vertex $p^{(j)} \in R^q$ of the box Π is an active vertex associated with V(x) if

$$\max_{x \in \Phi_1} \Lambda(x,p^{(j)}) = \Lambda_1$$

Denote (as previously) by ζ_j the unit vector at which the form QF_j, corresponding to the active vertex $p^{(j)}$, reaches the value Λ_1. The function $\Lambda(x,p^{(j)})$ attains the value Λ_1 at all points $x^{(j)} \in R^n$ such that $x^{(j)} = \alpha \, \zeta_j$, where α is an arbitrary, different from zero number.

Definition 2. A point $x^{(j)} = \alpha \, \zeta_j$ is called an active point associated with the vertex $p^{(j)}$ of the parallelepiped Π. We say that the active vertex $p^{(j)}$ associated with V(x) is reached by a trajectory of the system, if the trajectory passes at some instant t* by an active point $x^{(j)}$, associated with the vertex $p^{(j)}$ and at this instant the uncertainty realization takes the value of the vertex, $p_r(x^*,t^*) = p^{(j)}$.

Definition 3. We say that monitoring data are complete if during the monitoring all the active vertices associated with V(x) have been reached.
Further discussion will be carried out under the Assumption A (see Section: Some Definitions and Lemma), and:

Assumption B. During identification of optimal Lyapunov function (OLF) the monitoring data are complete.

We are ready to proceed with the identification of the OLF. We are given the set Z of the points x_i, $i = 1,\ldots,s$ of the trajectories of the system (1) and we are able to calculate the values of expressions (16)-(19) at these points. At the same time by virtue of the completeness of data we know that for the system (1) the maximum value Λ_1 of the function $\Lambda(x,p(x,t))$ is reached at some of the points x_i belonging to Z and only at these points. Thus in search of the better Lyapunov function the Lemma may be used.

Under the Assumption A the number s^* of points at which the maximum of $\Lambda(x,p(x,t))$ is reached is not larger than 2^q. Denote these points by $x^{(j)*}$, $k = 1,...,s^*$ and the associated vertices (mutually different by virtue of the Assumption A) by $p^{(j)*}$.

As it was done in the Sec. IV we consider the function $V + \epsilon\, \Delta V$ assuming sufficiently small ϵ. The expressions $\Delta QF_j(x^{(j)*})$ representing the Lyapunov derivative of the function ΔV at the points $x^{(j)*}$ corresponding to the active vertices $p^{(j)*}$ are obtained from (19). The expressions $\Delta QF_j(x^{(j)*})$ are solely linear functions of the elements of the matrix ΔS of the function ΔV. Thus the inequalities (6), with the points ζ_j replaced by the points $x^{(j)*}$ represent the conditions under which the function $V + \epsilon\, \Delta V$ is better than the function V for the system (1). Application of the substitution (7) leads to the:

Theorem 2. Let be an un-identified system with structured uncertainties (1), let the monitoring data be complete. Let the quadratic Lyapunov function be given and Assumptions A and B fulfilled. If there exists a solution to the inequality

$$\mathbf{B\,u} < 0 \tag{20}$$

then the function $V + \epsilon\, \Delta V$ is for sufficiently small ϵ better than the function V.

IDENTIFICATION ALGORITHM

Out of a number of possible options an on-line-like algorithm was selected. The limited memory storage as well limited data processing speed were assumed, which prompts the restriction on the number of state space points acquired and processed. Following are the principal steps of the algorithm to identify the optimal Lyapunov function.

1. Put $S_o = I$ or accept the other appropriate zero-approximation of Lyapunov function matrix.
2. Set up the initial value of Λ_1. Assign the number $\delta > 0$ defining the width of the band of acceptable values of $\Lambda(x_i,p_r(x_i,t_i))$ as well as the maximal number s^* of points to be considered simultaneously.
3. Start monitoring. Acquire a pair of points x_i, x_{i+1} belonging to the same trajectory. Use (16) and (17) to calculate $\Lambda(x_i,p_r(x_i,t_i))$.
4. If

   $$\Lambda(x_i,p_r(x_i,t_i)) < \Lambda_1 - \delta \tag{21}$$

   discard x_i and return to the step 3. Else store the point and the value of the function for processing.
5. If

   $$\Lambda(x_i,p_r(x_i,t_i)) > \Lambda_1$$

   set the new value of Λ_1.

   $$\Lambda_1 = \Lambda(x_i,p_r(x_i,t_i))$$

   Discard these stored points for which the condition (21) is fulfilled. Proceed (go to the step 3) till all the monitored points are examined.
6. Use (18) to calculate the values of the vector $f(x_i,p_r(x_i,t_i))$ for the remaining stored points. Use (19) to calculate the matrix B.
7. Solve the inequality (20), determine the corresponding matrix ΔS. If (20) has no solution go to step 10.
8. Create a new Lyapunov function $V + \epsilon\, \Delta V$. Check is it better than V, if not adjust the constant ϵ.
9. Go to the step 3.
10. The function is optimal with accuracy to the accepted algorithm.

EXAMPLES OF IDENTIFICATION

To verify the method the procedure of identification of optimal Lyapunov function was carried out utilizing a numerical integration of solutions to uncertain systems of the form (1). For simulation the most unfavourable realizations of uncertainties were assigned in the form

$$p_j(x,t) = \max(\mathrm{abs}(p_j^-,p_j^+))\ \mathrm{sign}(x^T(SA_j)x)$$

Such a form of realizations maximizes the value of the derivative of the Lyapunov function.

In each of the Examples 5-9 about 4500 solutions were simulated with randomly selected (with uniform probability distribution) initial conditions at each example. Some initial Lyapunov function was selected at each example and the function optimization based on the generated solutions to the system was carried out. Together with the function, its performance measure based on the generated solutions was determined. It must be noted that the value of the performance measure calculated on the sole basis of the generated solutions may differ from the value calculated by the definition (2).

The identified function was not expected to be optimal since the monitoring data might be not complete. However to claim the procedure effective the performance measure (calculated according to (2)) of the identified function (called further the suboptimal function) should either be significantly better than the one of the initial Lyapunov function or be reasonably close to this of the optimal Lyapunov function.

To make the comparison the optimal Lyapunov function and the corresponding optimal performance measure λ_{opt} were determined according to (2), independently of the identification process, by the optimization procedure as described in the Sec. IV. Also the performance measures of the initial and suboptimal functions were so determined. The following notation has been used:

λ_{in} - the performance measure of the initial Lyapunov function, calculated by (2),
λ_{subopt} - the performance measure of the suboptimal Lyapunov function, calculated by (2),
λ^*_{subopt} - the performance measure of the suboptimal Lyapunov function, calculated on the basis of the monitoring data.

The following examples of identification were carried out.

Ex. 5. Identification of the system (9) with no uncertainties, $p_1(x,t) = 0$.
Ex. 6. Identification of the system (9) with $|p_1(x,t)| \le \sqrt{3}/2$.
Ex. 7. Identification of the system (9) with $|p_1(x,t)| \le \sqrt{3}/2$, with the initial conditions in the first quadrant only, $x_{oi} \ge 0$, i = 1,2.
Ex. 8. Identification of the system (12).
Ex. 9. Identification of the system (12), with the initial conditions fulfilling $x_{oi} \ge 0$, i = 1,..,4.
Ex.10. Identification of the 24-dimensional system.

The matrices of initial Lyapunov function were in all examples assumed as identity matrices. Table 1 lists the results of the identification.

Discussion of Results. The suboptimal performance measures obtained in Examples 5,6,8 represent significant improvement with regard to the initial values λ_{in}, and are reasonably close to the optimal values λ_{opt}.

At the Examples 7 and 9 the restriction was imposed on the initial conditions. The suboptimal function appears to be worse than the initial function ($\lambda_{subopt} > \lambda_{in}$). This effect results from the fact that during the optimization of the Lyapunov function the generated solutions are not reaching all the active points of the function. Still, the suboptimal function should have the better performance measure than the initial function when considering the monitoring data only. This should be also true for the points fulfilling the restrictions. Indeed in the Example 7 when the performance measures λ_{subopt}, λ_{in} are found according to (2), but with restricting the state x to belong to the first quadrant we have $\lambda_{subopt} = -0.3531$, $\lambda_{in} = 0.3229$. Thus the reversed inequality occurs $\lambda_{subopt} < \lambda_{in}$; moreover this value of λ_{subopt} is less than λ_{opt} calculated for x belonging to the entire R^2.

Table 1. Results of identification of optimal Lyapunov function

Example #	λ_{in}	λ_{subopt}	λ_{opt}	λ^*_{subopt}
5	0.000	-0.8766	-1.000	-0.8766
6	0.3229	0.0195	0.000	0.0195
7	0.3229	0.9879	0.000	-0.3353
8	1.868	-0.2300	-0.4333	-0.2313
9	1.868	0.7704	-0.4333	-0.1214
10	5.1182	-4.9822	-10.81	-10.1126

The Example 10 illustrates how high dimensionality of the problem may affect identification. The results listed in the Table 1 were obtained by randomly generating about 6,000 solutions. The suboptimal Lyapunov function obtained at this stage of the procedure is relatively close to the optimal one (i.e. λ_{subopt} close to λ_{opt}). In the second phase of identification, after next 4,000 simulations the measure λ_{subopt} of the new function V^{**} increased dramatically to the value 35.4749 while the measure λ^*_{subopt} decreased to the value -10.1170. For the next 2,000 simulations no significant changes in the measures were observed.

To explain the second-phase effect the eigenvalues of the derivative of the Lyapunov function V^{**} were calculated at the vertices of the uncertainty (i.e. the eigenvalues of the forms QF_j, j = 1,2). It was found that while the three largest eigenvalues were equal to +40.9024, -1.4497, -7.2118, the other 45 eigenvalues were all less than -10.1170. Thus the probability of randomly selecting the initial condition for which the measure λ_{subopt} could be improved was very low. While such an initial condition was assigned (by selecting the eigenvector corresponding to the maximal eigenvalue +40.9024), the solution simulated and the procedure applied, the instant change in λ_{subopt} was observed.

CONCLUSION

The identification procedure of the optimal Lyapunov function has been presented here under Assumptions A and B. It should be noted that the Assumption A may be removed at the cost of a significant increase in calculations complexity. Since the matrices with all eigenvalues mutually different are distributed densely in the space of symmetric matrices, it seems that, excluding pathological cases, Assumption A is always fulfilled.

The Theorem 2 formulated at the paper provides a basis for a recursive identification procedure. The procedure discloses typical properties of the learning scheme at which the results of the previous trial are used to improve the function at the next trial. The data completeness is the necessary condition of determination of the optimal Lyapunov function. The completeness of data requirement may be modified in such applications, where initial conditions are restricted to some subset of R^n.

Apart from solving the identification problem the procedure may be modified to determine Lyapunov function for some class of dynamic systems with restricted initial conditions.

Example 2 shows that quadratic Lyapunov functions are not always best suited for robust stability investigation. Example 3 shows that present methods of robust controller design are still conservative. Example 4 shows that the procedure of the controller design is effective for large-scale systems, the area where conservatism of estimates often makes any analysis impossible. Examples 5 - 10 relate to the identification procedure and show that it is effective. The results of Examples 7 and 9 imply that identification of the suboptimal Lyapunov function is possible for systems with restricted initial conditions, and that the suboptimal Lyapunov function for solutions so restricted may be better than the optimal Lyapunov function for the system without the restriction.

The Example 10 shows that the procedure worked well for multi-dimensional systems. During both phases of the identification the performance measure λ^*_{subopt} was decreasing. For the cases when there is no restrictions imposed on the initial conditions the second-phase effect, i.e. the increase of the performance measure λ_{subopt} in the second phase is unacceptable. The occurrence of the effect may be eliminated by:

(i) ensuring that the monitoring data are complete. This may be done either by sufficiently long monitoring or by training - i.e. by running the system with initial conditions which are known to be critical,

(ii) providing sufficiently large memory storage to avoid discarding the points of the state space as well as providing sufficient speed for data processing.

REFERENCES

[1] D.Siljak, "Parameter Space Methods for Robust Control Design: A Guided Tour," *IEEE Trans. Automat. Contr.*, Vol. AC-34, pp. 674-688, July 1989.

[2] S.N. Singh and A.A.R. Coelho, "Nonlinear Control of Mismatched Uncertain Linear Systems and Applications to Control of Aircraft," *ASME Journal of Dynamic Systems, Measurement and Control*, Vol. 106, pp. 203-210, 1984.

[3] D. Siljak, *Large Scale Dynamic Systems: Stability and Structure*, North-Holland, Amsterdam, The Netherlands, 1978.

[4] R.V. Patel and M. Toda, "Quantitative Measures of Robustness of Multivariable Systems," *Proc. JACC*, San Francisco, TP8-A, 1980.

[5] R.K. Yedavalli and Z. Liang, "Reduced Conservatism in Stability Robustness Bounds by State Transformation," *IEEE Trans. Automat. Contr.*, Vol. AC-31, pp. 863-866, 1986.

[6] N. Becker and W.M. Grimm, "Comments on Reduced Conservatism in Stability Robustness Bounds by State Transformations," *IEEE Trans. Automat. Contr.*, Vol. AC-33, pp. 223-224, 1988.

[7] B. Radziszewski, "O. Najlepszej Funkcji Lapunowa," IFTR Reports, No. 26, 1977.

[8] A. Olas, "On Robustness of Systems with Structured Uncertainties," *Lecture Notes in Control and Information Science*, Vol. 151, Springer Verlag 1991, pp. 295-310.

[9] A. Olas, "Optimal Quadratic Lyapunov Functions for Robust Stability Problem," *Dynamics and Control*, Vol. 2, 1992, pp. 265-279.

[10] A. Olas and F. Ahmadkhanlou, "Globally Optimal Quadratic Lyapunov Functions for Robust Stability of systems with Structured Uncertainty," submitted for print.

[11] Y.H. Chen and J.S. Chen, "Robust Control of Uncertain Systems with Time-Varying Uncertainty: A Computer-Aided Setting," *Lecture Notes in Control and Information Sciences*, Vol. 151, pp. 97-114, Springer-Verlag, 1991.

[12] K.S. Narendra and S.S. Tripathi, "Identification and Optimization of Aircraft Dynamics," *J. Aircraft*, Vol. 10, pp. 193-199, 1973.

[13] A. Olas, "Construction of Optimal Lyapunov Function for Systems with Structured Uncertainties", *IEEE Trans. Automat. Contr.*, to appear Oct. 1993.

[14] Ahmadkhanlou F., "Globally Optimal Lyapunov Functions for Robust Stability Problem", *Ph.D. Thesis*, Oregon State University, June 1992.

A NEW TOOL FOR ROBUST CONTROL

Argiris G. Soldatos[1], Chingyei Chung[2], and David M. Auslander[1]

[1]Department of Mechanical Engineering
University of California
Berkeley, CA 94720
[2]Department of Automatic Control Engineering
Feng–Chia University
Tai–Chung City, Taiwan

ABSTRACT

We present a new tool for the analysis and design of uncertain linear systems. A robustness measure is developed based on the characteristic equation root locations. This measure has a useful geometric interpretation.

INTRODUCTION

In recent years, much effort has been devoted to the problem of obtaining some desired behavior from a linear, time invariant, uncertain system. In much of this research and in this paper, the uncertainties involved are modeled deterministically rather than stochastically. For example, typical system representations contain unknown parameters pertinent to the system description (parameter uncertainty) as well as incomplete model information (unmodeled dynamics). Hence, the task is to control the system so as to achieve the desired objective in the presence of these uncertainties. This process leads to robust control with respect to the aforementioned contingencies.

In the classical control literature, uncertainty and robustness issues are often dealt with by using the phase margin, the gain margin and the resonance peak M_p indicators (Kuo [12], Dorf [8]). Classical gain margins of at least 6db and phase margins of 30° to 80° have traditionally led to the assertion of possessing good robustness properties. In MIMO systems (e.g. Letomaki et al. [14]) it has been shown that a system with significant gain and phase margins may not be robust. Here, it is demonstrated that for a SISO system having sufficiently large margins and small resonance peak M_p, small changes in the characteristic polynomial coefficients may lead to instability.

Mechanics and Control, Edited by R.S. Guttalu
Plenum Press, New York, 1994

Recent investigators have formulated the uncertainty in the system representation in a number of ways. For example, using statistical considerations (e.g. Astrom [1]) the uncertain quantities involved are assumed to behave according to some probabilistic patterns. In the deterministic approach some researchers (e.g. Leitmann [13]) assume that the uncertain quantities are described by functions whose values may range in prescribed sets. The robustness results are also targeted differently. To mention a few, the H_∞-optimization techniques (e.g. Zames [18]) minimize the weighted infinity norm of an appropriate closed loop transfer function or combination of transfer functions. In the state space formalism Kautsky et al. [11], for example, minimize the sensitivity of the closed loop poles due to perturbations in the system and gain matrices.

Subsequently, we will provide some motivation for the type of uncertainty and robustness problems under investigation. Loosely speaking, we will be analyzing the behavior of the closed loop poles of a linear system when some parameters vary within predefined bounds. The consideration of the closed loop characteristic equation seems natural in this setting.

An exposition of the already existing characteristic polynomial literature as it relates to uncertainly would be rather long. However, some key results are reported for comparison since this paper does not follow the usual framework.

In 1978, V.L.Kharitonov introduced a stability theorem for a class of polynomials, e.g., see Barmish and Khargonekar [2] for a review of the continuous-time case and Bose and Zeheb [5] for the discrete-time case. These polynomials are defined by letting each coefficient vary independently in an arbitrary but prespecified interval. This would be typically the case when considering e.g. the effect of parameter uncertainly in the characteristic polynomial. The theorem states that the whole class is Hurwitz if and only if four special, well defined polynomials are Hurwitz. Various extensions of this result have been proposed. Among those, multidimensional system stability (e.g. Bose [6]), stability in the presence of both structured perturbations and unmodeled dynamics (Barmish and Khargonekar [2]), stability in the presence of structured perturbations and time delays (Mori and Kokame [15]), and root locus in the presence of structured perturbations (Barmish and Tempo [3]).

A relatively simple performance evaluation and design methodology is presented here. It is efficient in computation time and suitable for hand calculation. In addition, it has a "visual" interpretation that makes it attractive, as we hope to show.

Only the characteristic equation of the closed loop system needs to be considered together with some knowledge related to the uncertainty. At first, the different types of uncertainty of interest are modeled in a unifying manner. This takes the form of an analytic function with "unknown" coefficients. Then, the roots of that function are derived using some bounds on the unknown coefficients. This forms the basis for evaluating the performance of the closed loop system. Similarities and differences with the other approaches reported in this introduction will be evident during the proposed development. The efficacy of the results is illustrated in a number of examples.

UNCERTAINTY REPRESENTATION

We will initially be concerned with some performance and stability issues arising in connection with linear, time invariant, continuous-time systems. A typical open loop transfer function of such a system may be written as

$$G(s) = \frac{N(s)}{D(s)}, \tag{1}$$

where $N(s)$ and $D(s)$ are polynomials with real coefficients in the complex variable s, of degrees m and n respectively, $n \geq m$ and $D(s)$ is a monic polynomial. The closed loop poles are the n roots of the characteristic equation

$$C(s) \triangleq D(s) + N(s) = 0, \tag{2}$$

where it may be considered , without loss of generality, that $C(s)$ is a monic polynomial. Here, we focus our attention to uncertain systems whose closed loop characteristic equation is of the form

$$C(s) + \epsilon \hat{C}(s) = 0, \tag{3}$$

where $\hat{C}(s)$ is a polynomial of s with complex coefficients and ϵ is a constant. The degree, l, of $\hat{C}(s)$ depends on the uncertainty in the system representation. For $\epsilon = 0$, equation (3) reduces to the closed loop characteristic equation of the nominal (i.e. with no uncertainty) system. Equation (3) affords the following interpretation: The nominal system dynamics represented by the $C(s)$ of equation (2) are augmented by $\epsilon\hat{C}(s)$ to allow for possible parameter uncertainty, unmodeled dynamics and unknown delays in the system description. Equation (3) incorporates all the system information relevant to this study.

Parameter uncertainty of interest can be represented by some unknown but "small" variation in the coefficients of $C(s)$ in equation (2); i.e., by some $\epsilon\hat{C}(s)$ where $\hat{C}(s)$, is of degree at most $n-1$.

High frequency unmodeled dynamics can also be casted in the form of equation (3). Let an unknown high order unmodeled frequency, ω_1, be included in the representation. The new open loop transfer function is

$$G_F(s) = \frac{N(s)}{D(s)} \frac{m_1 s + \omega_1}{s + \omega_1}, \tag{4}$$

where $\omega_1, m_1 \in \mathbf{C}$ are unknown with $\mathrm{Re}(\omega_1) > 0$, $|\omega_1|$ "much greater" than the absolute value of both the roots of $D(s) = 0$ and $|m_1|$. The effect of the additional multiplicative term in the right hand side of equation (4) is such that the asymptotic Bode plot for this term has unity gain up to ω_1. The new closed loop characteristic equation is

$$D(s)(s + \omega_1) + N(s)(m_1 s + \omega_1) = 0,$$

i.e.,

$$C(s) + \frac{1}{\omega_1}[s(D(s) + m_1 N(s))] = 0. \tag{5}$$

This equation is in the same form as equation (3) with $\epsilon = 1/\omega_1$.

Introduction of more unmodeled frequencies can be effected by additional transfer function multiplications in equation (4). For example, a second unknown frequency, ω_2, yields the following closed loop characteristic equation:

$$D(s)(s + \omega_1)(s + \omega_2) + N(s)(m_1 s + \omega_1)(m_2 s + \omega_2) = 0,$$

i.e.,

$$C(s) + \frac{1}{\omega_1\omega_2}\left\{D(s)\left[s^2 + (\omega_1\omega_2)s\right] + N(s)\left[m_1 m_2 s^2 + (\omega_1 m_2 + \omega_2 m_1)s\right]\right\}. \tag{6}$$

The resulting closed loop characteristic equation (6) is in the form of equation (3) with $\epsilon = 1/\omega_1\omega_2$. The arguments can be extended to any number of unmodeled frequencies.

Transportation lag might be introduced in equation (1) as

$$G_T(s) = \frac{N(s)}{D(s)} e^{-Ts}, \tag{7}$$

where $T \geq 0$ is the dead time. T is usually unknown but "small". The closed loop characteristic equation is derived using equation (7).

$$D(s) + N(s)e^{-Ts} = 0 \quad \text{or} \quad D(s) + N(s) + \left(e^{-Ts} - 1\right) N(s) = 0,$$

i.e.,

$$C(s) + T\left(-s + \frac{Ts^2}{2!} - \frac{T^2 s^3}{3!} + \cdots\right) N(s) = 0. \tag{8}$$

Equation (8) is in the form of equation (3) with $\epsilon = T$. Given a desired degree of accuracy, a truncation of terms may take place in equation (8).

Hence, when uncertainties (parameter uncertainty, unmodeled dynamics or transportation lag) are introduced, the closed loop characteristic equation may be written in the form of equation (3).

To facilitate exposition, the unknown polynomial $\hat{C}(s)$ may be expressed more completely as

$$\hat{C}(s) = \hat{C}(s,q) \triangleq \sum_{i=0}^{d} \gamma_i(q) s^i, \tag{9}$$

where $q \in Q$ and $Q \subset \mathbf{C}^v$ with v a known positive integer. Q is assumed to be compact and known so that all possible uncertainties can be described by the elements of this set. The coefficients $\gamma_i(\cdot) : Q \to \mathbf{C}$ are assumed to be continuous functions of their argument. Hence, $\hat{C}(s)$ belongs to a one parameter family, F, of polynomials indexed by q, i.e.,

$$F \triangleq \left\{\hat{C}(\cdot, q) \mid q \in Q\right\}. \tag{10}$$

Due to the nature of the aforementioned uncertainties, the coefficients of the unknown polynomial $\epsilon\hat{C}(s)$ are "small" and bounded by known constants.

A ROBUSTNESS MEASURE

We would like to quantify the effect of the uncertainties on the closed loop poles; i.e., the roots of equation (3). It turns out that in the presence of uncertainty, all the poles of the nominal system (obtainable from equation (2)) change location in the complex plane. The perturbation of each root under the proposed uncertainties can be calculated to a desired degree of accuracy by summing up a sufficiently large number of terms in a series (see Lemma 1). Hence, given the "configuration" of the nominal poles in the complex plane one is able to obtain the neighborhood of all possible excursions for each root. This is particularly useful in the case of the dominant closed loop poles.

Lemma 1: Let $C(s)$ be a polynomial of degree n with real coefficients in the complex variable s, and λ_r be one isolated root of $C(s) = 0$. Then, for sufficiently small ϵ, one root of equation (3) is

$$\lambda_r(\epsilon) = \lambda_r - \epsilon\frac{\hat{C}(\lambda_r)}{C'(\lambda_r)} + \epsilon^2 \frac{2C'(\lambda_r)\hat{C}'(\lambda_r)\hat{C}(\lambda_r) - C''(\lambda_r)\hat{C}^2(\lambda_r)}{2[C'(\lambda_r)]^3} + \cdots, \tag{11}$$

where a prime indicates differentiation with respect to s.

Proof: Let λ_r be an isolated root of $C(s)$. One root $\lambda_r(\epsilon)$ of $C(s)+\epsilon\hat{C}(s)$ is (Wilkinson [16])

$$\begin{aligned}\lambda_r(\epsilon) &= \lambda_r + \sum_{k=1}^{\infty} p_k \epsilon^k \\ &= \lambda_r + \epsilon p_1 + \epsilon^2 p_2 + \epsilon^3 p_3 + \cdots\end{aligned} \tag{12}$$

This perturbation series converges for sufficiently small ϵ, that is, there is a $\sigma > 0$ such that for all $|\epsilon| < \sigma$ this series converges. Because $\lambda_r(\epsilon)$, $r = 1, 2, \cdots, n$, are the perturbed roots, they satisfy

$$\begin{aligned}0 &= C(\lambda_r(\epsilon)) + \epsilon\hat{C}(\lambda_r(\epsilon)) \\ &= C(\lambda_r + \epsilon p_1 + \epsilon^2 p_2 + \cdots) + \epsilon\hat{C}(\lambda_r + \epsilon p_1 + \epsilon^2 p_2 + \cdots),\end{aligned}$$

where $r = 1, 2, \cdots, n$. Because $C(s)$ and $\hat{C}(s)$ are polynomials, they can be expanded as Taylor series with respect to the unperturbed roots λ_r. So, equation (3) can then be written as

$$\begin{aligned}0 &= C(\lambda_r) + \frac{C'(\lambda_r)}{1!}(\epsilon p_1 + \epsilon^2 p_2 + \cdots) + \frac{C''(\lambda_r)}{2!}(\epsilon p_1 + \epsilon^2 p_2 + \cdots)^2 \\ &\quad + \cdots + \epsilon\left[\hat{C}(\lambda_r) + \frac{\hat{C}(\lambda_r)}{1!}(\epsilon p_1 + \epsilon^2 p_2)\right] + \mathcal{O}(\epsilon^3).\end{aligned}$$

Collecting the like powers of ϵ terms and noting that $C(\lambda_r) = 0$, we obtain

$$0 = \epsilon\left\{p_1 C'(\lambda_r) + \hat{C}(\lambda_r)\right\} + \epsilon^2\left\{C'(\lambda_r)p_2 + \frac{C''(\lambda_r)}{2!}(p_1 + \hat{C}'(\lambda_r))p_1\right\} + \mathcal{O}(\epsilon^3).$$

The above equation is valid for all $\epsilon \in (-\sigma, \sigma)$ and $\lambda_r \in \mathbf{C}$. Also, from the uniqueness theorem for analytic functions, the coefficients of the like powers of ϵ should vanish. Thus, we obtain equations of the form

$$\begin{aligned}&p_1 C'(\lambda_r) + \hat{C}(\lambda_r) = 0, \\ &C'(\lambda_r)p_2 + \frac{C''(\lambda_r)}{2!}p_1^2 + \hat{C}'(\lambda_r)p_1 = 0, \\ \cdots\end{aligned}$$

Therefore,

$$\begin{aligned}p_1 &= -\frac{\hat{C}(\lambda_r)}{C'(\lambda_r)}, \\ p_2 &= \frac{2C'(\lambda_r)\hat{C}'(\lambda_r)\hat{C}(\lambda_r) - C''(\lambda_r)\hat{C}^2(\lambda_r)}{2[C'(\lambda_r)]^3}.\end{aligned}$$

Substituting these into equation (12), we obtain equation (11). For the multiple roots case, consult the Appendix (Wilkinson [16]). ■

Equation (11) shows that, under the influence of the assumed uncertainties, λ_r will be shifted to a new location in the complex plane. The incurring dislocation is governed by the additional terms containing the uncertainties. It is evident that the

$C'(\lambda_r)$, $r = 1, 2, \cdots, n$, play an important role when evaluating the contribution of each additional term. Therefore, a convenient sensitivity measure discerning the effect of the uncertainties considered is provided by $|C'(\lambda_r)|$. The larger $|C'(\lambda_r)|$ is, the lesser the effect of the uncertainties on the corresponding root. This is particularly useful when considering a dominant closed loop pole location.

Definition: For an uncertain linear system with nominal characteristic polynomial $C(s)$ and corresponding poles λ_i, indexed by i, the *conjugate sensitivity measure* for each of these roots is $|C'(\lambda_i)|$.

We are interested in interpreting the conjugate sensitivity measure, $|C'(\lambda_i)|$, geometrically. This visualization will be instructive throughout the rest of this paper.

Lemma 2: Given the $C(s)$ of equation (2),

$$|C'(\lambda_i)| = \prod_{j=1, j\neq i}^{n} |\lambda_i - \lambda_j|, \tag{13}$$

where $i = 1, 2, \cdots, n$ and λ_k, $k = 1, 2, \cdots, n$ is the kth root of equation (2).

Proof: Since $C(s)$ is a monic characteristic polynomial, we have

$$C(s) = (s - \lambda_1)(s - \lambda_2) \cdots (s - \lambda_n). \tag{14}$$

Upon differentiation with respect to s,

$$\begin{aligned} C'(s) &= (s - \lambda_2)(s - \lambda_3) \cdots (s - \lambda_n) + (s - \lambda_1)(s - \lambda_3) \cdots (s - \lambda_n) \\ &\quad + \cdots + (s - \lambda_1)(s - \lambda_2) \cdots (s - \lambda_{n-1}). \end{aligned} \tag{15}$$

Substituting λ_i into this equation, we obtain equation (13). ■

This leads to a pictorial interpretation in the complex plane. The absolute value of the derivative, $|C'(\lambda_i)|$, evaluated at the specific pole, λ_i, is equal to the product of the distances between that and each of the remaining poles. Accordingly, a "large" $|C'(\lambda_i)|$ corresponds to a pole that is at a distance from the rest. Clustered poles yield a "small" conjugate sensitivity measure and are very sensitive to uncertain perturbations.

Remark 1: When $C(s)$ is a first-order polynomial, the conjugate sensitivity measure is always equal to one. This means that when there is a constant perturbation, the variation of the root is this constant itself (see equation (11)).

Remark 2: Note that if the degree of $\epsilon\hat{C}(s)$ in equation (3) is higher than that of the nominal characteristic polynomial, $C(s)$, then there will be some extra poles generated. These extra poles are called *singular perturbed poles* (Bender and Orszag [4]). In this paper, we assume that these singular perturbed poles are stable. This is true for minimum phase, high-frequency unmodeled dynamics.

Remark 3: From the robust stability viewpoint, nondominant roots can tolerate more variation than dominant ones. Therefore, it is more interesting to consider the conjugate sensitivity measure for the dominant roots.

Consider a system with a dominant closed loop pole, λ_d. To account for uncertainty in the system description, define the first order approximate root, $\lambda_{d,app}(\epsilon)$, as

$$\lambda_{d,app}(\epsilon) \triangleq \lambda_d - \frac{\epsilon \hat{C}(\lambda_d)}{C'(\lambda_d)}, \tag{16}$$

where $\lambda_{d,app}(\epsilon)$ is an approximation to $\lambda_d(\epsilon)$.

Let δ be the distance of the dominant root, λ_d, from the stability boundary. For a stable nominal system in continuous-time, $\delta = -\mathrm{Re}(\lambda_d)$ and in discrete-time, $\delta = |1 - |\lambda_d||$.

Considering the first-order approximation in equation (16), one can show:

Proposition: If $\delta|C'(\lambda_d)| \geq |\epsilon \hat{C}(\lambda_d)|$ then $\lambda_{d,app}(\epsilon)$ is stable.

Proof: Without loss of generality, consider the continuous-time system for which the above inequality implies

$$\left|\mathrm{Re}\left[\frac{\epsilon \hat{C}(\lambda_d)}{C'(\lambda_d)}\right]\right| \leq \left|\frac{\epsilon \hat{C}(\lambda_d)}{C'(\lambda_d)}\right| \leq \delta. \tag{17}$$

Then, from equations (16) and (17), we conclude that

$$\begin{aligned} \mathrm{Re}(\lambda_{d,app}(\epsilon)) &= \mathrm{Re}(\lambda_d) + \mathrm{Re}\left[-\frac{\epsilon \hat{C}(\lambda_d)}{C'(\lambda_d)}\right] \\ &= -\delta + \mathrm{Re}\left[-\frac{\epsilon \hat{C}(\lambda_d)}{C'(\lambda_d)}\right] \\ &\leq -\delta + \left|\mathrm{Re}\left[-\frac{\epsilon \hat{C}(\lambda_d)}{C'(\lambda_d)}\right]\right| \\ &\leq 0. \end{aligned} \tag{18}$$

Then, $\lambda_{d,app}(\epsilon)$ is stable. The discrete-time case is analogous. ■

Remark 4: The first-order approximate root, $\lambda_{d,app}(\epsilon)$, is an approximation for the actual perturbed root, $\lambda_d(\epsilon)$ with $|\lambda_d(\epsilon) - \lambda_{d,app}(\epsilon)| = \mathcal{O}(\epsilon^2)$, so it is possible that $\lambda_d(\epsilon)$ is unstable even though $\lambda_{d,app}(\epsilon)$ is stable.

Definition: For a system with nominal characteristic polynomial $C(s)$ and dominant pole λ_d, the *robustness measure* for the stability of the first-order approximation root is $\delta|C'(\lambda_d)|$.

This measure should be considered as in the proposition and with the understanding that $\lambda_{d,app}(\epsilon)$ is only an approximation to $\lambda_d(\epsilon)$. Note that the geometric interpretation of the proposed robustness measure is similar to that given for the conjugate sensitivity measure. Here, there is one more distance, δ, to be included in the product.

EXAMPLES

The system analyst often wants to know if the polynomials in the family described by equation (10) have their roots in some prespecified set $D \subset \mathbf{C}$. In the continuous time case for example, the polynomials exhibit stable roots if D is the left-half of the complex plane.

Another possible use of the results in the previous sections is to determine if the roots of equation (3) are within some set D which contains the roots of equation (2). We will consider in more detail the case where the $\mathcal{O}(\epsilon^2)$ and higher order terms in equation (11) are not significant.

Example 1

Consider the following plant transfer functions with unity feedback

$$G_1(s) = \frac{5s^2 - 3s + 10}{s^3 - 4.4s^2 + 3.12s - 9.992}, \quad G_2(s) = \frac{3s^2 + 5.4s + 1}{s^3 + 4.2s^2 + 6s + 1}. \tag{19}$$

The closed-loop systems for the above two transfer functions are stable and with the same dominant root at -0.2. However, checking the closed-loop poles, we find that they are clustered for $G_1(s)$ and well conditioned for $G_2(s)$. The robustness measure for the dominant pole, $\lambda_d = -0.2$, for $G_1(s)$ is $\delta|C'(-0.2)| = 4 \times 10^{-11}$ and for $G_2(s)$ is $\delta|C'(-0.2)| = 1.732$. Thus, we can predict that the system $G_2(s)$ is more robust than $G_1(s)$ when considering the same "perturbations". Actually, introduction of a "small" parametric and unmodeled dynamics uncertainty in the closed-loop characteristic equation (equation (3)), leads the first system to instability. The parametric uncertainty may be introduced by adding perturbations in either the numerator or the denominator of the above transfer functions. The nominal and perturbed closed loop poles are shown in Table 1.

Table 1. Closed loop transfer function dynamics (poles in italics are unstable).

Characteristic equation	*Closed loop system poles for $G_1(s)$*	*Closed loop system poles for $G_2(s)$*
Nominal system	$-0.2 \pm 0.00001i$, -0.19999	$-5.0, -2.0, -0.2$
With added perturbation $0.02s^2 + 0.04s - 0.04$	*$0.0019 \pm 0.28673i$*, -0.58380	-5.01808, -2.00734, -0.19458
With added perturbation $0.025s - 0.05$	*$0.00109 \pm 0.31035i$*, -0.58380	-4.98779, -2.01853, -0.19368
With multiplicative unmodeled dynamics $\frac{0.65s + 25}{s + 25}$	*$0.02233 \pm 0.10541i$*, -0.74407, -23.1506	-23.64376, -5.31814, -0.20001, -1.98809

The results given in Table 1 demonstrate that the closed loop system for $G_1(s)$ can not tolerate even a 1% change in the coefficients of the denominator or the numerator polynomials. Such a difference may be introduced by e.g. measurement error. On the other hand, given the robustness measure one is able to determine, using the Proposition, how big a perturbation can be tolerated.

Although our emphasis is on continuous-time systems, the same results can be used in the discrete-time analyses. The unit disc of the complex plane is of importance for stable behavior as shown in the following example.

Example 2

Consider the unity feedback digital systems with stable open loop transfer functions

$$G_3(z) = \frac{1.7z^2 + 1.43z + 0.229}{z^3 + z^2 + z + 0.5}, \quad G_4(z) = \frac{0.8z^2 + 0.6209z + 0.22981}{z^3 + z^2 + z + 0.5}. \tag{20}$$

Notice that the closed-loop systems for $G_3(z)$ and $G_4(z)$ are stable and with the same dominant pole which is ill-conditioned for $G_3(z)$. Considering the robustness measure for the dominant pole, $\lambda_d = -0.9$, we obtain $\delta|C'(-0.9)| = 10^{-11}$ for $G_3(z)$ and $\delta|C'(-0.9)| = 0.273$ for $G_4(z)$. By analogy to Example 1, we construct Table 2.

Table 2. Closed loop poles for the discrete-time systems (poles in italics are unstable).

Characteristic equation	*Closed loop system poles for $G_3(z)$*	*Closed loop system poles for $G_4(z)$*
Nominal system	-0.9, $-0.89999 \pm 0.00001i$	-0.9, $-0.45 \pm 0.78i$
With added perturbation 0.01	-1.11544, $-0.79228 \pm 0.18658i$	-0.91217, $0.44392 \pm 0.78357i$
With added perturbation $-0.01z$	-1.1240, $-0.7880 \pm 0.16623i$	-0.91110, $-0.44445 \pm 0.77684i$
With added perturbation $0.01z^2$	-1.13433, $-0.78783 \pm 0.14828i$	-0.91010, $-0.44995 \pm 0.77424i$
With added perturbation $0.001z^2 - 0.001z + 0.001$	-1.04646, $-0.82727 \pm 0.11495i$	-0.90334, $-0.44883 \pm 0.77946i$

It is worth noting that the system $G_3(z)$ is so sensitive that it can not tolerate even a 0.1% parameter change.

Letomaki et al. [14] have shown that in MIMO systems, gain and phase margin considerations do not guarantee robustness with respect to system parameter variations. The following example demonstrates that gain and phase margins are not necessarily the best robustness measures for SISO systems as well.

Example 3

Consider the continuous transfer function (parameters are entered with 5 decimal digits for accuracy):

$$G(s) = \frac{0.37(s^4 + 0.13429s^3 + 2.90987s^2 + 0.20526s + 2.08937)}{s^6 + 0.23s^5 + 4.03443s^4 + 0.61943s^3 + 5.28324s^2 + 0.40927s + 2.24320}. \tag{21}$$

This plant possesses a 3° phase margin. We compensate this system with a P-D controller $G_c = s + 1.042398$. The compensated system has 58° phase margin, infinity gain margin, and a resonance peak $M_p < 1.2$. The compensated open loop transfer function can be written as

$$G_cG(s) = \frac{0.37s^5 + 0.128446s^4 + 1.19824s^3 + 0.85223s^2 + 0.85223s + 0.80584}{s^6 + 0.23s^5 + 4.03443s^4 + 0.619434s^3 + 5.28324s^2 + 0.409268s + 2.243197}.$$

According to common practice, one should expect this system to possess superior "relative stability" properties (Kuo [12], Dorf [8]), i.e., the system should be able to resist a certain degree of uncertainty. However, slight parameter perturbations indicate the opposite. Consider, for example, the following three cases:

(1) Additive perturbation of $0.006s^2+0.006s$ in the denominator and $0.004s^2+0.004s$ in the numerator of $G_cG(s)$.

(2) Additive perturbation of $0.005s + 0.005$ in the denominator and $0.005s + 0.005$ in the numerator of $G_cG(s)$.

(3) Additive perturbation of $0.0035s^5 - 0.0035s^4 + 0.0035s^3 + 0.0035s^2 + 0.0035s + 0.0014$ in the denominator and $0.001s^5 - 0.001s^4 + 0.001s^3 + 0.001s^2 + 0.0004$ in the numerator of $G_cG(s)$.

The Nyquist plots for the nominal and the perturbed systems are shown in Figures 1-3.

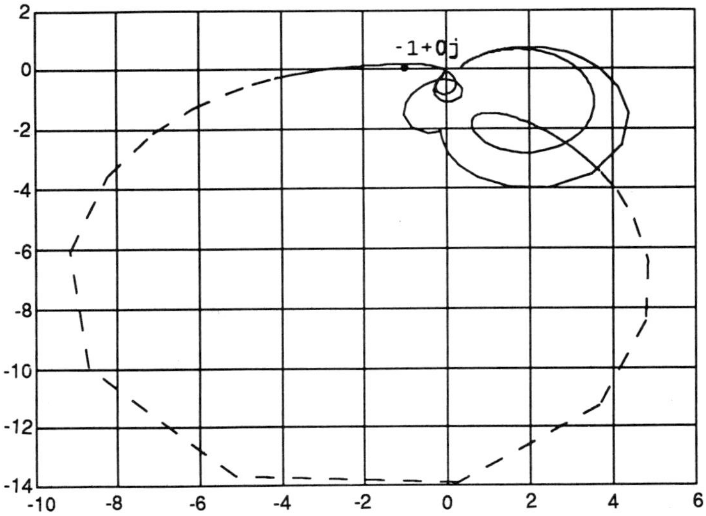

Figure 1. Perturbed and nominal Nyquist plots for case 1; – – perturbed, —- nominal.

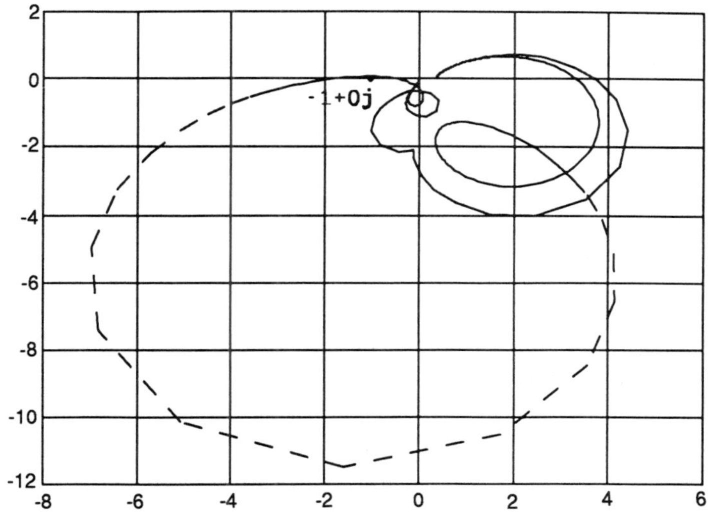

Figure 2. Perturbed and nominal Nyquist plots for case 2; – – perturbed, —- nominal.

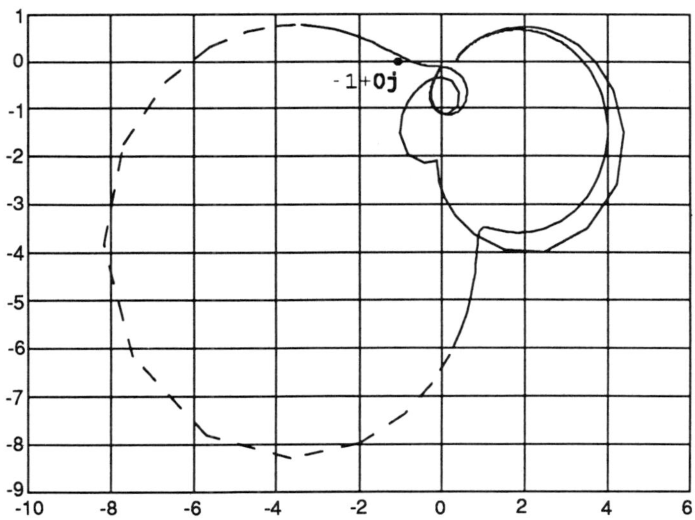

Figure 3. Perturbed and nominal Nyquist plots for case 3; – – perturbed, —- nominal.

Notice that in the above three cases, the perturbed open loop transfer functions are stable and correspond to minimum phase systems. It is obvious that there are dramatic changes under the influence of the above perturbations. The results can be explained using the robustness measure and its geometric interpretation. Here, $|C'(-0.11509 \pm 1.20976i)| = 0.0036$ is relatively low.

ROBUST REGULATOR DESIGN

Consider a system whose closed loop poles (see equation (2)) belong to a known set $D \subset \mathbf{C}$. Let D be such that, when the poles are contained in it, the system stability and performance are assured. It was shown earlier that a slight perturbation in the characteristic polynomial (typified by equation (3)) may drastically change the system dynamics by moving poles out of the set D. It is desired to design a compensator, such that the performance specifications are met under any admissible perturbation. In the above setting, this amounts to pole placement so that any possible perturbation leaves the poles (and places any new ones) in D. This is possible using equation (11) so that each pole λ_i is placed in such a way that under the worse uncertainty, $\lambda_i(\varepsilon) \in D$.

We revisit Examples 1 and 2 using a compensator design based on the Diophantine equation (see Chen [7]).

Example 4

Consider the plant transfer function in equation (19) for $G_1(s)$. The unity feedback system is stable. However, it is very sensitive to parameter changes and unmodeled dynamics.

Considering the robustness measure, we design the second-order compensator (Chen [7])

$$G_c(s) = \frac{4.53795s^2 + 4.3695s + 53.2142}{s^2 + 1.1102s + 51.8092}, \tag{22}$$

which yields the closed loop characteristic polynomial

$$\begin{aligned} C(s) &= (s + 0.2 \pm 0.6i)(s + 1.5 \pm 0.1i)(s + 16) \\ &= s^5 + 19.4s^4 + 58.26s^3 + 63.864s^2 + 34.586s + 14.464. \end{aligned} \tag{23}$$

The closed loop poles have been placed in such a way that the condition of the proposition is satisfied for the dominant pole under the assumed perturbations.

Table 3 and Figures 4-5 show the changes in the characteristic roots due to different perturbations.

Table 3. Compensated and uncompensated system dynamics under perturbations (poles in italics are unstable).

Characteristic equation	*Closed loop system poles without compensator*	*Closed loop system poles with compensator*	*Approximate dominant poles for compensated system from equation (11)*
Nominal system	$-0.2 \pm 0.00001i$, -0.19999	-16.0, $-1.5 \pm 0.1i$, $-0.2 \pm 0.6i$	
With added perturbation in the numerator of $G_1(s)$ $0.02s^2 + 0.04s - 0.04$	-0.5838, *$0.0019 \pm 0.2867i$*	-15.9032, $-1.5648 \pm 0.3190i$, $-0.1383 \pm 0.6245i$	$-0.1296 \pm 0.6267i$
With added perturbation in the numerator of $G_1(s)$ $0.025s - 0.05$	-0.6021, *$0.0011 \pm 0.3103i$*	-16.0098, $-1.5530 \pm 0.3993i$, $-0.1421 \pm 0.6291i$	$-0.1326 \pm 0.6314i$
With added multiplicative unmodeled dynamics in $G_1(s)$ $\dfrac{0.65s + 25}{s + 25}$	-23.1505, -0.7441, *$0.0223 \pm 0.1054i$*	$-16.7559 \pm 12.5369i$, $-1.34452 \pm 0.94504i$, $-0.1288 \pm 0.5377i$	$-0.1308 \pm 0.5139i$

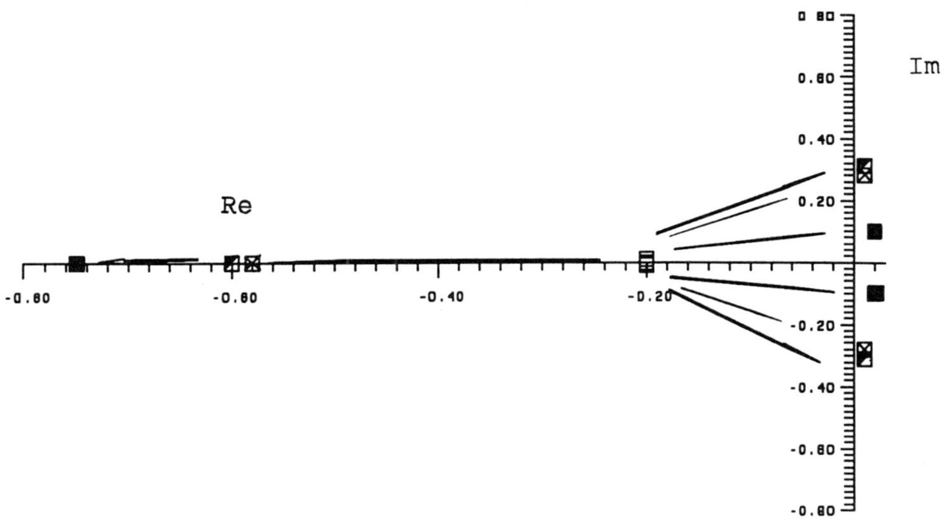

Figure 4. Nominal and perturbed poles for Example 4 (without compensator).

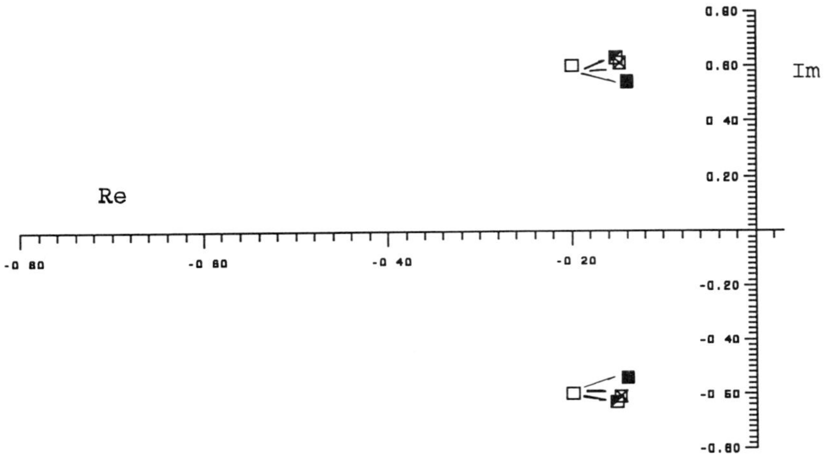

Figure 5. Nominal and perturbed poles for Example 4 (with compensator). Distant poles are not shown.

Example 5

A unit feedback system with an open loop transfer function given by equation (20) for $G_3(z)$ is stable. However, this system is very sensitive to parameter perturbations since it cannot tolerate a 0.1% coefficient changes.

Using the robustness measure, a compensator

$$G_c(z) = \frac{0.67288z^2 - 0.2794z + 0.38467}{z^2 - 0.54389z - 0.11834} \tag{24}$$

is designed (Chen [7]) such that the closed loop characteristic polynomial is

$$\begin{aligned} C(z) &= (z+0.9)(z-0.1 \pm 0.35i)(z+0.45 \pm 0.2i) \\ &= z^5 + 1.6z^4 + 0.825z^3 + 0.24625z^2 + 0.095806z + 0.028918 \end{aligned} \tag{25}$$

The improvement is shown in Table 4 and Figures 6-7.

DISCUSSION

In this paper, we define the conjugate sensitivity measure which equals the product of the distances from a specific pole to all the other poles of a system. A control system possesses a poor robustness property when the closed loop poles are clustered. This measure is rather simple to apply for controller designs in the time domain. Effective designs can be achieved by pole placement techniques. Both continuous and discrete-time systems can be treated.

For a more general controller design, one might first choose the dominant poles to satisfy requirements such as bandwidth, damping ratios, settling time, etc. Then, the other poles could be assigned to meet the robustness measure requirements for the dominant poles. This method is the subject of current research.

Table 4. Compensated and uncompensated system dynamics under perturbations (poles in italics are unstable).

Characteristic equation	*Closed loop poles without compensator*	*Closed loop poles with compensator*	*Approximate dominant poles for compensated system from equation (11)*
Nominal system	$-0.9, -0.89999 \pm 0.0001i$	$-0.9, -0.1 \pm 0.35i, -0.45 \pm 0.2i$	
With added perturbation in the numerator of $G_3(z)$ 0.01	*−1.1154*, $-0.7923 \pm 0.1866i$	$-0.9381,\ 0.0920 \pm 0.3446i,\ -0.4229 \pm 0.2315i$	−0.9434
With added perturbation in the numerator of $G_3(z)$ $-0.01z$	*−1.1240*, $-0.7880 \pm 0.1662i$	$-0.9359,\ 0.0990 \pm 0.3533i,\ -0.4310 \pm 0.2091i$	−0.9391
With added perturbation in the numerator of $G_3(z)$ $0.01z^2$	*−1.1343*, $-0.7878 \pm 0.1483i$	$-0.9337,\ 0.1012 \pm 0.3500i,\ -0.4394 \pm 0.2005i$	−0.9352
With added perturbation in the numerator of $G_3(z)$ $0.001z^2 - 0.001z + 0.001$	*−1.0465*, $-0.8272 \pm 0.1149i$	$-0.9114,\ 0.0992 \pm 0.3498i,\ -0.0440 \pm 0.2045i$	−0.9118

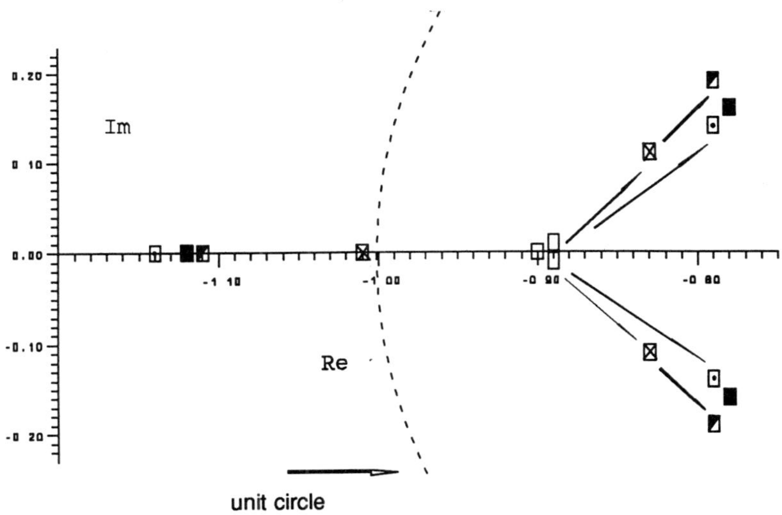

Figure 6. Nominal and perturbed poles for Example 5 (without compensator).

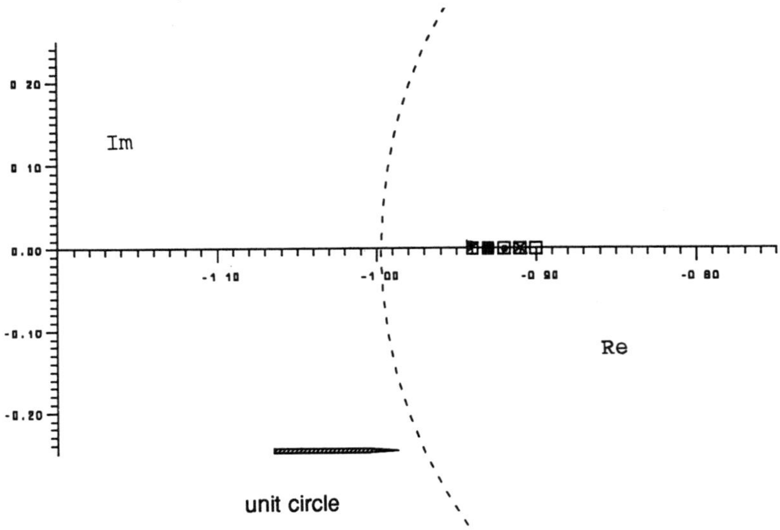

Figure 7. Nominal and perturbed poles for Example 5 (with compensator).

ACKNOWLEDGEMENTS

We would like to thank Nirmala and Ramesh Guttalu for their useful discussions on this work.

The first author would also like to acknowledge the suggestions of Professor George Leitmann.

REFERENCES

1. K. J. Astrom, *Introduction to Stochastic Control Theory*, Academic Press, New York, 1970.

2. B. R. Barmish and P.P. Khargonekar, "Robust stability of feedback control systems with uncertain parameters and unmodeled dynamics," *Proceedings of American Control Conference*, Atlanta, Georgia, 1988.

3. B. R. Barmish and R. Tempo, "The robust root locus," *Automatica*, Vol. 26, pp. 283-292, 1990.

4. C. M. Bender and S. A. Orszag, *Advanced Mathematical Methods for Scientists and Engineers*, McGraw-Hill, New York, 1989.

5. N. K. Bose and E. Zeheb, "Kharitonov's theorem and stability test of multidimensional digital filters," *IEE Proceedings-G*, Vol. 133, pp. 187-190, 1986.

6. N. K. Bose, "Robust multivariate scattering Hurwitz interval polynomials," *Linear Algebra and its Applications*, Vol. 98, pp. 123-136, 1988.

7. C. T. Chen, *Introduction to Linear Systems Theory*, Holt, Rinehart and Winston, New York, 1970.

8. R. C. Dorf, *Modern Control Systems*, Addison-Wesley, Reading, Mass., 1992.

9. L. R. Fletcher, J. Kautsky, G. K. G. Kolka and N. K. Nichols, "Some necessary and sufficient conditions for eigenstructure assignment," *Int. J. Cont.*, Vol. 42, pp. 1457-68, 1985.

10. G. F. Franklin, J. D. Powell and A. Emami-Naeini, *Feedback Control of Dynamic Systems*, Addison-Wesley, Reading, Mass., 1991.

11. J. Kautsky, N. K. Nichols and P. van Dooren, "Robust pole assignment in linear state feedback," *Int. J. Cont.*, Vol. 41, pp. 1129-55, 1985.

12. B. C. Kuo, *Automatic Control Systems*, Prentice-Hall, Englewood Cliffs, New Jersey, 1975.

13. G. Leitmann, "Deterministic control of uncertain systems," *Acta Astronautica*, Vol. 7, pp. 1457-61, 1980.

14. N. A. Letomaki, N. R. Snadell and M. Athans, "Robustness results in linear quadratic Gaussian based multivariable control design," *IEEE Trans. Autom. Cont.*, Vol. 26, pp. 75-93, 1981.

15. T. Mori and H. Kokame, "An extension of Kharitonov's theorem and its application," *Proceedings of American Control Conference*, Minneapolis, Minnesota, 1987.

16. J. H. Wilkinson, "The evaluation of zeros of ill-conditioned polynomials," *Num. Math.*, Vol. 1, pp. 150-180, 1959.

17. J. H. Wilkinson, *Rounding Errors in Algebraic Processes*, Prentice-Hall, Englewood Cliffs, New Jersey, 1963.

18. G. Zames, "Feedback and optimal sensitivity: Model reference transformations, multiplicative seminorms, and approximate inverses," *IEEE Trans. Autom. Cont.*, Vol. 26, pp. 301-20, 1981.

APPENDIX

Multiple roots case for Lemma 1.

If λ_r is a root of multiplicity m, then from Wilkinson [16], the perturbation series can be expressed as

$$\lambda_r(\epsilon) = \lambda_r + \epsilon^{\frac{1}{m}} b_1 + \epsilon^{\frac{2}{m}} b_2 + \epsilon^{\frac{3}{m}} b_3 + \cdots \tag{26}$$

Substituting equation (26) into equation (3), taking the Taylor series expansion and using the fact that the coefficients of like powers of ϵ should vanish, we obtain, for example,

$$b_1 = \left[\frac{m!\, \hat{C}(\lambda_r)}{C^m(\lambda_r)}\right]^{\frac{1}{m}}.$$

Finally, we have

$$\lambda_r(\epsilon) = \lambda_r + \left[\frac{-2\hat{C}(\lambda_r)}{C''(\lambda_r)}\right]^{\frac{1}{2}} \epsilon^{\frac{1}{2}} + \frac{C'''(\lambda_r)\hat{C}(\lambda_r) - 3C''(\lambda_r)\hat{C}'(\lambda_r)}{3\left[C''(\lambda_r)\right]^2}\epsilon + \cdots$$

NEW DIRECTIONS IN THE CONTROL OF NONLINEAR MECHANICAL SYSTEMS

Firdaus E. Udwadia[1] and Robert E. Kalaba[2]

[1]Professor of Mechanical and Civil Engineering & Decision Systems
[2]Professor of Biomedical and Electrical Engineering & Economics
University of Southern California
Los Angeles
CA 90089-1453

ABSTRACT

The newly developed equations of motion for constrained mechanical systems are used as a means for the explicit determination of the control forces needed to cause mechanical systems to satisfy given constraints, such as motion along a given curve in phase space. Two simple examples are described utilizing this new approach to the control of nonlinear mechanical systems.

INTRODUCTION

Consider a nonlinear mechanical system, consisting of n particles, described in a rectangular inertial frame of reference by the equation

$$M\ddot{\mathbf{x}} = \mathbf{F}(\mathbf{x},\dot{\mathbf{x}},t). \tag{1}$$

Here the 3n-vector of position is denoted by the vector $\mathbf{x}$, and M is a constant, positive definite, diagonal 3n by 3n matrix. The dots refer to differentiation with respect to time. The "given force" $\mathbf{F}$ and the matrix M are assumed known.

We now desire to control this mechanical system so that, in addition, it satisfies the m consistent equations given by

$$\varphi_i(\mathbf{x},\dot{\mathbf{x}},t) = 0,\ i = 1, 2, \ldots, m, \tag{2}$$

which need not be functionally independent. This set of equations (2) may then be thought of as an additional set of *constraint* equations which the mechanical system defined by equation (1) is further required to satisfy.

On differentiating equations (2) we obtain the equation

$$A(\mathbf{x},\dot{\mathbf{x}},t)\ddot{\mathbf{x}} = \mathbf{b}(\mathbf{x},\dot{\mathbf{x}},t), \tag{3}$$

where the elements of the m by 3n matrix A as well as those of the m-vector $\mathbf{b}$ are known functions of $\mathbf{x}$, $\dot{\mathbf{x}}$ and t.

Mechanics and Control, Edited by R.S. Guttalu
Plenum Press, New York, 1994

Given the set of initial conditions $\mathbf{x}(t_0)$ and $\dot{\mathbf{x}}(t_0)$ which satisfy the constraint equations (2), the equation of motion of the constrained mechanical system (described jointly by equations (1) and (3)) can now be explicitly expressed as (see [1])

$$M\ddot{\mathbf{x}} = \mathbf{F}(\mathbf{x},\dot{\mathbf{x}},t) + M^{1/2}C^{+}(\mathbf{b} - A\mathbf{a}) \tag{4}$$

where the 3n-vector $\mathbf{a} = M^{-1}F(\mathbf{x},\dot{\mathbf{x}},t)$ and the matrix $C = A(\mathbf{x},\dot{\mathbf{x}},t)M^{-1/2}$. The superscript "+" denotes the Moore-Penrose inverse of the matrix C. We note that the vector $\mathbf{a}$ is simply the acceleration corresponding to the unconstrained system as obtained by using equation (1).

EXPLICIT DETERMINATION OF THE CONTROL FORCE

Noting that the presence of the constraints brings into play a set of constraint forces, we can express the equation of motion of the constrained system as

$$M\ddot{\mathbf{x}} = \mathbf{F}(\mathbf{x},\dot{\mathbf{x}},t) + \mathbf{F}^{c}(\mathbf{x},\dot{\mathbf{x}},t)\ , \tag{5}$$

where $\mathbf{F}^{c}(\mathbf{x},\dot{\mathbf{x}},t)$ is the force of constraint brought into play by Nature so that the constraints (2) are exactly satisfied. Comparing equations (4) and (5) we find that the force of constraint can be expressed explicitly as

$$\mathbf{F}^{c}(\mathbf{x},\dot{\mathbf{x}},t) = M^{1/2}C^{+}(\mathbf{b} - A\mathbf{a}). \tag{6}$$

We may alternately think of this force as the "control force" exerted by Nature so that the unconstrained (uncontrolled) system given by equation (1) is forced to satisfy the constraint set (2) in the presence of the 'given' forces. Thus we have obtained an explicit expression for a control force which would cause the system to satisfy the constraints. We next give two examples where this control philosophy is illustrated.

EXAMPLES

(1) Let a particle of unit mass be located in a Cartesian coordinate frame of reference at x=0, y= L at time t =0. The particle is subjected to the downward force of gravity acting along the positive y direction. We want this particle to follow the given trajectory

$$x^2 + y^2 = L^2, \tag{7}$$

where L is a given constant. We are desirous of finding a control force that will do this for us, given that the particle's initial velocity at time t =0 is $\dot{x}(0) = \dot{x}_0$ and $\dot{y}(0) = 0$. (Note that our initial conditions satisfy equation (7)).

Since the equations of motion of the unconstrained (uncontrolled) system can be written as

$$\begin{aligned} \ddot{x} &= 0 \\ \ddot{y} &= g, \end{aligned} \tag{8}$$

the matrix M = I, and the acceleration $\mathbf{a}$ of the uncontrolled system is simply the vector $[0 \quad g]^T$. Differentiating equation (7) twice we obtain the equation

$$[x \quad y]\begin{bmatrix}\ddot{x}\\ \ddot{y}\end{bmatrix} = -(\dot{x}^2+\dot{y}^2) \tag{9}$$

so that the matrix $C = A = [x \quad y]$, and the vector **b** is simply the scalar $-(\dot{x}^2+\dot{y}^2)$. The Moore-Penrose inverse of C is given by

$$C^+ = \frac{1}{x^2+y^2}\begin{bmatrix}x\\ y\end{bmatrix} = \frac{1}{L^2}\begin{bmatrix}x\\ y\end{bmatrix} \tag{10}$$

so that the "control force" that Nature would use to control this system would be simply given by equation (6) as

$$\mathbf{F}^c(\mathbf{x},\dot{\mathbf{x}},t) = -\frac{1}{L^2}\begin{bmatrix}x\\ y\end{bmatrix}(\dot{x}^2+\dot{y}^2+gy). \tag{11}$$

It must be remembered that there are a large number of "control forces" which will cause the particle to follow the trajectory specified by equation (7). *One* such set of control forces, based on our observation of the way Nature operates, is given by equation (11). These control forces, in the presence of the external influence of gravity, will ensure that the particle will follow the desired trajectory. Hence the equation of motion of the "controlled" system is given by

$$\begin{bmatrix}\ddot{x}\\ \ddot{y}\end{bmatrix} = \begin{bmatrix}0\\ g\end{bmatrix} - \frac{1}{L^2}\begin{bmatrix}x\\ y\end{bmatrix}(\dot{x}^2+\dot{y}^2+gy); \quad x(0)=0,\ y(0)=L,\ \dot{x}(0)=\dot{x}_0,\ \dot{y}(0)=0. \tag{12}$$

This is nothing other than the equation of motion of a pendulum of length L, as can be easily verified!

(2) Consider a particle of mass m subjected to the forces F_x, F_y, and F_z, in the x, y, and z Cartesian coordinate directions respectively. These forces are given functions of $x,\dot{x},y,\dot{y},z,\dot{z}$ and t. Let us require that in the presence of these forces, the particle be controlled so that it follows a trajectory along which

$$\dot{y}(t) = z(t)\dot{x}(t) + \sin(\alpha t) \tag{13}$$

where α is a given constant. The initial conditions of the particle at time t=0 are given by $x(0)=a,\ y(0)=b,\ z(0)=c,\ \dot{x}(0)=d,\ \dot{y}(0)=cd,\ \dot{z}(0)=e$.

A control force which would be capable of doing this can be explicitly found as follows. The unconstrained (uncontrolled) motion of the particle is given by

$$m[I]\begin{bmatrix}\ddot{x}\\ \ddot{y}\\ \ddot{z}\end{bmatrix} = \begin{bmatrix}F_x(x,\dot{x},y,\dot{y},z,\dot{z},t)\\ F_y(x,\dot{x},y,\dot{y},z,\dot{z},t)\\ F_z(x,\dot{x},y,\dot{y},z,\dot{z},t)\end{bmatrix}. \tag{14}$$

Differentiating equation (13) once we get

$$[-z \quad 1 \quad 0]\begin{bmatrix}\ddot{x}\\ \ddot{y}\\ \ddot{z}\end{bmatrix} = \dot{z}\dot{x} + \alpha\cos(\alpha t) \tag{15}$$

so that the matrix $A = [-z \quad 1 \quad 0]$, and the vector $\mathbf{b}$ is the scalar $\dot{z}\dot{x} + \alpha\cos(\alpha t)$. Using equation (6) we get the control force to be

$$\mathbf{F}^c = \frac{1}{1+z^2}\begin{bmatrix}-z\\ 1\\ 0\end{bmatrix}\left\{m\dot{z}\dot{x} + m\alpha\cos\alpha t + zF_x - F_y\right\}. \tag{16}$$

This control force, in the presence of the given forces F_x, F_y, and F_z will cause the particle to follow a trajectory that satisfies equation (13), given that it has the prescribed initial conditions. The equation of motion of the controlled dynamic system will then be

$$m[I]\begin{bmatrix}\ddot{x}\\ \ddot{y}\\ \ddot{z}\end{bmatrix} = \begin{bmatrix}F_x(x,\dot{x},y,\dot{y},z,\dot{z},t)\\ F_y(x,\dot{x},y,\dot{y},z,\dot{z},t)\\ F_z(x,\dot{x},y,\dot{y},z,\dot{z},t)\end{bmatrix} + \frac{1}{1+z^2}\begin{bmatrix}-z\\ 1\\ 0\end{bmatrix}\left\{m\dot{z}\dot{x} + m\alpha\cos\alpha t + zF_x - F_y\right\}. \tag{17}$$

CONCLUSIONS

In this paper we have presented a new approach to the control of nonlinear mechanical systems through the use of the newly developed equation of motion for constrained systems. These ideas are only preliminary; as such, a great deal more research needs to be done. In particular, issues such as the uniqueness, the existence and the stability of equation (4) need to be attended to before we can go much further. We leave these issues for future communications.

REFERENCE

1. Udwadia F. E. and R. E. Kalaba, "A New Perspective on Constrained Motion, " *Proceedings of the Royal Society of London*, vol. 439, pp. 407-410, 1992.

PARTIALLY SINGULAR CONTROL TRAJECTORIES REVISITED

David J. Bell

Department of Mathematics
UMIST
P.O.Box 88
Sackville Street
Manchester, M60 1QD
U.K.

ABSTRACT

Some necessary conditions for optimality are known at any point where a nonsingular subarc is joined to a singular subarc to form an optimal control trajectory. However, certain results still await verification. This paper presents those results which are known and reports on a further contribution in the investigation of McDanell's conjecture.

INTRODUCTION

The existence of partially singular optimal control trajectories has been known for three decades [9]. Since the early 1960's the singular subarcs which form part of these control trajectories have been investigated by many researchers. Both necessary and sufficient conditions for optimality are now well-established [3]. Necessary conditions have also been obtained at junctions where a singular subarc joins a nonsingular one. The earliest attempt at obtaining such junction conditions was carried out by Kelley, Kopp and Moyer [6] and generalised later by McDanell and Powers [8]. Further work remains to be done particularly in resolving the question of McDanell's conjecture. This paper draws attention to a further contribution [1] towards the final resolution of this latter problem. The known results on junction conditions are also reviewed.

THE FUNDAMENTAL PROBLEM

The optimal control problem to be considered is that of controlling a system governed by the vector equation

$$\dot{x}(t) = f(x(t)) + u(t)\ g(x(t)) \tag{1}$$

where $x(t) \in \mathbb{R}^n$, $x(t_0) = x_0$, $u(t) \in \mathbb{R}$, $|u(t)| \le 1$, $t_0 \le t \le t_f$, t_0, t_f specified, so that a performance index

$$G[x(t_f)] + \int_{t_0}^{t_f} [L_0(x(t)) + u(t)\ L_1(x(t))]dt$$

is minimised.

The switching function for this problem is given by

$$\phi(t) = \frac{\partial H}{\partial u}(x(t), \lambda(t))$$

where the Hamiltonian function H is defined by

$$H(x,u,\lambda) = L_0(x) + \lambda^T f(x) + [L_1(x) + \lambda^T g(x)]u$$

and the adjoint vector $\lambda(t)$ satisfies the vector differential equation

$$\dot{\lambda}(t) = -\frac{\partial H}{\partial x}(x(t), \lambda(t), u(t)).$$

On a singular subarc the control is such that $\phi(t) = 0$ for all t, whereas on a nonsingular subarc the control is determined by Pontryagin's Principle as

$$u(t) = -\operatorname{sgn} \phi(t) .$$

Since the control in (1) is scalar it will first appear explicitly in an even time derivative of the switching function, say

$$\frac{d^{2q}}{dt^{2q}}\left(\frac{\partial H}{\partial u}\right)(x,\lambda,u) = A(x,\lambda) + B(x,\lambda)u .$$

The positive integer q is the *order* of the singular problem. One of the necessary conditions for optimality of a singular subarc is the Generalised Legendre Clebsch condition which, for all t on the subarc, is

$$(-1)^q A(x(t), \lambda(t)) \geq 0 .$$

Let

$$\alpha(t) = A(x(t), \lambda(t)),$$
$$\beta(t) = B(x(t), \lambda(t)),$$

$u^{(r)}$($r \geq 0$) be the lowest time derivative of u which is discontinuous at a junction, and $\beta^{(m)}$($m \geq 0$) the lowest time derivative of β which is nonzero at that junction.

JUNCTION CONDITIONS

The control is either piecewise analytic in a neighbourhood of a junction or else the junction is nonanalytic. An example of a nonanalytic junction has been given by Fuller [5]. At first it was thought that no singular problem of odd order could have a junction at which the control was continuous. However, a counterexample to this was given by McDanell and Powers [8]. Also, there was a period of time during which it was thought that all singular problems of even order have nonanalytic junctions. Again this was proved false, this time by Maurer [7].

The following important junction condition was obtained by McDanell and Powers [8]. If a control is piecewise analytic in a neighbourhood of a junction and if the Generalised Legendre Clebsch condition is satisfied in strengthened form at the junction, then the integer q + r is odd.

McDANELL'S CONJECTURE

This conjecture can be expressed in the following way. If $m > 0$, then

(i) $q + r + m$ is an odd integer when $m \leq r$

(ii) $- \operatorname{sgn} [\beta^{(m)}(t_c^{+}) \, \beta^{(m)}(t_c^{-})] = (-1)^{q+r+m}$ when $m > r$,

where t_c is the time at a junction.

Bell and Boissard [2] showed that this conjecture is true for $m = 1$ and $q = 1$ while Bortins [4] showed that it was true for $m = 1$ and $q > 1$. However, Pan and Bell [10] have given a counterexample to (ii). The validity or otherwise of (i) for $m > 1$ is not yet known but Bell [1] has presented some evidence to suggest that this part of the conjecture is also untrue in general.

CONCLUSIONS

Optimality conditions at junction points where nonsingular and singular subarcs of optimal trajectories are joined have been reviewed. The status of McDanell's conjecture is still not completely established.

REFERENCES

1. D.J. Bell, "Optimality conditions at junctions of singular and nonsingular controls", *J. Opt. Th. Applics.*, to appear, 1993.
2. D.J. Bell and M. Boissard, "Necessary conditions at the junction of singular and nonsingular control subarcs", *Int. J. Contr.*, vol. 29, pp. 981-990, 1979.
3. D.J. Bell and D.H. Jacobson, *Singular Optimal Control Problems.* Academic Press: London, 1975.
4. R. Bortins, "On joining nonsingular and singular optimal control subarcs", *Int. J. Contr.*, vol. 37, pp.867-872, 1983.
5. A.T. Fuller, "Study of an optimum nonlinear control system", *J. Electronics Contr.*, vol. 15, pp. 63-71, 1963.
6. H.J. Kelley, R.E. Kopp and H.G. Moyer, "Singular extremals", in *Topics in Optimization*, (G. Leitmann, ed.), pp.63-101. Academic Press: New York, 1967.
7. H. Maurer, "An example of a continuous junction for a singular control problem of even order", *SIAM J. Contr.*, vol. 13, pp. 899-903, 1975.
8. J.P. McDanell and W.F. Powers, "Necessary conditions for joining optimal singular and nonsingular subarcs", *SIAM J. Contr.*, vol. 9, pp. 161-173, 1971.
9. A. Miele, "Extremization of linear integrals by Green's Theorem", in *Optimization Techniques with Applications to Aerospace Systems*, (G. Leitmann, ed.) pp.69-98. Academic Press: New York, 1962.
10. G.L. Pan and D.J. Bell, "A counterexample to a conjecture of McDanell", *Syst. Contr. Letters*, vol. 9, pp.85-87, 1987.

APPLICATIONS TO AEROSPACE SYSTEMS

PURSUIT-EVASION DIFFERENTIAL GAMES APPLIED TO A 3-DIMENSIONAL MISSILE GUIDANCE PROBLEM USING THE LIAPUNOV APPROACH

Nigel John Conrad Greenwood

Department of Mathematics
University of Queensland
St. Lucia, Queensland 4072
Australia

ABSTRACT

In this paper we shall outline a technique using Liapunov functions and sufficiency conditions to estimate the winning region for a three-phase missile entering its coast phase. We shall use point-mass dynamics with variable mass to model the missile trajectories in three dimensions. We will then use this estimated winning region to provide a procedure whereby we can judge the robustness of a particular three dimensional guidance law when one or more of its parameters has bounds imposed upon it. This technique effectively establishes a benchmark against which guidance laws with different bounds upon their parameters can be compared. We shall then give an example using a pursuit-evasion scenario between an AMRAAM AIM-120A missile and a F-15E "Strike Eagle" fighter.

INTRODUCTION

The Formulation of the Game

The study of differential games arose from the desire to provide a mathematical perspective through which to study systems involving conflict between two or more *players*. The system is described by a state vector $\mathbf{z}$, which can be influenced by the players using *control variables* $\mathbf{u}_i$. We shall restrict our attention to systems involving conflict between two players with mutually conflicting objectives.

The dynamics of the game can be represented mathematically by the state equation

$$\dot{\mathbf{z}} = \mathbf{f}(\mathbf{z}, \mathbf{u}_1, \mathbf{u}_2, t) \tag{1}$$

Mechanics and Control, Edited by R.S. Guttalu
Plenum Press, New York, 1994

where the state vector $\mathbf{z}(t)$ varies for $t \geq t_0$ along the solution trajectories

$$\boldsymbol{\varphi}(\mathbf{z}^0, t_0, t)$$

in a specified open bounded playing set Δ within the state space $\mathbb{R}^N$, with the control vectors $\mathbf{u}_i$, $i = 1, 2$ varying in specified compact sets $\mathcal{U}^i$ of control constraints within the spaces of control values $\mathbb{R}^{r_i}$, $i = 1, 2$.

Differential games may be classified in either one of two categories, namely either *qualitative* or *quantitative* differential games. In the next two sections we shall consider each of these in turn. It should be noted that as well as qualitative or quantitative formulations, it is possible to use either deterministic or stochastic techniques (cf. [2]). In *deterministic games* the system of ordinary differential equations is usually describing the dynamic behaviour of both players. In *stochastic games* the governing equation includes some Gaussian noise modelling the uncertainties associated with states and/or unmodelled dynamics.

Qualitative Differential Games

The study of qualitative differential games addresses itself to the issue of whether there exist trajectories $\boldsymbol{\varphi}$ issuing from a particular point $\mathbf{z}^0 \in \Delta$ such that a particular qualitative objective Q(i) for player i is attainable on $\boldsymbol{\varphi}$. A typical qualitative objective for a pursuer in an air combat pursuit-evasion gane would be "interception of the evader before the evader escapes to a region of safety". The emphasis in a qualitative game is to establish a "map of the Game of Kind" in the sense of Isaacs, [13]; i.e. to dissect Δ into subsets based upon whether all of the points for a particular subset can be used to achieve the specified qualitative objective Q(i). Consider the following two definitions, taken from Skowronski and Stonier, [24].

Definition 1 (i-Controllability) The system is *i-controllable* at $(\mathbf{z}^0, t_0)$ for Q(i) iff there exist control programs $\mathbf{p}_*^1, \mathbf{p}_*^2$ such that all of the solution trajectories to

$$\dot{\mathbf{z}} \in \left\{ \mathbf{f}(\mathbf{z}^0, \mathbf{u}_1^*, \mathbf{u}_2^*, t_0) \,\middle|\, \mathbf{u}_1^* = \mathbf{p}_*^1(\mathbf{z}(t)),\ \mathbf{u}_2^* = \mathbf{p}_*^2(\mathbf{z}(t)) \right\} \tag{2}$$

exhibit Q(i). We call the set of all such $\mathbf{z}^0$ the *region of i-controllability* Δ_q^i.

Definition 2 (Strong i-Controllability) The system is called *strongly i-controllable* at $(\mathbf{z}^0, t_0)$ for Q(i) iff there exists a control program $\mathbf{p}_\star^i$ such that all of the solution trajectories to

$$\dot{\mathbf{z}} \in \left\{ \mathbf{f}(\mathbf{z}^0, \mathbf{u}_i^\star, \mathbf{u}_j, t_0) \,\middle|\, \mathbf{u}_i^\star = \mathbf{p}_\star^i(\mathbf{z}(t)),\ \mathbf{u}_j \in \mathcal{U}^j \right\} \tag{3}$$

exhibit Q(i) regardless of any strategies of player j, $j \neq i$. We call the set of all such $\mathbf{z}^0$ the *region of strong i-controllability* Δ_Q^i, and we call $\mathbf{p}_\star^i(.)$ the *winning strategy.*

Note that an analysis of the game of kind often does not address itself to determining the actual winning strategy; indeed, the problem of calculating such a strategy is often far more difficult than calculating the region for which it exists.

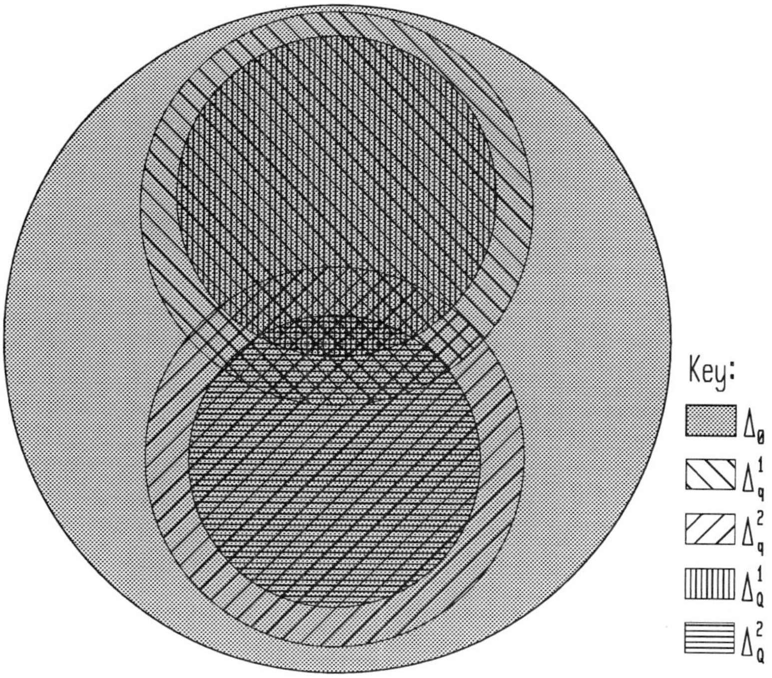

Figure 1. The generic map of the Game of Kind.

We can use the sets Δ_q^i and Δ_Q^i, $i = 1, 2$ to establish the general form for the map of the Game of Kind, illustrated in Figure 1. For our pursuit-evasion game we shall denote the evader as player "1" and the pursuer as player "2".

Consistent with the previous definitions, the map of the Game of Kind can be dissected as follows:

	Δ_q^1	is the *1 – Controllable Region.*
	Δ_q^2	is the *2 – Controllable Region.*
	Δ_Q^1	is the *Region of 1 – Strong Controllability.*
	Δ_Q^2	is the *Region of 2 – Strong Controllability.*
	Δ_n	$\triangleq \Delta_0 \setminus (\Delta_q^1 \cup \Delta_q^2)$ is the *Dead Zone.*
W_1	$\triangleq$	$\Delta_Q^1 \setminus \Delta_Q^2$is the *1 – Winning Region.*
W_2	$\triangleq$	$\Delta_Q^2 \setminus \Delta_Q^1$is the *2 – Winning Region.*
W_{12}	$\triangleq$	$\Delta_Q^1 \cap \Delta_Q^2$ is the *Mutual Kill Region.*
RNGO	$\triangleq$	$(\Delta_q^1 \cap \Delta_q^2) \setminus (\Delta_Q^1 \cup \Delta_Q^2)$ is the *Region of No Guaranteed Outcome.*

Now we shall turn our attention to quantitative differential games.

Quantitative Differential Games

The second form of game is the *quantitative game*, otherwise known as the *Game of Degree.* This formulation addresses itself specifically to the problem of determining an optimal strategy to achieve a particular objective. It does not address the question of whether the desired objective is actually attainable, but tacitly assumes that this is the case and then proceeds to calculate a control program which, when implemented, will optimise some payoff function (such as time before interception, miss-distance, fuel or energy).

Most of the literature on differential games appears to be dedicated to the analysis of the Game of Degree formulation of air combat, with relatively little work done on the Game of Kind. This is largely due to the mathematical intractability of the traditional approach originally proposed by Isaacs. This shall be discussed further in the next section.

Literature Review: Air Combat as a Differential Game

Methods based on differential games approach have been applied to the guidance and control problems since the mid-fifties. The particular problem was to solve the mathematical model of aerial combat between two or more fighter-aircraft. This also can be easily re-formulated as a problem of missile guidance.

Isaacs [13] originally developed the basic differential game definitions and tools in the 1950's while working for RAND Corporation on mathematical modelling of aerial combat. Since then his methods have been continually used to determine winning regions or suitable tactics (aircraft manoeuvres) for simplified models of one-to-one air combat. Although his techniques for the Game of Kind have been widely used in the literature, they are extremely dependent upon geometrical arguments; in particular, upon the existence of "barriers" and other "singular surfaces" in the state space. This dependence upon geometry and the requirement to retro-integrate solution trajectories from singular surfaces have until relatively recently rendered calculations of the map of the Game of Kind intractable for any realistic models of air combat, as such models require a state space of up to twelve dimensions and involve highly non-linear O.D.E's. Two of the better examples of the Isaacs approach to the map of the game by analysis of surfaces using necessary conditions were performed by Getz and Pachter, [8] and [18], where they analysed simplied models of air combat confined to a plane, using both circular and fan-shaped targets.

A great deal of subsequent work has concentrated instead upon the analysis of the Game of Degree. This approach directly confronts the problem of optimal strategies and guidance laws. Anderson [1] presented a comparison between designs of guidance laws based on optimal control and differential games, showing that the major advantage of the latter is reduced sensitivity to errors in the estimates of target manoeuvre.

Recently Moritz, Polis and Well [17] have analysed aircraft tactics combined with the simulation of missiles operating under pure proportional navigation. Their analysis is based on zero- and non-zero sum differential games with the solution of the latter defined as Nash equilibrium. Their work is an extension of problems previously described by Järmark [14] and Herbst [12].

Merz [16] discussed the use of optimal differential games for rôle determination in the aerial combat. Davidovitz and Shinar [5] also analysed air combat as a zero-sum differential game, although they investigated a duel between aircraft of different speed and manoeuverability (a slower, but very manoeuverable Harrier and a faster conventional fighter aircraft) equipped with "all-aspect"missiles.

Breitner et al [3] concentrated on an air combat between a medium range air-to-air missile and a high performance aircraft. The air combat described in [3] was based on the available data on the thrust profile and drag coefficients for a F-15E *Strike Eagle* fighter and for the medium range air-to-air missile of the type of *AMRAAM AIM-120A*.

Criticism of quantitative analysis in recent years has restored attention to the study of qualitative analysis. There exist two major criticisms of the quantitative approach.

1. The first is of the computational complexity associated with optimization. The calculation of an optimal strategy for a missile attempting interception of an aircraft using full dynamics in three dimensions involves solving a Two Point Boundary Value Problem (TPBVP) in twelve dimensions. Although recent developments have been made which may make such calculations feasible for aeroplanes (cf. [15]), they are unlikely to do so for on-board computations in missiles.

2. The second criticism is that the qualitative approach provides no guarantee of success. Considering the expense of a modern air-to-air missile, we want some idea of regions from which interception is guaranteed.

Developments in recent years have made the calculation of winning regions more feasible, largely through the use of Liapunov functions. Such work has been performed with various degrees of complexity by a number of authors, for example Skowronski, [23], Stonier, [24], Getz and Leitmann, [7] and Greenwood [9]. A detailed discussion of the Liapunov technique shall be provided later. At this stage it is sufficient to say that the inherent advantages of this approach render it an attractive tool to use. They are

1. Once a Liapunov function is designed, it is used to define estimated winning regions using sufficiency conditions. For this reason, regions stipulated as winning by the use of a Liapunov function are guaranteed to be so. This is in contrast to the Isaacs approach, which was based upon necessary conditions. It also implies that any estimates for winning regions will err on the conservative side. For this reason, when we are talking about estimated winning regions we shall denote them as $\widehat{W}_i$, to distinguish them from the actual winning regions W_i. Note that $\widehat{W}_i \subseteq W_i$.

2. Calculations using the Liapunov method do not require calculation of solution trajectories at any stage. The qualitative behaviour of trajectories can be determined in a computationally simple way, by determining for what values of the control and state vectors the derivative of the Liapunov function is negative. This means that systems of large dimension and highly non-linear behaviour may be readily analysed.

However, there exist two main disadvantages with such work.

1. The objective of this approach to the "Game of Kind" (in the sense of Isaacs, [13]) is to establish a guaranteed winning region for the pursuer.

   Expressed mathematically, it seeks to determine the region $W_2 \subseteq \mathbb{R}^N$ for which there exists a control program $p^2_\star : \mathbb{R}^N \to \mathbb{R}$ so that *regardless* of any strategy played by the evader, the pursuer's objectives are attained. Provided the pursuer remains within his winning region and plays his optimal strategy $p^2_\star$, then he is guaranteed successful interception of his quarry.

Unfortunately, this approach does not necessarily specify the actual strategy $p_\star^2$ to be used- it merely identifies the regions for which it exists.

2. The other problem is that it is much more difficult to use the Liapunov technique for non-autonomous systems than for autonomous ones. Although this is no problem when considering the map of the game for systems which can be approximated by an autonomous model, such as that of two aircraft engaged in combat, it becomes significant when we come to analyse inherently time-dependent systems such as those for a three-phase air-to-air missile attempting interception of an aircraft.

Guidance Law Verification by Analysis of Winning Regions

For this paper we shall be using the map of the game in a rather unusual fashion. If we have a guidance law which we wish to test for efficiency, we can calculate a winning region $\widehat{W}_2$ for which there exists some guaranteed winning strategy for interception of an evader. We can then run simulations for a missile starting from points within $\widehat{W}_2$, relying upon this guidance law to hit its target. If there exist positions within $\widehat{W}_2$ for which the missile fails to strike its evader, then there exists a superior guidance law which the missile could have used instead of the implemented law.

The Guidance Law: Proportional Navigation

The guidance law under discussion is the celebrated **Proportional Navigation Law**. Developed during the Second World War independently in Germany and in the US (cf. [28]), it is still being implemented in many advanced missiles.

The aim of the PN guidance is to produce the lateral acceleration to re-direct the missile on the predicted collision course with the target. The magnitude of the lateral acceleration a_L is proportional to the line-of-sight rate $\dot{\lambda}$ and the magnitude of the closing (relative) velocity $\mathbf{v_c}$:

$$a_L = N\dot{\lambda}v_c \tag{4}$$

where $3 \leq N \leq 5$.

While the above formula is generally accepted, the possible implementations of the PN law may still differ depending on coordinate system under which the guidance law is derived. When the lateral acceleration is applied normal to the line-of-sight (i.e. with respect to the missile-body coordinates) the resulting PN is referred to as *True* Proportional Navigation (TPN) (cf. Guelman, [10]). When the lateral force is applied orthogonally to the missile velocity vector it is referred to as *Pure* Proportional Navigation (PPN), while in the case of applying the correcting acceleration in a direction normal to the relative velocity vector the resulting guidance law is called *Ideal* Proportional Navigation (IPN) (see Yuan, [26]).

It has been shown in literature that the implementation of the guidance law (effectively the coordinates system under which the missile operates) is one of the factors which determines the envelope of the missile. While the True PN is perhaps easiest to implement it is not very robust and the resulting envelope (or capture volume) of the missile is smaller than under Pure or Ideal PN. While the line-of-sight (LOS) and the LOS-rate are readily available (given by the seeker), information about the velocity vector is produced by the missile's Inertial Navigation System (INS). Since not all missiles could be equipped with the full INS [1], the guidance must rely only on information available in real time from the seeker. It is claimed that the Ideal PN gives the largest capture region (synonymous with the winning region in the parlance of differential games), but the full information on target dynamics is required for the succesful implementation of ideal PN. This particular requirement makes the ideal PN rather difficult to implement. The real-time estimation (viz. Kalman filtering techniques) can give some information on target manoeuvres, but it is subject to the usual pitfalls of linear estimation techniques.

This paper will present the investigation of winning regions of missiles using the Pure Proportional Navigation law for their guidance, where the missile flight is modelled using point-mass dynamics with variable mass in a three-phase trajectory. The trajectories are simulated in three dimensional Cartesian space. Not many unclassified papers discuss 3D guidance laws and resulting 3D trajectories of missiles. Authors are usually content with the linear approximation of the dynamics and 2-dimensional simulation. The reason for the three dimensional simulation is to include the effect of the realistic target manoeuvre on the missile guidance (particularly the component of the target acceleration normal to the instantaneous missile flyplane - see the discussion in [6] on three-dimensional homing, pp. 234-235).

DESCRIPTION OF THE PROBLEM

Objectives

The objective of this paper is to provide a technique whereby the robustness of a guidance law can be tested when bounds are imposed upon the line-of-sight rate observable by the missile. It can be regarded as divided into two components.

1. First of all we outline a technique whereby winning regions for the pursuer can be calculated using a Liapunov function and sufficiency conditions, rather than by the use of necessary conditions. This is a technique which is computationally simple (although sometimes algebraically complicated for multiple objectives) and which provides the existence of a strategy which guarantees success. We then give the example of a particular winning region $\widehat{W}_2$ estimated this way. This winning region can be regarded as a missile target in its own right, to be reached at the beginning of the final phase of missile flight.

2. Secondly, we shall test different versions of the same guidance law upon $\widehat{W}_2$, in order to ascertain the effect of placing bounds upon particular parameters. In this case we have chosen to study the line-of-sight rate. The same technique could easily be applied to studying the navigation constant or other parameters in a particular guidance law. This is the component of the paper which requires computer simulation. As the simulation model uses an accelerating target flying through a realistic atmosphere with point-mass aircraft dynamics, this should provide some useful insights as regards acceptable values for these bounds.

[1]It usually consists of triads of orthogonal gyroscopes and accelerometers. The output of gyroscopic instruments gives the angular rates, and (after integration) the orientation (or attitude) angles. The output of accelerometers gives the specific forces, which after integration produces the velocity and position vectors.

The Fighter-Missile Scenario

The scenario which we studied is as follows:

There exists an unarmed high-performance fighter, designated the *evader* (in this case an F-15E Strike Eagle) which is engaging in an evasive manoeuvre from another aircraft, designated the *launch platform*. This second aircraft is armed with a medium range air-to-air missile, designated the *pursuer* (here, an AMRAAM AIM-120A), which is launched at time t=0. This missile is programmed with a PPN guidance law, and will attempt to intercept the evader before its energy expires.

In order to avoid the missile the F-15 is is flying at full throttle, and is engaged in a shallow dive. Once it commences this manoeuvre it follows an open loop program and does not respond to any further actions of the pursuer.

The missile has a three-phase trajectory. Following the terminology of Breitner et al, [3], the phases are described as "boost", "march" and "coast" phases and occur for the time intervals $[0,3), [3,15)$ and $[15,\infty)$ seconds respectively. As fuel is consumed the missile loses weight until it reaches the coast phase, when the fuel is completely exhausted and the missile depends upon its accumulated energy to ensure interception.

We shall assume the missile is modelled by a point mass in space, which is at the centre of a hypothetical sphere of radius 100 metres. This sphere represents the missile's "kill" zone. During the encounter the missile must pass within 100 metres of the F-15 for a successful interception.

After missile launch at t=0 the launch platform shall be completely ignored for the rest of the game.

The Coordinate System and State Equations

We shall assume a Cartesian geodetic coordinate system, with an arbitrary but fixed Origin on the ground. The X and Y ground coordinates are taken relative to this Origin. The horizontal heading (*yaw*) angles are taken to be anticlockwise from the X axis. Using the convention that the subscript i=1 denotes the evader, i=2 denotes the pursuer, we have the equations of motion for the system as follows:

$$\dot{x}_i = v_i \cos \chi_i \cos \gamma_i \quad (5)$$

$$\dot{y}_i = v_i \sin \chi_i \cos \gamma_i \quad (6)$$

$$\dot{h}_i = v_i \sin \gamma_i \quad (7)$$

$$\dot{v}_i = \frac{[\sigma_i \, T_M(M_i, h_i, t) - D_i(M_i, n_i, t)]}{m_i(t)} - g \sin \gamma_i \quad (8)$$

$$\dot{\gamma}_i = \frac{g}{v_i}[n_i \cos \theta_i - \cos \gamma_i] \quad (9)$$

$$\dot{\chi}_i = \frac{g \, n_i \sin \theta_i}{v_i \cos \gamma_i} \quad (10)$$

where

x_i, y_i are the Cartesian ground coordinates for the ith aircraft,
h_i is the altitude,
v_i is the aircraft speed,

χ_i is the horizontal flight path angle,
γ_i is the vertical flight path angle,
m_i is the mass,
^iT_M is the maximum thrust available,
σ_i is the throttle,
θ_i is the bank angle,
and n_i is the load factor.

This system of equations is commonly found in the literature; see Moritz et al [17], Breitner et al [3] and Greenwood [9].

The controls are taken to be the load factor n_i, the bank angle θ_i and (in the case of the F-15 evader) the throttle σ_1. The throttle for the missile is taken to be fixed at $\sigma_2 = 1$ for the boost and march phases; during the coast phase it is irrelevant, as all fuel has been expended.

Atmospheric and Aerodynamic Data

We shall assume that the missile weight is a function of time, as fuel is expended, and that the rate of expenditure depends on the thrust phase.

$$m_2 = m_2(t) \tag{11}$$

We assume the F-15's change of weight during the encounter is negligable, so $\dot{m}_1 = 0$ $\forall t > 0$.

If we denote Mach number by $M_i(h_i, v_i)$, the reference wing surface area by S_i and the lift coefficient by ^iC_L, then the aerodynamic lift L_i for both the pursuer and the evader is given by the equation

$$\begin{aligned} L_i(n_i, t) &= n_i(t) m_i(t)\, g \\ &= {}^iC_L(n_i, M_i, t) \frac{S_i\, \rho(h_i)\, v_i^2}{2}. \end{aligned} \tag{12}$$

Furthermore, the drag D_i is given by

$$D_i(n_i, h_i, v_i, t) = \frac{S_i\, \rho(h_i)\, v_i^2}{2}\, {}^iC_D(n_i, M_i, t), \tag{13}$$

where the drag coefficient ^iC_D is assumed to be of the form of a parabolic drag polar, so that

$${}^iC_D(n_i, M_i, t) = {}^iC_{D_0}(M_i) + {}^iC_L(n_i, M_i, t)^2\, {}^iC_{D_I}(M_i), \tag{14}$$

where $^iC_{D_0}$ is the *zero-lift drag coefficient* and $^iC_{D_I}$ is the *induced drag coefficient.*

The equations for the atmospheric data shall be taken from Breitner et al, [3]. In that paper, the atmospheric density $\rho(h)$ is given by the equation

$$\rho(h) = \exp[r_6 + r_5\, h + \exp(r_4\, h^4 + r_3\, h^3 + r_2\, h^2 + r_1\, h + r_0)] \tag{15}$$

and the speed of sound a(h) is given by the expression

$$a(h) = \sqrt{a_3\, h^3 + a_2\, h^2 + a_1\, h + a_0} \tag{16}$$

According to [3], the relative error between equation (15) and the US standard atmosphere varies between -6% and $+2\%$, whereas that for equation (16) and the speed of sound for the standard atmosphere varies between -1% and $+6\%$.

Once we have an expression for the speed of sound we can calculate the Mach number, defined to be

$$M_i(h_i,\ v_i) = \frac{v_i}{a(h_i)}. \tag{17}$$

Analytic approximations for the aeronautical coefficients of the F-15E are also provided in [3]. The following functions are used:

$$^1C_{D_0} = \frac{b_4\ M_1^4 + b_3\ M_1^3 + b_2\ M_1^2 + b_1\ M_1 + b_0}{\bar{b}_4\ M_1^4 + \bar{b}_3\ M_1^3 + \bar{b}_2\ M_1^2 + \bar{b}_1\ M_1 + \bar{b}_0} \tag{18}$$

$$^1C_{D_I} = \frac{c_4\ M_1^4 + c_3\ M_1^3 + c_2\ M_1^2 + c_1\ M_1 + c_0}{\bar{c}_4\ M_1^4 + \bar{c}_3\ M_1^3 + \bar{c}_2\ M_1^2 + \bar{c}_1\ M_1 + \bar{c}_0} \tag{19}$$

$$T_1 = \sigma_1\ {}^1T_M \tag{20}$$

$$= \sigma_1 \sum_{i=0}^{5}\sum_{j=0}^{5} d_{ij}\ h_1^i\ M_1^j \tag{21}$$

It should be noted that the denominators of the rational approximations (18) and (19) have no real zeros in the range $M_1 \in [0,\ 5.4]$.

Similarly the drag coefficients for the AIM-120A are approximated by the following functions:

$$^2C_{D_0} = e_3\ \ln\{1 + \exp[\frac{e_2}{e_3}\ (3 - M_2)]\} + e_1 \tag{22}$$

$$^2C_{D_I} = \bar{e}_1 \tag{23}$$

Consistent with our previous comments, the thrust is given by

$$^2T_M = \begin{cases} {}^2T_b, & t \in [0,3) \\ {}^2T_m, & t \in [3,15) \\ 0, & t \in [15,\infty). \end{cases} \tag{24}$$

The PPN Guidance Law

We shall use for our PPN missile guidance laws versions of the guidance law presented in Moritz et al [17], using the load factor n_2 and the bank angle θ_2 as controls. From [17] the guidance law for these controls under PPN has the following derivation:

We shall denote the displacement between the pursuer and evader by

$$\mathbf{r_{12}} = (\ x_1 - x_2,\ y_1 - y_2,\ h_1 - h_2\ )^T, \tag{25}$$

so that the vector of closing velocity is

$$\mathbf{v_c} = (\dot{x}_1 - \dot{x}_2, \dot{y}_1 - \dot{y}_2, \dot{h}_1 - \dot{h}_2)^T, \tag{26}$$

$$= \mathbf{v_1} - \mathbf{v_2}, \tag{27}$$

where the evader and pursuer velocity vectors are given by the equations

$$\mathbf{v_1} = (\ \dot{x}_1,\ \dot{y}_1,\ \dot{h}_1)^T, \quad |\ \mathbf{v_1}\ | = v_1, \tag{28}$$

$$\mathbf{v_2} = (\ \dot{x}_2,\ \dot{y}_2,\ \dot{h}_2)^T, \quad |\ \mathbf{v_2}\ | = v_2, \tag{29}$$

and the lateral acceleration vector (normal to the velocity vector) is chosen to be (using the notation of equation (4),

$$\begin{aligned}\mathbf{a_L} &= (\,a_{L1},\, a_{L2},\, a_{L3})^T &(30)\\ &= N\,\dot{\lambda}\,v_c\,\hat{\mathbf{w}} &(31)\end{aligned}$$

where $\hat{\mathbf{w}}$ is a unit vector normal to the velocity vector. This acceleration vector is taken to be

$$\mathbf{a_L} = N\,\frac{\mathbf{v_2} \times (\,\mathbf{v_c} \times \mathbf{r_{12}})}{|\,\mathbf{r_{12}}\,|^2}. \tag{32}$$

If we follow the assumption of [17] and neglect the term $-(\cos\gamma_2)/v_2$ in $\dot{\gamma}_2$ we obtain the control laws

$$\begin{aligned}n_2^* &= \min(\,|\,\mathbf{a_L}\,|\,/g,\, {}^2n_M) &(33)\\ \sin\theta_2^* &= \frac{(\,a_{L2}\cos\chi_2 - a_{L1}\sin\chi_2)}{|\,\mathbf{a_L}\,|} &(34)\\ \cos\theta_2^* &= \frac{a_{L3}}{(|\,\mathbf{a_L}\,|\;\cos\gamma_2)} &(35)\end{aligned}$$

where ${}^2n_M$ denotes the load factor limit for the missile.

We shall use these equations for the missile control laws in our simulation of PPN. We are interested in establishing a technique for comparing guidance laws to compare their effectiveness when bounds are placed on their mathematical parameters. In this context, the physical constraint imposed by the maximum load factor limit in (33) for the particular missile used provides an unnecessary complication and shall be ignored; i.e. we shall assume that the PPN control law for load factor never saturates to the AMRAAM AIM-120A's limit in the course of the interceptions simulated here.

The basic (parameter-unconstrained) guidance law to be used will be denoted PPN1, and is given by the equations

$$\begin{aligned}n_2^I\sin\theta_2^I &= \frac{(\,a_{L2}\cos\chi_2 - a_{L1}\sin\chi_2)}{g}, &(36)\\ n_2^I\cos\theta_2^I &= \frac{a_{L3}}{g\;\cos\gamma_2}. &(37)\end{aligned}$$

If we place constraints upon the observable line-of-sight rate $\dot{\lambda}$, we get the second guidance law, which is given by the equations

$$n_2^{II}\sin\theta_2^{II} = \begin{cases} (\,a_{L2}\cos\chi_2 - a_{L1}\sin\chi_2)/g, & \dot{\lambda} \le \kappa,\\ 0, & \dot{\lambda} > \kappa.\end{cases} \tag{38}$$

$$n_2^{II}\cos\theta_2^{II} = \begin{cases} a_{L3}/(g\;\cos\gamma_2), & \dot{\lambda} \le \kappa,\\ 0, & \dot{\lambda} > \kappa.\end{cases} \tag{39}$$

We shall consider this second guidance law for two values of κ,, namely $\kappa = 0.5$ and $\kappa = 0.1$. We shall refer to the guidance law provided by the first of these cases as PPN2, and the second as PPN3.

Calculation of the Winning Region Using a Liapunov Function

In order to calculate an estimated winning region $\widehat{W}$ for the pursuer we shall use the following theorem for autonomous systems, a version of which appears in Skowronski, [23].

Theorem 3 Given the sets $\mathcal{T}$ (target set) and $\Delta \subseteq \mathbb{R}^N$, then the system

$$\dot{\mathbf{z}} = \mathbf{f}(\mathbf{z}, \mathbf{u}_1, \mathbf{u}_2, t)$$

is strongly controllable on Δ for collision with $\mathcal{T}$ if there exists a program $p_\star^2(.)$ and a $\mathcal{C}^1$ Liapunov function

$$V(.) : \overline{\Delta \setminus \mathcal{T}} \to \mathbb{R}, \tag{40}$$

such that, given constants

$$\nu^+ \triangleq \sup V(\mathbf{z}) \mid \mathbf{z} \in \partial\Delta \tag{41}$$
$$\nu^- \triangleq \inf V(\mathbf{z}) \mid \mathbf{z} \in \partial\mathcal{T} \tag{42}$$

the function V satisfies the following conditions:

1. $V(\mathbf{z}) > \nu^-, \quad \forall \mathbf{z} \notin \mathcal{T}$
2. $V(\mathbf{z}) \leq \nu^+, \quad \forall \mathbf{z} \in \Delta \setminus \mathcal{T}$
3. $\forall \mathbf{u}_2^\star \in p_\star^2(.) \quad \exists T_C$ such that

$$\int_{t_0}^{t_0+T_C} \nabla V(\mathbf{z})^T.\mathbf{f}(\mathbf{z},\ \mathbf{u}_1,\ \mathbf{u}_2^\star,\ \tau)d\tau \leq -(\nu^+ - \nu^-), \quad \forall \mathbf{u}_1 \in \mathcal{U}^1.$$

Proof. The proof is indirect. Non-trivially, assume $\mathbf{z}^0 \notin \mathcal{T}$. Suppose there exists some trajectory $\boldsymbol{\varphi}(\mathbf{z}^0, t_0, t)$ which does not enter $\mathcal{T}$. Then condition 1 implies that

$$V\left(\boldsymbol{\varphi}(\mathbf{z}^0, t_0, t)\right) = V(\mathbf{z}^0) + \int_{t_0}^{t} \dot{V}(\mathbf{z}(\tau))d\tau > \nu^- \quad \forall t \geq t_0. \tag{43}$$

Since $V(\mathbf{z}^0) \leq \nu^+$ we have that

$$\int_{t_0}^{t_0+T_C} \dot{V}(\mathbf{z}(\tau))d\tau > -(\nu^+ - \nu^-) \tag{44}$$

which contradicts condition 3. Hence no such trajectory exists, and the system is strongly controllable on Δ for collision with $\mathcal{T}$.

Corollary 4 The previous theorem on collision holds if condition 3 is replaced by the condition

4. $\forall \mathbf{u}_2^\star \in p_\star^2(.)$ we have

$$\nabla V(\mathbf{z})^T.\mathbf{f}(\mathbf{z},\ \mathbf{u}_1,\ \mathbf{u}_2^\star,\ t) < 0 \quad \forall \mathbf{u}_1 \in \mathcal{U}^1.$$

Proof. Consider a trajectory $\boldsymbol{\varphi}(\mathbf{z}^0, t_0, t)$ which does not enter $\mathcal{T}$. As in the previous proof, condition 1 implies that

$$V\left(\boldsymbol{\varphi}(\mathbf{z}^0, t_0, t)\right) = V(\mathbf{z}^0) + \int_{t_0}^{t} \dot{V}(\mathbf{z}(\tau))d\tau > \nu^- \quad \forall t \geq t_0.$$

By condition 4, $\dot{V}(\mathbf{z}) < 0$ generates an outflux of $V(\mathbf{z}(t))$. This implies that $\exists T_C > 0$ sufficiently large for $t = t_0 + T_C$ to imply that $V\left(\boldsymbol{\varphi}(\mathbf{z}^0, t_0, t)\right) = \nu^-$, contradicting equation (43). Therefore no such trajectory exists, and the system is strongly controllable on Δ for collision with $\mathcal{T}$.

If we attempt to visualise the function V(.) we can regard it as a "basin" in state space, with the target lying at the bottom of the basin. Condition 4 is then equivalent to saying that the strategy $p_\star^2(.)$ forces all solution trajectories to the equation

$$\dot{\mathbf{z}} = \mathbf{f}(\mathbf{z},\ \mathbf{u}_1,\ \mathbf{u}_2^\star,\ t) \tag{45}$$

down the slopes of this basin - into the target.

NUMERICAL RESULTS

The Navigation Constant

Throughout these simulations we have taken the value of the navigation constant N to be N = 4.

Initial Conditions at Missile Launch and Coast Phase

As has already been noted, although the Liapunov approach theoretically works for both autonomous and non-autonomous systems, in practice it is far more difficult to implement for non-autonomous systems. Because of this, and because of the design of the actual Liapunov function used in this paper, we require the launch platform to have superior specific energy to the evader at time t=0. (The *specific energy* of an aircraft is the ratio of its total energy (potential and kinetic) over its weight at a given time). It also means that we must restrict our attention to the coast phase of the missile, i.e. $t \in [15, t_f)$, where t_f is the time of game termination (taken to be ∞ if the pursuer fails in its interception). For this time interval the system is effectively autonomous.

We shall assume that the launch platform is at a sufficient distance from the evader at time $t = 0$ so that successful interception does not occur before the pursuing missile enters the coast phase at time $t = 15$ seconds. Secondly, we assume that by $t = 15$ the missile has reached speeds close to those corresponding to its maximum Mach number.

We assume the position (in metres) of the evader at the beginning of the pursuer's coast phase is $x_1(15) = 0$, $y_1(15) = 0$, at a height $h_1(15) = 10\ 000$ metres and travelling at 450 ms^{-1} along the X-axis with zero yaw and azimuth angles, i.e $\chi_1(15) = 0$, $\gamma_1(15) = 0$. After ten seconds the evader has accelerated to a speed of 475 ms^{-1} and has flown 4.63 km along the X-axis, losing 130 metres in altitude in the process.

Equations for $\widehat{W}_2$

Using the aerodynamic coefficients of the models for the F-15E and AIM-120A and a suitable Liapunov function, we obtained an estimated winning region in the horizontal plane at an altitude $h_1(t^*)$, $t^* \in [15, t_f)$ of

$$\widehat{W}_2(t^*) = \Bigg\{ \mathbf{z} \in \mathbb{R}^N \mid \Big[[x_2(t^*) - x_1(t^*)]^2 + [y_2(t^*) - y_1(t^*)]^2 \Big]^{\frac{1}{2}} < \frac{1}{v_2(t^*) + {}^1v_D(t^*)} \left\| \frac{0.295\ \left[(v_2^0)^2 - (v_1^0)^2\right]}{g} + 0.59\ \left(h_2^0 - h_1^0\right) \right\| \times \left[\frac{v_2(t^*) D_2(t^*)}{m_2(t^*)} - \frac{{}^1v_D}{2m_1} \rho\left({}^1h_D(\mathbf{z}(t^*))\right) S_1\ {}^1C_D\left({}^1v_D(\mathbf{z}(t^*)),\ {}^1h_D(\mathbf{z}(t^*))\right) \right] \Bigg\} \tag{46}$$

where

v_i^0 is the speed of the ith aircraft at time t=0,
h_i^0 is the height of the ith aircraft at time t=0,
$m_2(t^*)$ is the (constant) missile mass during coast phase,

and where ${}^1h_D\,(\mathbf{z}(t^*))$, ${}^1v_D\,(\mathbf{z}(t^*))$ denote the "best possible" values for height and speed which the evader could possibly achieve in its future trajectory, starting from the current position $\mathbf{z}(t^*)$, in order to maximize drag before interception. In this way the pursuer deliberately over-estimates the evader's capabilities to ensure the evader has no option of escape.

For t = 15 seconds and with the stated conditions at this time, the winning region becomes

$$\widehat{W}_2(15) = \Bigg\{ \mathbf{z} \in \mathbb{R}^N \mid \Big[x_2(15)^2 + y_2(15)^2 \Big]^{\frac{1}{2}} < \frac{1}{v_2(15) + {}^1v_D\,(\mathbf{z}(15))} \Big\| \frac{0.295\,\left[(v_2^0)^2 - (v_1^0)^2\right]}{g} + 0.59\,(h_2^0 - h_1^0) \Big\| \times \left[\frac{v_2(15) D_2(15)}{m_2(15)} - \frac{{}^1v_D\,(\mathbf{z}(15))\,\rho\Big({}^1h_D\,(\mathbf{z}(15))\Big)}{2m_1} S_1\,{}^1C_D\Big({}^1v_D(\mathbf{z}(15)),\ {}^1h_D(\mathbf{z}(15))\Big) \right] \Bigg\} \tag{47}$$

At the time of missile launch the launch platform is taken to have an advantage in specific energy over the evader of 300 metres, so we take

$$\frac{(v_2^0)^2 - (v_1^0)^2}{2\,g} + h_2^0 - h_1^0 = 300.$$

Equation (47) effectively defines a target in the XY plane at an altitude of $h_1(15)$ metres, which the pursuer wishes to reach at the commencement of coast phase. Once this is done successful interception of the evader is assured, provided the missile guidance law is adequate.

This region can be regarded as a set of nested concentric circles centred at the evader's position at t=15 (see Figure 2). We can regard the radius of the circle $\partial\widehat{W}_2$ as a monotonic function of pursuer speed $v_2(15)$. Typical values are given in Table 1. Note that the shape of the estimated winning region is dictated by the Liapunov function used in the calculations, rather than by the shape of the actual winning region. This is because we are using sufficiency conditions rather than necessary conditions.

Verification of PPN2, PPN3

Now we have established the estimated winning region we shall proceed to examine our guidance laws with $\dot{\lambda}$ constrained. It is known through simulation that PPN1 provides winning strategies for all points within $\partial\widehat{W}_2$. We shall examine PPN2 and PPN3 using the following procedure:

1. Firstly we shall take $v_2(15) = 900\text{ms}^{-1}$. This gives us a radius for $\partial\widehat{W}_2$ in Figure 2 of 1831 metres. (See Table 1).

Table 1. Table showing typical values of the radius of $\partial\widehat{W}_2$ in the XY plane as a function of $v_2(15)$.

speed $v_2(15)$, m s^{-1}	radius of $\partial\widehat{W}_2$, metres
900	1831
875	1767
850	1702
825	1634
800	1563

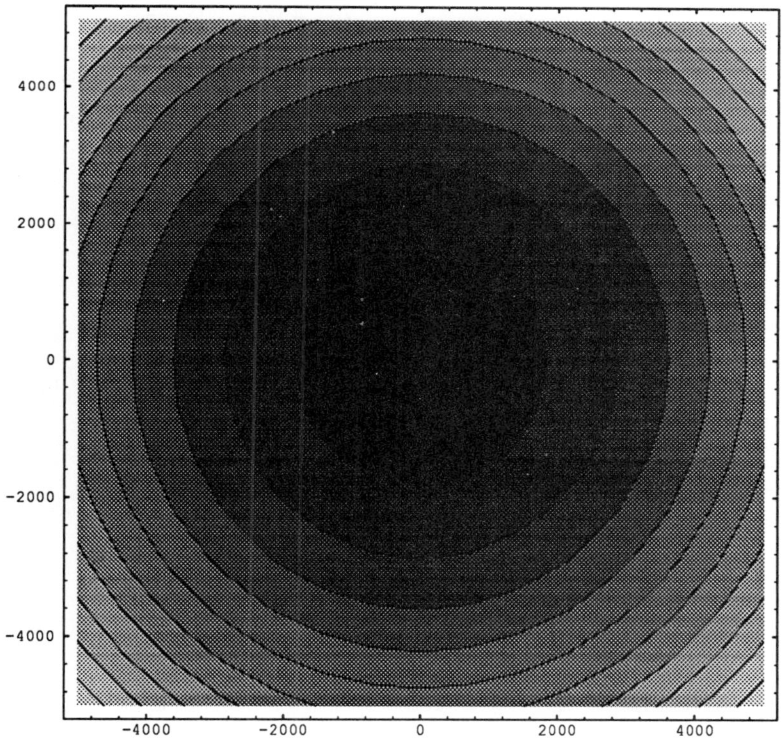

Figure 2. Contour plot of the winning region $\widehat{W}_2$ in the XY plane at height $h_1(15)$ metres above sea level at time $t = 15$, located with its centre at the evader's position at that time. Different values for the radius of $\partial\widehat{W}_2$ correspond with different values of $v_2(15)$. Distances along X and Y axes are in metres.

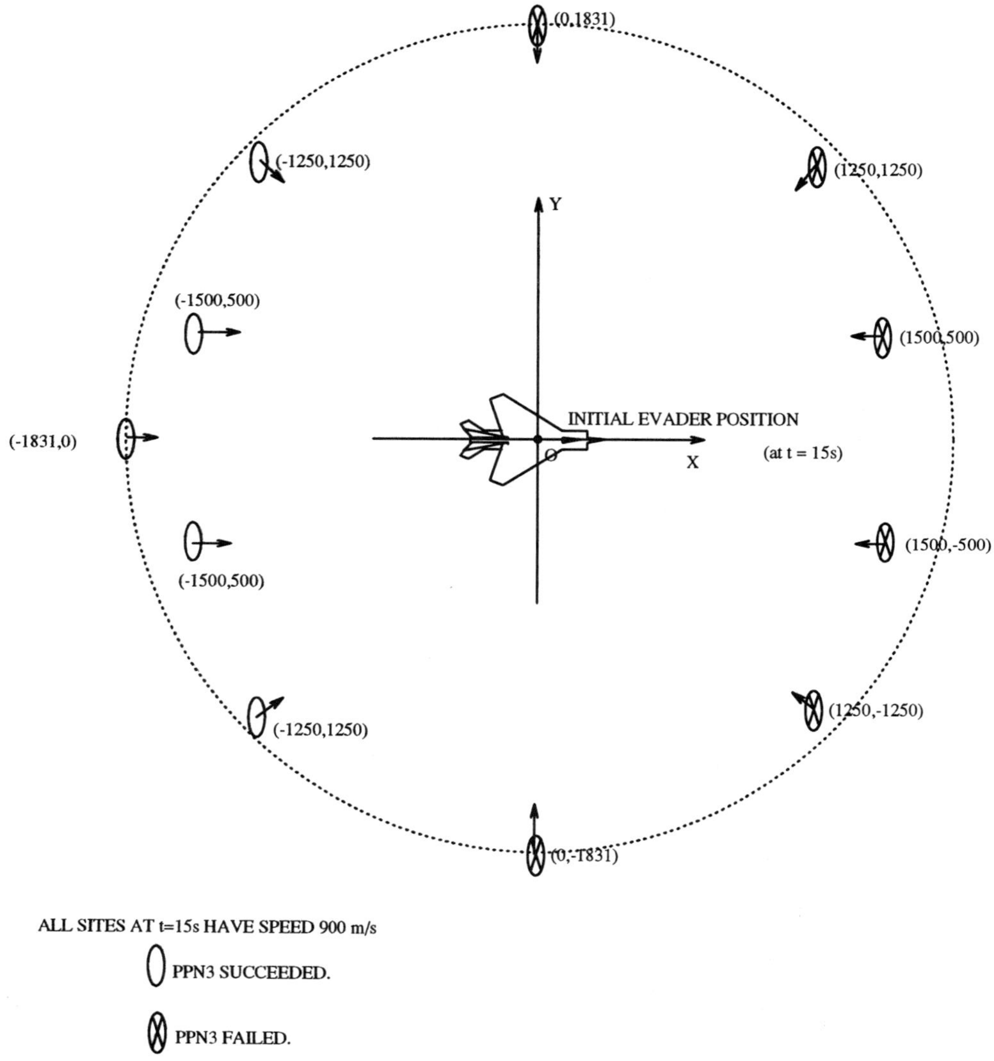

Figure 3. Results of attempted interceptions from sites within $\partial\widehat{W}_2$ using PPN3. Note the existence of failed attempts.

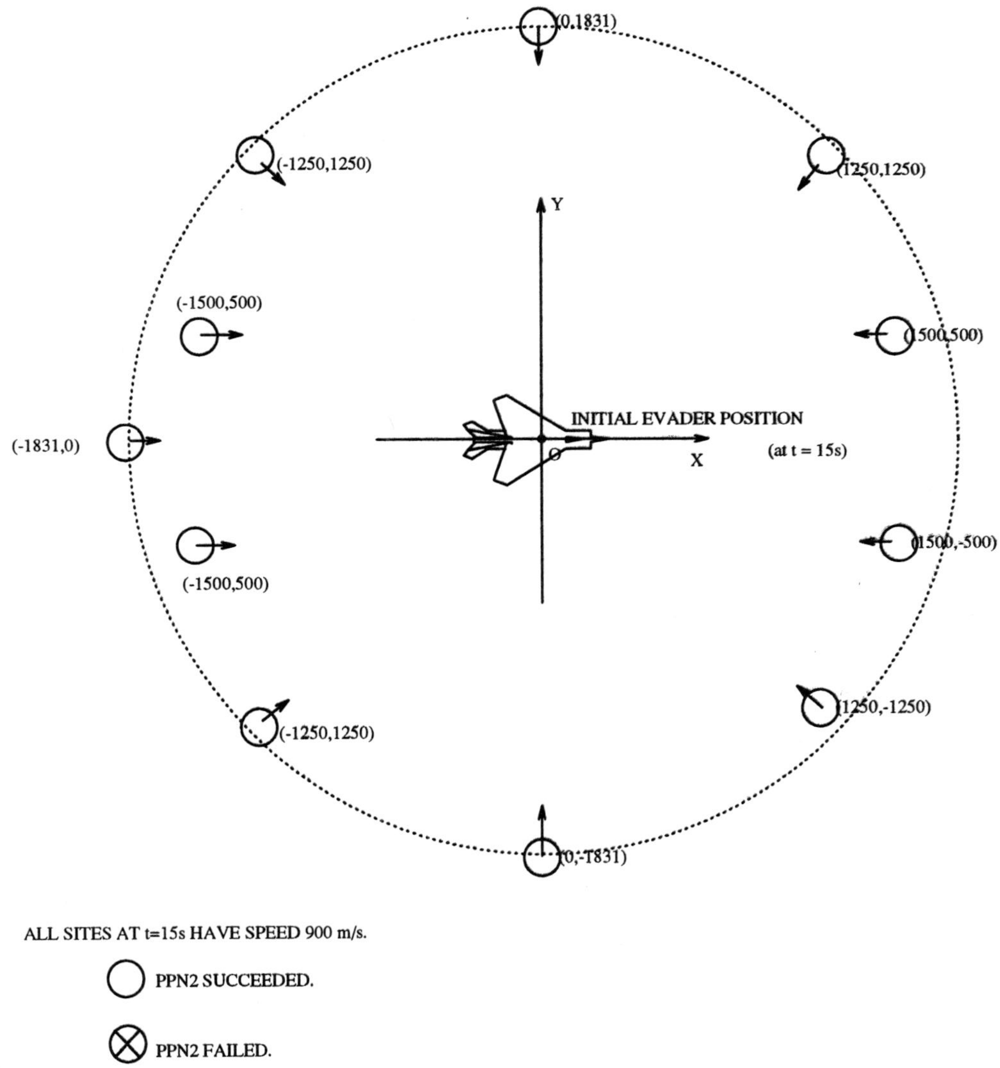

Figure 4. Results of attempted interceptions from sites within $\partial\widehat{W}_2$ using PPN2. Note the success of all attempts.

2. Secondly, we select a guidance law and simulate attempted interceptions from locations around the perimeter of $\partial\widehat{W}_2$. These locations shall be referred to as *sites*, and represent positions (and associated headings) consistent with the missile entering $\partial\widehat{W}_2$ at t= 15 seconds.

3. Thirdly, having done this for both of our guidance laws for sites spaced regularly around the perimeter of $\partial\widehat{W}_2$, we compare the results to determine the robustness of the guidance law when $\dot{\lambda}$ is constrained.

Figures 3 and 4 show the results of these simulations. It can be readily seen that PPN2 (where we have set $\kappa = 0.5$) is an acceptable guidance law for this winning region, as interception from all sites were successful. However, a significant number of sites failed to yield successful interception using PPN3 (where we set $\kappa = 0.1$), indicating that this law provides distinctly inferior results.

Thus we can establish a test for the acceptability of a guidance system. Comparisons can be made between guidance laws of the same form but different bounds upon their parameters, to determine the most desirable guidance law in terms of the percentage of sites within $\widehat{W}_2$ which yield successful interceptions.

CONCLUSIONS

In conclusion, we have outlined the technique of using a Liapunov function to estimate the winning region for a missile entering its coast phase. Once this was done we then used this region to provide a technique whereby we could judge the robustness of a particular guidance law when one or more of its parameters had bounds imposed upon it.

The winning region was calculated for a pursuit-evasion scenario between an AMRAAM AIM-120A missile and a F-15E "Strike Eagle" fighter. This region was then used to test a guidance law based on Pure Proportional Navigation (PPN) when the parameter $\dot{\lambda}$ was constrained. It was found that a constraint of $\kappa = 0.5$ was acceptable in providing winning strategies for the missiles; however, when this constraint was changed to $\kappa = 0.1$ a significant number of attempted interceptions from 2-winning sites failed.

We have established a simple test for ascertaining the acceptability of a particular guidance system, by the use of a benchmark set of sites corresponding with guaranteed interception. It should be noted that this procedure could also be used for non-military aerospace applications.

ACKNOWLEDGEMENTS

This work was supported in part by a grant from the Australian Defence Science and Technology Organisation (DSTO). The author would also like to thank Michael Breitner, Technischen Universität München, Germany, for the provision of technical data.

REFERENCES

1 G.M. Anderson, "Comparison of Optimal Control and Differential Game Intercept Missile Guidance Laws", *J. Guidance, Control and Dynamics*, Vol. 4, No. 2, 1981.

2 A. Blaquière, Gérard, and G. Leitmann, *Qualitative and Quantitative Differential Games*, Academic Press, 1975.

3 M.H. Breitner, H.J. Pesch, and W. Grimm, *Complex Differential Games of Pursuit-Evasion Type with State Constraint, Part 1 : Necessary Conditions for Optimal Open-Loop Strategies; Part 2 : Numerical Computation of Optimal Open-Loop Strategies.* Mathematisches Institute, Technische Universität München, Reports 314/315, 1992.

4 A.E. Bryson and Y.C. Ho, *Applied Optimal Control*, Hemisphere Publishing Corp., Washington DC, 1975.

5 A. Davidovitz, J. Shinar, "Eccentric Two-Target Model for Qualitative Air-Combat Game Analysis", *J. Guidance, Control and Dynamics*, Vol. 8, No. 3, 1985.

6 P. Garnell, *Guided Weapon Control Systems*, (Second Edition), Pergamon Press, 1980.

7 W.M. Getz and G. Leitmann, "Qualitative Differential Games with Two Targets", *Journal of Mathematical Analysis and Applications*, vol. 68(2), 1979, pp.421-430.

8 W.M. Getz and M. Pachter, "Two-Target Pursuit-Evasion Differential Games in the Plane", *JOTA*, Vol. 34, No. 3, pp. 383-403, July 1981.

9 N.J.C. Greenwood, "A Differential Game in Three Dimensions: The Aerial Dogfight Scenario", *J.Dynamics and Control*, vol. 2(2), pp.161-200, 1992.

10 M. Guelman, "A Qualitative Study of Proportional Navigation", *IEEE Transactions on Aerospace and Electronic Systems*, Vol. AES-7, No. 4, pp. 637-643, 1972.

11 M. Guelman, J. Shinar, and A. Green, "Qualitative Study ofa Planar Pursuit Evasion Game in the Atmosphere", *Proc. of AIAA Guidance, Navigation and Control Conference*, Paper No. 88-4158-CP, Minneapolis, Minnesota, 1988.

12 W. Herbst, "Dynamics of Air Combat", *Journal of Aircraft*, Vol. 20, pp. 594-598, 1983.

13 R. Isaacs, *Differential Games*, John Wiley and Sons, New York 1965.

14 B. Järmark, "A Missile Duel Between Two Aircraft", *J. Guidance, Control, and Dynamics*, Vil. 8, No. 3, 1985, pp. 508-513.

15 P.K.A. Menon and E.L. Duke, "Time-Optimal Pursuit-Evasion with a Weapon Envelope Constraint", *Proc. of the 1990 American Control Conference*, San Diego, pp. 2337-2342, 1990.

16 A.W. Merz, "To Pursue or to Evade - That is the Question" , *J. Guidance, Control and Dynamics*, Vol 8, No 2, pp. 161-166, 1985.

17 K. Moritz, R. Polis, and K.H. Well, "Pursuit-Evasion in Medium-Range Air-Combat Scenarios", *Comp. Math. Applic.*, Vol. 13, No. 1-3, pp 167-180, 1987.

18 M. Pachter and W.M. Getz, "The Geometry of the Barrier in the Game of Two Cars", *Optimal Control Applications and Methods*, Vol. 1, pp. 103-118, 1980.

19 R.A. Poulter and G.M. Anderson, "A Guidance Concept for Air-to-Air Missiles Based on Nonlinear Differential Game Theory", *Proc. NAECON*, pp. 605-609, 1976.

20 J. Shinar, and S. Gazit, "Optimal No-Escape Envelopes of Guided Missiles", *Proc. of AIAA Guidance, Navigation and Control Conference*, Paper 85-1960, Snowmass, Colorado, 1985.

21 J. Shinar and S. Gutman, "Three-dimensional Optimal Pursuit and Evasion with Bounded Control", *IEEE Trans. on Automatic Control*, Vol AC-25, pp. 492-496, 1980.

22 J.M. Skowronski, "The Barrier in a Pursuit-Evasion Game with Two Targets", *Comput. Math. Applic.*, vol. 13(1-3), pp.37-45, 1987.

23 J.M. Skowronski, "Winning Controllers for Nonlinear Air Combat Game with Reduced Dynamics", *Proc. AIAA Guidance, Navigation and Control Conference*, Minneapolis, pp.866-873, 1988.

24 J.M. Skowronski and R.J. Stonier, "Two-Person Qualitative Differential Games with Two Objectives", *Computers Math. Applic.*, vol 18(1-3), Pergamon Press 1989, 133-150.

25 R. De Villiers, C.J. Wright and Y. Yavin, "A Stochastic Pursuit-Evasion Differential Game in 3-D : Optimal Evasion Strategies", *Comp. Math. Applic.*, Vol. 13, No. 7, pp. 623-630, 1987

26 P.-J. Yuan and J.-S. Chern, "Ideal Proportional Navigation", *J. Guidance, Control and Dynamics*, Vol. 15, No. 5, 1992.

27 P.-J. Yuan and J.-S. Chern, "Solutions of True Proportional Navigation for Manuevering and Nonmanuevering Targets", *J. Guidance, Control and Dynamics*, Vol. 15, No. 1, pp. 268-271, 1992.

28 P. Zarchan, "Tactical and Strategic Missile Guidance", *Progress in Astronautics and Aeronautics*, Vol 124, AIAA Tactical Missile Series, Washington DC, 1990.

GAME THEORETIC REGULAR PERTURBATION GUIDANCE STRATEGIES FOR AEROASSISTED PLANE CHANGE MISSIONS

Marc R. Ilgen[1], and Jason L. Speyer[2]

[1]The Aerospace Corporation
El Segundo, CA 90245-4691
[1]School of Engineering and Applied Science
University of California
Los Angeles, CA 90024-1592

ABSTRACT

The calculus of variations regular perturbation methodology is extended to solve a class of differential game problems and to apply this methodology to the development of robust approximate optimal guidance laws for aeroassisted plane change missions. The class of differential game problems examined arises from the treatment of system parameter uncertainties as deterministic control variables controlled by Nature as an intelligent adversary. When the parameter uncertainties are introduced using the concept of fictitious feedback loops, the optimal solutions to the resulting formulation are sought from the class of solutions satisfying first and second order necessary conditions for a game theoretic saddle point. The solution to the zeroth order problem (the problem resulting from setting a small parameter ϵ to zero in the equations of motion) is found analytically while first and higher order corrections are found from a set of numerical quadratures. The required calculations are quite modest in number and can easily be performed in real time. Impressive numerical results indicate the robust behavior of this guidance law.

REFERENCES

1. M. R. Ilgen and J. L. Speyer, "Robust Approximate Optimal Plane Change Guidance Using Differential Game Theoretic Methods," AIAA Paper 92-4453-

Mechanics and Control, Edited by R.S. Guttalu
Plenum Press, New York, 1994

CP, *AIAA Guidance, Navigation and Control Conference*, Hilton Head, South Carolina, August 10-12, 1992.

2. M. R. Ilgen, J. L. Speyer and C. T. Leondes, "Robust Approximate Optimal Guidance Strategies for Aeroassisted Plane Change Missions: A Game Theoretic Approach," in *Control and Dynamic Systems*, (C. T. Leondes, ed.), Vol. 52, 1992.

MINIMUM-FUEL TRANSFER BETWEEN COPLANAR ELLIPTIC ORBITS - GLOBAL RESULTS USING GREEN'S THEOREM

Kenneth D. Mease[1] and Anil V. Rao[2]

Department of Mechanical and Aerospace Engineering
Princeton University
Princeton, NJ 08544

ABSTRACT

The use of Green's theorem to determine the minimum-fuel transfer between coplanar elliptic orbits in the time-free, orientation-free case is reviewed and extended to the consideration of aeroassisted transfers. In addition to the two-impulse Hohmann transfer, there are many other transfers between a given pair of elliptic orbits that satisfy the local optimality conditions. Green's theorem offers a straightforward and effective means of proving that most of these transfers cannot be globally optimal. Using Green's theorem, it is shown that the minimum-fuel aeroassisted transfer belongs to one of two one-parameter families of transfers. Direct numerical comparison of the fuel consumption is used to determine the optimal value of the parameter for each family. There remains a best aero-elliptic transfer and a best aero-parabolic transfer which must be compared with the Hohmann and bi-parabolic all-propulsive transfers to determine the minimum-fuel transfer for each pair of initial and final elliptic orbits.

INTRODUCTION

The problem of minimum-fuel transfer of a spacecraft between coplanar elliptical orbits in an inverse square gravity field has received much attention [6, 13]. Only the time-free, orientation-free case is considered in the present paper. Hohmann [8] proposed that between two coplanar circular orbits the minimum-fuel transfer is via an elliptical orbit that intersects both orbits tangentially at points 180° apart. Under the assumption that the on board rocket(s) can produce a thrust impulse and effect an instantaneous velocity change, the Hohmann transfer requires two impulses. With the number of impulses restricted to two, Lawden [10], using the calculus of variations, Marec [13], using a hodograph analysis, Battin [2], using a Lagrange multiplier approach, and Palmore [20] and Prussing [21], using ordinary calculus, have proved the optimality of the Hohmann transfer. If the circular orbits are sufficiently far apart, however, Shternfeld [22], Hoelker and Silber [9], and Edelbaum [5] have shown that a three-impulse bi-elliptic transfer outperforms the Hohmann. In fact, when the bi-elliptic transfer outperforms the Hohmann transfer, it is always best to use a bi-parabolic transfer, the limit of the bi-elliptic transfer as the intermediate apoapsis distance goes to infinity.

---

[1] Assistant Professor

[2] Graduate Student

Mechanics and Control, Edited by R.S. Guttalu
Plenum Press, New York, 1994

For transfer between coplanar elliptic orbits, Barrar [1] proved that the Hohmann transfer is the minimum-fuel two-impulse transfer. Ting [23] showed that any transfer involving four or more impulses could be replaced by a more fuel-efficient two-impulse or three-impulse transfer. Moyer [19] confirmed Ting's result for circle to ellipse transfers, showing that the minimum-fuel n-impulse transfer is either the Hohmann or the bi-parabolic. Hazelrigg [7] used Green's theorem to prove the suboptimality of all multi-impulse transfers between coplanar elliptical orbits but the Hohmann transfer and the one-parameter family of bi-elliptic transfers. He determined the global optimum from this drastically reduced set by direct numerical comparison of the fuel consumption. For a given set of initial and final orbits, either the Hohmann or the bi-parabolic transfer, a limiting case of the bi-elliptic transfer, was shown to be the global optimum. Vinh and Marchal [24] determined the switching structure by a judicious choice of state and independent variables and the analytic integration of the state and adjoint equations. They used a separatrix to determine global optimality. Vinh and Marchal also considered the use of aerodynamic braking, and showed that besides the propulsive Hohmann and bi-parabolic transfers, there are two aeroassisted transfers – the aero-elliptic transfer and the aero-parabolic transfer – that may be the global optimum depending on the initial and final orbits.

In honor of Professor Miele's pioneering work on the application of Green's theorem to flight path optimization [14,15] and his more recent research on aeroassisted orbital transfer [16-18], we review the Green's theorem approach to the problem of minimum-fuel transfer between coplanar elliptical orbits and extend this approach to the consideration of aeroassisted transfer. Our results for aeroassisted transfer are consistent with the previous results of Vinh and Marchal [24].

EQUATIONS OF MOTION AND OPTIMALITY CONDITIONS

We consider the problem of finding the thrust program to transfer a spacecraft between coplanar initial and final elliptic orbits denoted by Θ_I and Θ_F, respectively, such that the total fuel mass expended is a minimum. The equations of motion of a spacecraft are given as

$$\dot{\vec{r}} = \vec{V}$$
$$\dot{\vec{V}} = \vec{g} + \vec{T} / m \tag{2.1}$$

where $\vec{r}$ is the position of the spacecraft, $\vec{V}$ is the velocity of the spacecraft, $\vec{g}$ is the gravitational acceleration, $\vec{T}$ is the thrust, and m is the mass of the spacecraft. The mass varies according to the equation

$$\dot{m} = -T / (g_s I_{sp}) \tag{2.2}$$

where T is the thrust magnitude, g_s is the reference gravitational acceleration and I_{sp} is the specific impulse.

The ejection velocity (i.e. $g_s I_{sp}$) of the burned fuel is assumed constant. Under this assumption minimizing the characteristic velocity, given by the expression

$$C = \int_0^t \frac{T}{m} dt \geq 0 \tag{2.3}$$

is equivalent to minimizing the consumed fuel mass. Therefore, the objective is to minimize the total characteristic velocity for the transfer. We further assume that the planet is surrounded by a spherical atmosphere of radius R, that the duration of the transfer is unrestricted (the final time is free), that the orientation of the target elliptic orbit is unspecified, and that the propulsion system is capable of delivering a thrust impulse, i.e. the rockets can produce instantaneous changes in velocity. Noting that Marec [13] and Ting

[23] have shown that, in transferring a spacecraft between coplanar orbits, it is suboptimal leave the plane of the orbits, we restrict our attention to planar transfers.

The state of the vehicle at any given time t is defined by (r_p, r_a, ω, C) where r_p is the periapsis distance, r_a is the apoapsis distance, and ω is the longitude of periapse. The first three state variables describe the osculating ellipse along which the spacecraft is moving at time t. The control is given by

$$\eta = (\theta, \phi) \tag{2.4}$$

where θ is the true anomaly and ϕ is the thrust direction relative to the local horizontal. Since the state of the vehicle changes only when thrust is applied, it is convenient to use C as the independent variable [3,4]; this choice also ensures that the state variable rates are bounded even though the thrust is unbounded. The state equations [24] are

$$\frac{dr_p}{dC} = -\frac{2r_p^2}{nb(r_p + r_a)}\left[\sin\theta\sin\phi + \frac{(r_a - r_p)\cos^2\theta + 2(r_a + r_p)\cos\theta - (3r_a + r_p)}{(r_a + r_p) + (r_a - r_p)\cos\theta}\cos\phi\right] = f_1 \tag{2.5a}$$

$$\frac{dr_a}{dC} = \frac{2r_a^2}{nb(r_p + r_a)}\left[\sin\theta\sin\phi + \frac{(r_a - r_p)\cos^2\theta + 2(r_a + r_p)\cos\theta + (r_a + 3r_p)}{(r_a + r_p) + (r_a - r_p)\cos\theta}\cos\phi\right] = f_2 \tag{2.5b}$$

$$\frac{d\omega}{dC} = -\frac{4r_a r_p}{nb(r_a^2 - r_p^2)}\left[\cos\theta\sin\phi - \frac{2(r_a + r_p) + (r_a - r_p)\cos\theta}{(r_a + r_p) + (r_a - r_p)\cos\theta}\sin\theta\cos\phi\right] = f_3 \tag{2.5c}$$

$$\frac{dC}{dC} = 1 = f_0 \tag{2.5d}$$

where

$$b = \sqrt{r_a r_p} \quad , \quad n = \sqrt{8\mu / (r_a + r_p)^3} \tag{2.6}$$

and μ is the gravitational parameter for the inverse square field. The initial and final conditions are given as

$$\begin{aligned} &C(0) = 0, \quad r_a(0) = r_{a0} = a_0(1 + e_0), \quad r_p(0) = r_{p0} = a_0(1 - e_0), \quad \omega(0) = \omega_0 \\ &r_a(t_F) = r_{aF}, \quad r_p(t_F) = r_{pF}, \quad \omega(t_F) \text{ free} \end{aligned} \tag{2.7}$$

The necessary conditions for the minimum-fuel problem are as follows [24]. We define the adjoint vector $\vec{\lambda} = (\lambda_0, \lambda_1, \lambda_2, \lambda_3)$ and write the Hamiltonian for the system as

$$H = \lambda_0 + \lambda_1 f_1 + \lambda_2 f_2 + \lambda_3 f_3 \tag{2.8}$$

Since H is independent of ω and ω is unspecified at the final time, we have $\lambda_3 \equiv 0$. The stationary conditions for H with respect to the controls require that

$$\begin{aligned} \sin\theta &= \sin\phi = 0 \\ \cos\theta &= \pm 1 = \varepsilon_1 \\ \cos\phi &= \pm 1 = \varepsilon_2 \end{aligned} \tag{2.9}$$

which dictates that the impulses must always be applied tangentially at the apses. Substituting the optimal controls into Eqs. (2.5a,b) we have

$$\frac{dr_p}{dC}=-\frac{2r_p^2\varepsilon_2}{nb}\left[\frac{2(\varepsilon_1-1)}{(1-\varepsilon_1)r_p+(1+\varepsilon_1)r_a}\right] \tag{2.10a}$$

$$\frac{dr_a}{dC}=\frac{2r_a^2\varepsilon_2}{nb}\left[\frac{2(\varepsilon_1+1)}{(1-\varepsilon_1)r_p+(1+\varepsilon_1)r_a}\right] \tag{2.10b}$$

$$\frac{d\omega}{dC}=0 \tag{2.10c}$$

Therefore, there are four distinct types of extremal arcs corresponding to the four possibilities for the pair $(\varepsilon_1,\varepsilon_2)$ given as follows.

$\underline{(\varepsilon_1,\varepsilon_2)}$: (2.11)

(1,1):	Prograde tangential impulse at periapse to raise apoapse.
(1,-1):	Retrograde tangential impulse at periapse to lower apoapse.
(-1,1):	Prograde tangential impulse at apoapse to raise periapse.
(-1,-1):	Retrograde tangential impulse at apoapse to lower periapse.

The orientation of the ellipse, given by ω, remains constant at the initial value during the optimal transfer.

What remains is to determine the globally optimal transfer between specified initial and final orbits from the four types of extremal arcs. Vinh and Marchal [24] found the optimal transfers by analyzing the switching function and using a separatrix concept. We follow Hazelrigg [7] and use Green's theorem to rule out all but a few transfers and then determine the globally optimum transfers by direct comparison of the fuel consumption.

MINIMUM-FUEL PROPULSIVE TRANSFERS

We view the optimal transfer problem in the (r_p,X) plane, where $X=1/r_a$ (Fig. 1). The set of points corresponding to elliptic orbits is given by

$$E=\{(r_p,X): r_p>R, X>0, Xr_p\le 1\}$$

Points of the form $(r_p,0)$ correspond to parabolic orbits; points of the form $(r_p,1/r_p)$ correspond to circular orbits. In this plane, the four types of extremal arcs, defined in the previous section, correspond to downward vertical, upward vertical, rightward horizontal, and leftward horizontal motion, respectively. We think of each extremal arc as being traversed during a single impulse, but in fact the arc could be traversed over any number of orbits with the total impulse fractioned between them with no change in the total fuel consumption [7, 24]. For given initial and final elliptical orbits, any continuous curve between them, that is composed exclusively of vertical and horizontal segments, is an extremal transfer and a candidate for the globally optimal transfer. Note that it is sufficient to study the minimum-fuel, time-free, orientation-free transfer problem in this plane, since we have determined that ω remains constant during an extremal transfer and the dynamics of r_p and X are independent of ω. We allow segments of the line $X=0$ to be part of an extremal transfer between elliptical orbits, i.e., the transfer is allowed to involve parabolic orbits. Orbits with $r_p\le R$ are affected by aerodynamic force and are treated in the next section.

Using the inverse of Eq. (2.10a) with $\varepsilon_1=-1$, the characteristic velocity for changing periapse from r_{pi} to r_{pj} by a tangential impulse applied at apoapse is given by

$$C_p = \int_{r_{pi}}^{r_{pj}} \frac{\partial C\,(\varepsilon_2)}{\partial r_p} dr_p, \quad \varepsilon_2 = sign\left(r_{pj} - r_{pi}\right) \tag{3.1}$$

Similarly, using the inverse of Eq. (2.10b) with $\varepsilon_1 = 1$, the characteristic velocity for changing apoapse by a tangential impulse applied at periapse is given by

$$C_a = \int_{X_i}^{X_j} \frac{\partial C\,(\varepsilon_2)}{\partial X} dX, \quad \varepsilon_2 = sign\left(X_j - X_i\right) \tag{3.2}$$

The quantities $\dfrac{\partial C(\varepsilon_2)}{\partial r_p}$ and $\dfrac{\partial C(\varepsilon_2)}{\partial X}$ are given respectively by

$$\frac{\partial C(\varepsilon_2)}{\partial r_p} = \varepsilon_2 X \sqrt{\frac{\mu}{2r_p}} (1 + X r_p)^{-3/2} \tag{3.3}$$

$$\frac{\partial C(\varepsilon_2)}{\partial X} = -\varepsilon_2 \sqrt{\frac{\mu r_p}{2}} (1 + X r_p)^{-3/2} \tag{3.4}$$

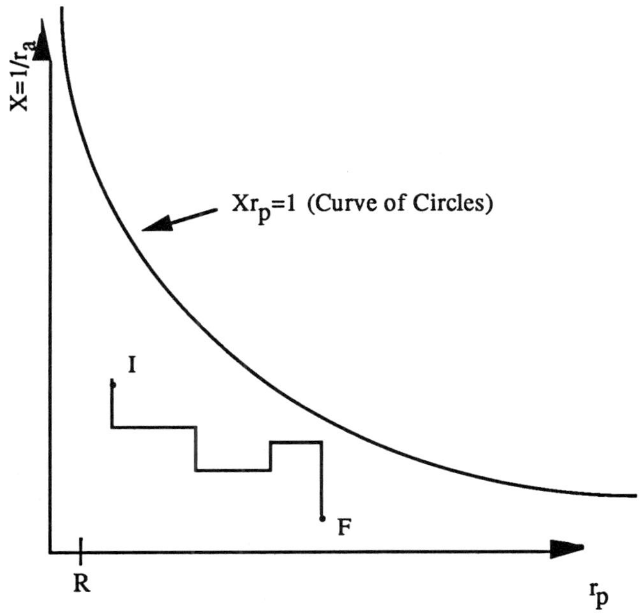

Figure 1. The X-r_p plane and an example of an extremal transfer from an initial orbit I to a final orbit F

The total cost of transfer from an initial orbit I to an final orbit F is given by the sum of the characteristic velocities for all the periapse and apoapse changes

$$C = \sum_{i=1}^{n} C_{a_i} + \sum_{j=1}^{m} C_{p_j} \tag{3.5}$$

Following Hazelrigg [7], we first use Green's theorem to determine the minimum characteristic velocity two-impulse transfer. Referring to Fig. 2, the only two extremal transfers between A and D that can be accomplished with two impulses are AED and ABD. Using the notation $\phi(\varepsilon_2) = \dfrac{\partial C(\varepsilon_2)}{\partial r_p}$ and $\psi(\varepsilon_2) = \dfrac{\partial C(\varepsilon_2)}{\partial X}$, the cost of transfer along path ABD is given by

$$C_{ABD} = \int_A^B \phi(1)dr_p + \int_B^D \psi(1)dX \tag{3.6}$$

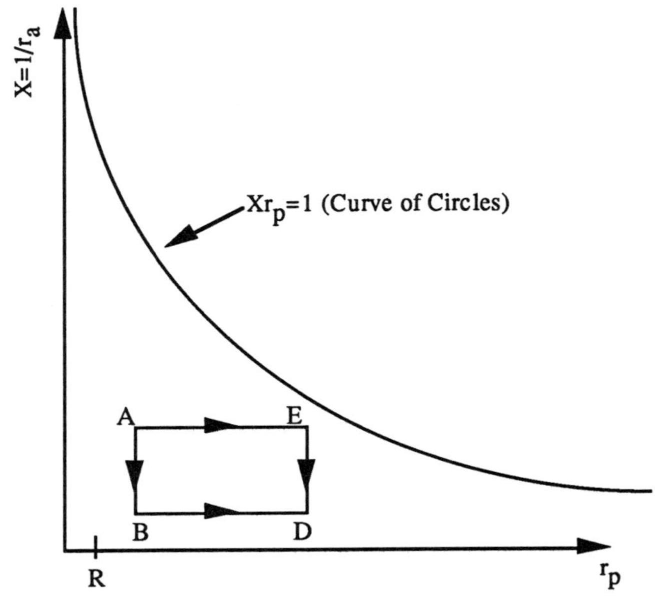

Figure 2. Examples of two-impulse transfers from I to F

Similarly, the cost of transfer along path AED is given by

$$C_{AED} = \int_A^E \phi(1)dr_p + \int_E^D \psi(1)dX \tag{3.7}$$

Subtracting equation (3.7) from equation (3.6) gives us

$$\Delta C = \int_A^B \phi(1)dr_p + \int_B^D \psi(1)dX - \int_A^E \phi(1)dr_p - \int_E^D \psi(1)dX \tag{3.8}$$

which can be expressed as the closed-path integral

$$\Delta C = \oint_{ABDEA} (\phi dr_p + \psi dX) \tag{3.9}$$

Since ϕ and ψ are continuous with continuous partial derivatives over the path of integration, we can use Green's theorem [14] to transform the line integral into an integral over the surface enclosed by the path ABDEA. This gives

$$\Delta C = \iint_{ABDEA} \left(\frac{\partial \psi}{\partial r_p} - \frac{\partial \phi}{\partial X}\right) dr_p dX \tag{3.10}$$

Hazelrigg [7] has shown that the sign of the integrand is negative at all points of the enclosed surface. Therefore, ΔC is negative, and ABD is the minimum-fuel two-impulse transfer. The analysis of the transfers E to B, D to A, and B to E proceeds similarly, and the following conclusions are obtained:

$$\begin{aligned} &(1)\ C_{ABD} < C_{AED} \\ &(2)\ C_{EDB} < C_{EAB} \\ &(3)\ C_{DBA} < C_{DEA} \\ &(4)\ C_{BDE} < C_{BAE} \end{aligned} \tag{3.11}$$

These results cover all the cases of two-impulse propulsive transfer between coplanar ellipses for the class of time-free, orientation-free transfers.

In the case of multiple-impulse transfer, Hazelrigg [7] has shown that there always exists either a two-impulse or three-impulse transfer of lower cost. The use of Green's theorem eliminates all but the two-impulse Hohmann and a one-parameter family of three-impulse transfers, namely the family of bi-elliptic transfers. Hazelrigg has shown numerically that in the cases where a bi-elliptic transfer is better than the Hohmann transfer, it is always best to let the intermediate apoapsis distance go to infinity ($X \to 0$). This limiting three-impulse transfer is the bi-parabolic transfer; the second impulse is infinitesimal. For example, for the initial and final orbits, denoted by I and F in Fig. 3, either the two-impulse Hohmann transfer I3F or the bi-parabolic transfer I123F is the minimum-fuel propulsive transfer.

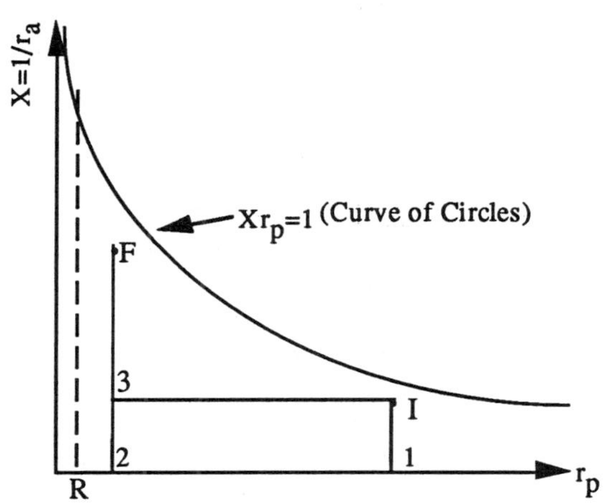

Figure 3. Possible optimal propulsive transfers from I to F

MINIMUM-FUEL AEROASSISTED TRANSFERS

In this section, we extend the Green's theorem approach to the consideration of aeroassisted orbital transfer. Similar to the all-propulsive case, the role of Green's theorem is to prove the suboptimality of all but two one-parameter families of transfers. Other analytical or numerical methods must then be used to determine the global optimum from this reduced set.

Suppose the spacecraft is in orbit about an attracting body with a spherical sensible atmosphere of radius R from the center of the attracting body. Consider an orbit transfer for which the initial apoapse is higher than the final apoapse, as shown in Fig. 4. In this case it may be advantageous to use atmospheric braking. The treatment of atmospheric braking is idealized in the following manner. We assume that with $r_p = R$, atmospheric braking allows apoapse lowering without fuel consumption. In the (r_p, X) plane, this corresponds to free transfer along the vertical line $r_p = R$ in the positive X-direction. Mease and Vinh [11] have shown that this idealization provides an accurate estimate of the characteristic velocity associated with an aerobraking maneuver. See [12] for a review of the work on aeroassisted orbital transfer optimization,

Consider an aeroassisted transfer from an initial orbit I to a final orbit F. Fig. 4 shows a number of representative extremal arcs that might be involved in the minimum-fuel transfer. Since the transfer is aeroassisted, there must be at least one arc on the line $r_p = R$ which is traversed in the positive X-direction, i.e., an apoapse lowering. The following lemmas and theorems determine the optimal path for reaching the atmosphere, the optimal path for leaving the atmosphere, and the optimal number of aerobraking arcs. We first define two important families of aeroassisted transfers.

Definition 1

An *aero-elliptic transfer* is composed of four arcs as shown in Fig. 4

(i) propulsive periapse lowering to the edge of the atmosphere (arc I45)
(ii) aerobraking to lower the apoapse to $X = X_A$ where $X_I < X_A \leq X_F$ (arc 5A)
(iii) propulsive raising of periapse to $r_p = r_{pF}$ (arc ABC)
(iv) propulsive lowering of apoapse (increasing X) to $X = X_F$ (arc CEF)

For given initial and final orbits, there is a family of aero-elliptic transfers parametrized by X_A. In the case where X_A=X_F, the aero-elliptic transfer is a three-arc transfer, the fourth arc degenerating to a point.

Definition 2

An *aero-parabolic transfer* is composed of five arcs as shown in Fig. 4

(i) propulsive raising of apoapse to infinity ($X = 0$) (arc I1)
(ii) propulsive lowering of periapse, with negligible fuel consumption, to the edge of the atmosphere (arc 1203)
(iii) aerobraking to lower apoapse to $X = X_A$ where $0 < X_A \leq X_F$ (arc 35A)
(iv) propulsive raising of periapse to $r_p = r_{pF}$ (arc ABC)
(v) propulsive lowering of apoapse (increasing X) to $X = X_F$ (arc CEF)

For given initial and final orbits, there is a family of aero-parabolic transfers parametrized by X_A. In the case X_A=X_F, the aero-parabolic transfer is a four-arc transfer, the fifth arc degenerating to a point.

Lemma 1

The minimum-fuel aeroassisted transfer reaches the atmosphere (the line $r_p = R$) either via arc (i) of the aero-elliptic transfer or arcs (i) and (ii) of the aero-parabolic transfer.

Proof

Lemma 1 is a direct consequence of the result that the minimum-fuel path from (r_{pI}, X_I) to (R, X_I) is either I45 or I12035 even in the absence of an atmosphere (Refer to Fig. 4). With aerobraking, the fuel consumption for the latter path is reduced. That paths like I4HCBA and I124HCBA offer no advantage follows from the two-impulse results (Eq. 3.11).

Lemma 2

The globally optimal transfer from I to F, if aeroassisted, proceeds from the line $r_p = R$ to the line $r_p = r_{pF}$ via a horizontal line, i.e., a single-impulse periapse raise.

Proof

Consider path ABCE versus path ABDE in Fig. 4. Using the 2-impulse propulsive results (Eq. 3.11), we have that BCE is better than BDE. Next consider path ABC versus ABGHC. Path BGHC is a bi-elliptic transfer between B and C. If it has a lower characteristic velocity than the direct path BC, the bi-parabolic transfer BG024HC has even lower cost. But either state K or state 0 was already achieved in reaching the atmosphere, according to Lemma 1. Since the round trip has cost fuel, it should have been avoided; aerobraking was not worthwhile. Moreover, the minimum-fuel all-propulsive transfer does not pass through either K or 0. Thus, it is only necessary to consider the horizontal path ABC.

Lemma 3

The globally optimal transfer from I to F, if aeroassisted, does not involve intermediate states with $r_p = R$ and $X > X_F$.

Proof

Suppose the aerobraking is continued to point 7 where point 7 is above the line $X = X_F$ in Fig. 4. Then from point 7 it would be better to proceed to point F via 76F than by 78F based on two-impulse results (Eq. 3.11). But the cost of traversing arc 76 could have been avoided by using the direct path 6F.

Lemma 4

It is never optimal to leave the line $r_p = R$ and then return to it.

Proof: To proceed from a point on the line $r_p = R$ to a higher point on that same line, the minimum-fuel path is along the segment of the line $r_p = R$ connecting the two points. By assumption, increasing X along the line $r_p = R$ is accomplished by aerobraking and requires no fuel consumption. Any excursion into the region where $r_p > R$ requires fuel consumption. Hazelrigg [7] has shown that Lemma 3 holds for propulsive transfer between points on vertical lines with $r_p > R$; in this case the minimum-fuel consumption is nonzero. Furthermore, Lemma 3 holds for propulsive transfer in the reverse direction [7].

Theorem 2

The minimum-fuel aeroassisted transfer is either an aero-elliptic transfer or an aero-parabolic transfer.

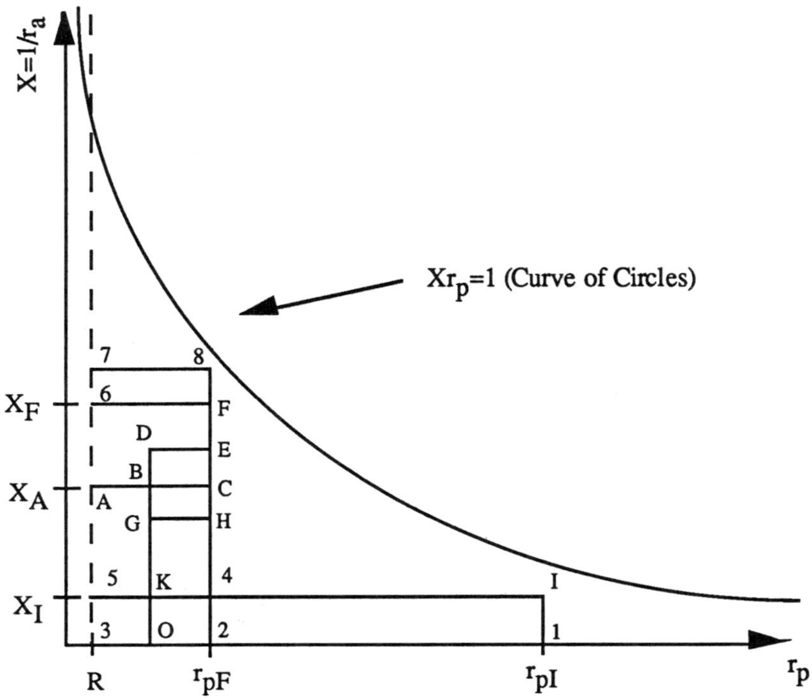

Figure 4. Extremal paths for aeroassisted transfer from I to F

<u>Proof</u>

The theorem is a straightforward consequence of Lemmas 1-4.

<u>Theorem 3</u>

The minimum-fuel aero-elliptic transfer and the minimum-fuel aero-parabolic transfer are each achieved with $X_A = X_F$.

<u>Proof</u>

Our approach to proving the theorem is to compare directly the fuel consumption. We only present results for one final orbit condition $(r_{pF}, X_F) = (1.5R, 1/(2.5R))$ and the atmosphere radius $R = 6498\ km$. The particular value of μ is inconsequential as it only changes the absolute, not the relative, fuel consumption. The characteristic velocity for the aero-parabolic transfer minus that for the all-propulsive bi-parabolic transfer is negative for all X_A such that $0 < X_A \leq X_F$. The difference in cost is plotted as a function of X_A in Fig. 5. The characteristic velocity for the aero-parabolic transfer decreases as X_A increases. The minimum is achieved when $X_A = X_F$, since Lemma 3 implies that the characteristic velocity will increase for $X_A > X_F$. Similar results have been obtained for other final orbit conditions and for the aero-elliptic transfer.

The two potentially optimal aeroassisted transfers - the aero-elliptic and the aero-parabolic, each with $X_A = X_F$ - along with the two potentially optimal all-propulsive transfers - the Hohmann and the bi-parabolic - must be compared to determine the minimum-fuel transfer for each initial/final orbit pair. Vinh and Marchal [24] have determined some conditions for the optimality of each transfer. Their results are repeated below for completeness.

(1) Hohmann: $\max(r_{pF}, r_{pI}) \le 11.938 \min(r_{pF}, r_{pI})$.

(2) Bi-parabolic: $\max(r_{pF}, r_{pI}) > 9 \min(r_{pF}, r_{pI})$ *and* $r_{pF} \ge 4R(1+RX_F)$.

(3) Aero-parabolic: $r_{pF} \le 4R(1+RX_F)$ *and* $r_{pI} > 4R(1+R/\min(X_I, X_F))$.

(4) Aero-elliptic: $X_I < X_F$, $r_{pI} \le 4R(1+RX_I)$, *and*

$$(4R - r_{pF}) + X_F R(4R + r_{pF}) + X_F^2 R^2 (R + r_{pF}) > 0$$

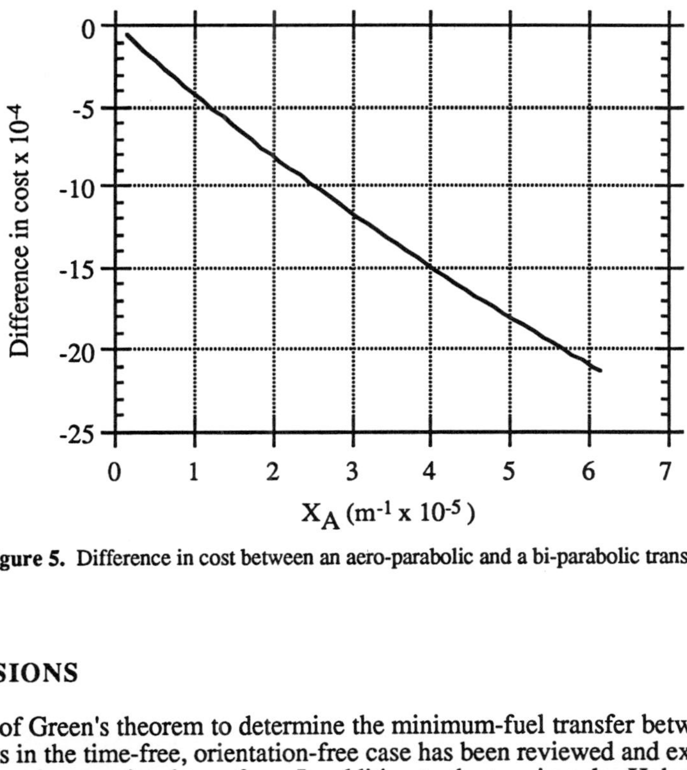

Figure 5. Difference in cost between an aero-parabolic and a bi-parabolic transfer

CONCLUSIONS

The use of Green's theorem to determine the minimum-fuel transfer between coplanar elliptic orbits in the time-free, orientation-free case has been reviewed and extended to the consideration of aeroassisted transfers. In addition to the two-impulse Hohmann transfer, there are many other transfers between a given pair of elliptic orbits that satisfy the local optimality conditions. Green's theorem offers a straightforward and effective means of proving that most of these transfers cannot be globally optimal. The new contribution in this paper is the proof, using Green's theorem, that the minimum-fuel aeroassisted transfer belongs to one of two one-parameter families of transfers. Direct numerical comparison of the fuel consumption was used to determine the optimal value of the parameter for each family. There remains a best aero-elliptic transfer and a best aero-parabolic transfer which must be compared with the Hohmann and bi-parabolic all-propulsive transfers to determine the minimum-fuel transfer for each pair of initial and final elliptic orbits.

REFERENCES

1. R.B. Barrar, "An Analytic Proof That the Hohmann-type Transfer is the True Minimum Two-Impulse Transfer," *Astronautica Acta*, 9 (1): 1 (1963).

2. R.H. Battin, "An Introduction to the Mathematics and Methods of Astrodynamics," AIAA Education Series, AIAA, New York (1987).

3. J.V. Breakwell, "Minimum Impulse Transfer," AIAA Paper 63-416 (1963).

4. P. Contensou, "Etude Theorique des Trajectories Optimales Dans un Champ de Gravitation. Application au Cas d'un Centre d'Attraction Unique," *Astronautica Acta*, 9: 134 (1963).

5. T.N. Edelbaum, "Some Extensions of the Hohmann Transfer Maneuver," *ARS Journal*, 29 (11): 864 (1959).

6. F.W. Gobetz and J.R. Doll, "A Survey of Impulsive Trajectories," *AIAA Journal*, 5 (7): 801 (1969).

7. G.A. Hazelrigg Jr., "Globally Optimal Impulsive Transfers via Green's Theorem," *Journal of Guidance, Control, and Dynamics* , 7 (4): 462 (1984).

8. W. Hohmann, "Die Erreichbarkeit der Himmelskorper," Oldenbourg, Munich 1925; also, NASA Translation TT-F-44 (1960).

9. R.E. Hoelker and R. Silber, "The Bi-Elliptic Transfer Between Circular Coplanar Orbits," Department of the Army, Army Ballistic Missile Agency, Redstone Arsenal, Ala., Technical Memo 2-59 (1959).

10. D.F. Lawden, "Optimal Trajectories for Space Navigation," Butterworths, London (1963).

11. K.D. Mease and N.X. Vinh, "Minimum-Fuel Aeroassisted Coplanar Orbit Transfer," *Journal of Guidance, Control, and Dynamics*, 8 (1): 134 (1985).

12. K.D. Mease, "Optimization of Aeroassisted Orbital Transfer: Current Status," *Journal of the Astronautical Sciences*, 36 (1/2): 7 (1988).

13. J.P. Marec, "Optimal Space Trajectories," Elsevier, Paris (1979).

14. A. Miele, "Problems of Minimum Time in the Nonsteady Flight of Aircraft," Atti della Accademia delle Scienze di Torino, Classe di Scienze Fisiche, *Matematiche e Naturali*, 85: 41 (1950-51) (in Italian).

15. A. Miele, "Extremization of Linear Integrals by Green's Theorem," *in: Optimization Techniques*, G. Leitmann, ed., Academic Press, New York (1962).

16. A. Miele and P. V. Venkataraman, "Optimal Trajectories for Aeroassisted Transfer," *Acta Astronautica*, 11 (7-8): 423 (1984).

17. A. Miele, V.K. Basapur, and W.Y. Lee, "Optimal Trajectories for Aeroassisted, Coplanar Orbital Transfer," *Journal of Optimization Theory and Applications*, 52 (1): 1 (1987).

18. A. Miele, V.K. Basapur, and W.Y. Lee, "Optimal Trajectories for Aeroassisted, Noncoplanar Orbital Transfer, Part I," *Acta Astronautica*, 15 (6/7): 399 (1987).

19. H.G. Moyer, "Minimum Impulse Coplanar Circle to Ellipse Transfers," *AIAA Journal*, 3 (4): 723 (1965).

20. J.I. Palmore, "An Elementary Proof of the Optimality of Hohmann Transfers," *Journal of Guidance, Control, and Dynamics*, 7 (5): 629 (1984).

21. J.E. Prussing, "A Simple Proof of the Global Optimality of the Hohmann Transfer," *Journal of Guidance, Control, and Dynamics*, 15 (4): 1037 (1992).

22. A. Shternfeld, "Soviet Space Science," Basic Books, Inc., New York (1959).

23. L. Ting,, "Optimum Orbital Transfer by Impulse," *ARS Journal*, 30 (11): 1013 (1960).

24. N.X. Vinh and C. Marchal, "Analytical Solutions of a Class of Optimum Orbit Modifications," *Journal of Optimization Theory and Applications*, 5 (3): 178 (1970).

PERIODIC OPTIMAL ENDURANCE CRUISE WITH VARIABLE CAMBER CONTROL

Gottfried Sachs[1] and Rainer Mehlhorn[2]

[1]Professor
[2]Research Assistant
Institute of Flight Mechanics and Flight Control
Technische Universität München
Arcisstr. 21
8000 München 2
Germany

ABSTRACT

Variable wing camber which is an efficient aerodynamic means for improving the lift/drag ratio of aircraft is used as a control in addition to thrust and angle of attack to maximize endurance. The potential of variable camber for increasing the maximum lift/drag ratio at each lift coefficient can be fully utilized in a dynamic (periodic) process optimally controlled.

It is shown that periodic optimal control of variable camber provides a significant endurance increase when compared to the best steady-state cruise. Quantitative results are presented using data of a high performance aircraft.

The control of thrust for an engine model without response delay shows a bang-bang type switching. Since this requires an infinite control rate, an expanded engine model accounting for a delay in thrust buildup is also considered in order to show which degradation in endurance performance results.

An optimization technique is used which is based on the minimum principle and on the method of multiple shooting. The computer code applied in the numerical investigation provides results of high accuracy.

NOMENCLATURE

C_D drag coefficient
C_L lift coefficient
D drag
g acceleration due to gravity
H Hamiltonian
h altitude

Mechanics and Control, Edited by R.S. Guttalu
Plenum Press, New York, 1994

J performance criterion
L lift
M Mach number
m mass
S reference area
$S_{1,2}$ switching functions
T thrust
t time
V speed
α angle of attack
γ flight path angle
δ_T thrust setting
δ_{VCi} variable camber setting, $i = 1, \ldots n$
ζ_T rate of thrust change
λ Lagrange multiplier
μ Lagrange multiplier
ϱ atmospheric density
σ fuel consumption factor
τ normalized time, $\tau = t/t_{cyc}$

INTRODUCTION

Aircraft cruise for maximizing either range or endurance represents a basic control problem in flight preformance optimization. To optimize endurance cruise, the flight time spent in the air is maximized while the distance travelled ist not of interest. An excellent treatment of this subject and the techniques for determining the best cruise performance is provided by Miele[1].

Aircraft cruise is usually considered as a steady-state flight at constant speed and/or altitude. Recent research in flight path optimization shows that periodic control may be a means for increasing range and endurance when compared with classical steady-state cruise[2-14]. A particular advantage may be possible for endurance performance[12,13]. According to classical theory, the best steady-state endurance cruise is a flight where fuel consumption per time is minimized. This steady-state endurance cruise may be regarded as the best compromise of operating conditions for aerodynamics and the propulsion system. On the contrary, periodic optimal cruise yields two alternating phases one of which can provide a better operating condition for the propulsion system in terms of increasing the energy added per fuel consumed and the other yields a better utilization of aerodynamics in terms of reducing the energy subtracted due to drag.

To achieve these alternating phases, two controls are usually considered in papers on periodic optimal cruise. One control is throttle setting and the other is the elevator (or lift coefficient, respectively). There has only recently been a paper where variable camber was considered as an additional control in periodic optimal cruise problems[14].

Variable camber which has received great interest for some time is an aerodynamic technique for improving the lift/drag ratio of aircraft by changing the profile of the wing[15-18]. This is illustrated in Fig. 1. Individual drag polars generated by an appropriate setting of the wing profile are shown. They may be used for different types of flight maneuvers, as also indicated in Fig. 1. The individual drag polars may be combined to yield an envelope which provides the best lift/drag ratio at each lift coefficient.

Variable camber setting is usually considered to be constant when performing a maneuver of the type indicated in Fig. 1. It is changed only when a transition from one maneuver to another takes place. In this paper, variable camber is considered as a non-constant control which is continuously changed during a maneuver which in itself

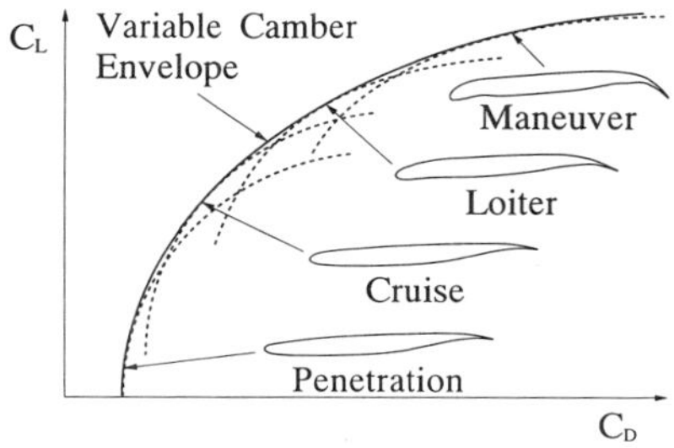

Figure 1. Effect of variable camber on lift/drag characteristics, from Ref. 16

is non-steady. Thus, the improvements in lift/drag ratio due to variable camber can be utilized at each instant during the non-steady phases of periodic optimal endurance cruise.

PROBLEM FORMULATION

The optimal control problem may be formulated as maximizing the flight time per fuel consumed. The performance criterion may be defined as

$$J = \frac{t_{\rm cyc}}{(m_{\rm f})_{\rm cyc}} \tag{1}$$

where $t_{\rm cyc}$ is the time of one period of the trajectory and $(m_{\rm f})_{\rm cyc}$ is the fuel consumed.

The performance criterion is subject to the equations of motion which are usually based on point mass dynamics for performance optimization problems. Using the normalized time $\tau = t/t_{\rm cyc}$ as independent variable, the equations of motion may be written as

$$\begin{aligned}
\frac{dV}{d\tau} &= t_{\rm cyc}\left(\frac{T}{m} - \frac{D}{m} - g\sin\gamma\right)\\
\frac{d\gamma}{d\tau} &= t_{\rm cyc}\left(\frac{L}{mV} - \frac{g}{V}\cos\gamma\right)\\
\frac{dh}{d\tau} &= t_{\rm cyc}\, V\sin\gamma\\
\frac{dm_{\rm f}}{d\tau} &= t_{\rm cyc}\, \dot{m}_{\rm f}(V,\, h;\, \delta_{\rm T})
\end{aligned} \tag{2}$$

The aerodynamic model for lift and drag is

$$\begin{aligned}
L &= C_{\rm L}(\varrho/2)V^2 S\\
D &= C_{\rm D}(\varrho/2)V^2 S
\end{aligned} \tag{3}$$

where

$$\begin{aligned}
C_{\rm L} &= C_{\rm L}(M;\, \alpha,\, \delta_{\rm VCi}), \quad i = 1,\ldots n\\
C_{\rm D} &= C_{\rm D}(M;\, \alpha,\, \delta_{\rm VCi}), \quad i = 1,\ldots n
\end{aligned} \tag{4}$$

Variable camber may be realized by a multiple set of individual control devices $\delta_{\rm VCi}$ up to a number of n for changing the profile of the wing. The actual number

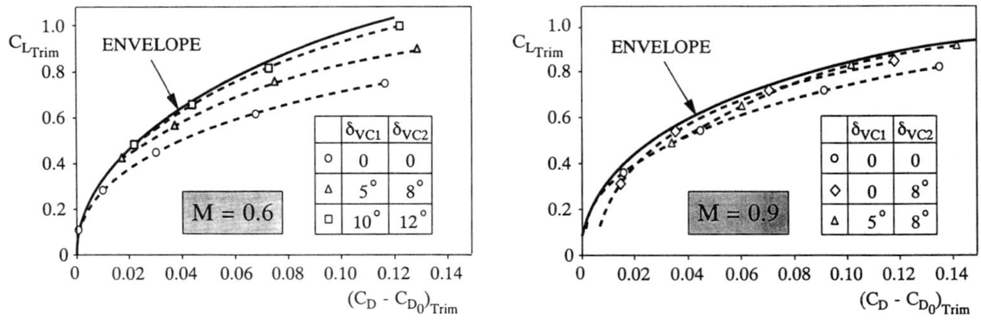

Figure 2. Effect of variable camber on drag characteristics, from Ref. 15

depends on the specifics of a technical design such as a segmentation of a wing[15]. In the numerical investigation of this paper, a combination of leading and trailing edges is considered (implying that $n = 2$). An example for the drag characteristics of such a variable camber design is shown in Fig. 2.

From the results presented it follows that the lift/drag ratio can be significantly improved by an appropiate camber setting as indicated by the envelope. Fig. 2 also illustrates the effect of Mach number. As may be seen, the effectiveness of variable camber is reduced at high subsonic Mach numbers.

Thrust and fuel consumption are considered to be dependent on speed, altitude and throttle setting. A realistic thrust model may be expressed as

$$T(V, h; \delta_T) = \delta_T T_{max}(V, h) \tag{5}$$

The model for fuel consumption characteristics is

$$\dot{m}_f(V, h; \delta_T) = \dot{m}_{f0}(h) + \delta_T \sigma(V) T_{max}(V, h) \tag{6}$$

The mass of the aircraft can be considered as constant for one period ($\tau = 1$) since the fuel consumed during such a time interval is small when compared to the total of the mass, i.e.

$$m_f(1) - m_f(0) \ll m \tag{7}$$

Periodicity of the flight path implies the following boundary conditions

$$V(0) = V(1), \quad \gamma(0) = \gamma(1), \quad h(0) = h(1) \tag{8a}$$

The initial condition for fuel mass can be written as

$$m_f(0) = 0 \tag{8b}$$

Control variables are angle of attack α, variable camber setting $\delta_{VCi} (i = 1, ..., n)$ and throttle setting δ_T which are subject to the following inequality constraints

$$\begin{aligned} \alpha_{min} \leq \alpha &\leq \alpha_{max} \\ (\delta_{VCi})_{min} \leq \delta_{VCi} &\leq (\delta_{VCi})_{max}, \quad i = 1, \ldots n \\ 0 \leq \delta_T &\leq 1 \end{aligned} \tag{9}$$

The atmospheric model which is used for describing the dependencies of air density, speed of sound and thrust on altitude agrees with the ICAO Standard Atmosphere[19].

The periodic control problem is to find the control histories α, δ_{VCi} $(i = 1, \ldots n)$ and δ_T, the initial states $(V(0), \gamma(0), h(0))$ and the optimal cycle length $\bar{t}_{cyc}$ which maximize the performance criterion $J = t_{cyc}/m_f(1)$ subject to the system described by Eq. (2), the boundary conditions given by Eqs. (8a,b) and the inequality constraints for the control variables, Eq. (9a,b).

OPTIMALITY CONDITIONS

Necessary conditions for optimality can be determined by applying the minimum principle. For this purpose, the Hamiltonian is defined as

$$H = t_{cyc}\left[\lambda_V\left(\frac{T}{m} - \frac{D}{m} - g\sin\gamma\right) + \lambda_\gamma\left(\frac{L}{mV} - \frac{g}{V}\cos\gamma\right) + \lambda_h V\sin\gamma + \lambda_f \dot{m}_f\right] \quad (10)$$

where the Lagrange multipliers $\lambda = (\lambda_V, \lambda_\gamma, \lambda_h, \lambda_f)^T$ have been adjoined to the system of Eq. (2). The Lagrange multipliers are determined by*

$$\frac{d\lambda_V}{d\tau} = t_{cyc}\left[\lambda_V\frac{D_V}{m} + \lambda_\gamma\frac{L - VL_V - mg\cos\gamma}{mV^2} - \lambda_h\sin\gamma - \lambda_f\delta_T\left(\sigma_V T_{max} + \sigma\frac{\partial T_{max}}{\partial V}\right)\right]$$

$$\frac{d\lambda_\gamma}{d\tau} = t_{cyc}\left(\lambda_V g\cos\gamma - \lambda_\gamma\frac{g}{V}\sin\gamma - \lambda_h V\cos\gamma\right)$$

$$\frac{d\lambda_h}{d\tau} = t_{cyc}\left[\lambda_V\frac{D_h}{m} - \lambda_\gamma\frac{L_h}{mV} - \lambda_f\left(\sigma\delta_T\frac{\partial T_{max}}{\partial h} + \frac{\partial \dot{m}_{f0}}{\partial h}\right)\right]$$

$$\frac{d\lambda_f}{d\tau} = 0 \quad (11)$$

with the following boundary conditions

$$\begin{aligned} \lambda_V(0) &= \lambda_V(1), & \lambda_\gamma(0) &= \lambda_\gamma(1) \\ \lambda_h(0) &= \lambda_h(1), & \lambda_f(1) &= -t_{cyc}/m_f^2(1) \end{aligned} \quad (12)$$

The optimal cycle time $\bar{t}_{cyc}$ can be determined by applying two further differential equations

$$\frac{dt_{cyc}}{d\tau} = 0\ , \quad \frac{d\lambda_t}{d\tau} = -\frac{H}{t_{cyc}} \quad (13)$$

subject to the boundary conditions

$$\lambda_t(0) = 0\ , \quad \lambda_t(1) = \frac{1}{m_f(1)} \quad (14)$$

The optimal controls $\alpha, \delta_{VCi}\,(i = 1,\dots n)$ and δ_T are such that H is minimized. For this reason, α is determined either by

$$C_{L_\alpha}\lambda_\gamma - \lambda_V V\frac{\partial C_D}{\partial \alpha} = 0 \quad (15)$$

(resulting from $\partial H/\partial\alpha = 0$) or by the constraining bounds of Eq. (9a).

The condition $\partial H/\partial\delta_{VCi} = 0$ for the optimal setting of variable camber yields

$$\lambda_\gamma\frac{\partial C_L}{\partial \delta_{VCi}} - \lambda_V V\frac{\partial C_D}{\partial \delta_{VCi}} = 0\ , \quad i = 1,\dots n \quad (16)$$

*Partial derivatives of D and L are denoted by subscripts, e.g. $D_V = \partial D/\partial V$.

This relation applies when δ_{VCi} is within its admissible set. Otherwise, the constraining bounds of Eq. (9a) become active.

With regard to throttle setting δ_{T}, the Hamiltonian shows a linear dependence. Accordingly, a bang-bang type control or singular arcs can exist. The bang-bang type control means that the throttle takes on its boundary values according to

$$\begin{aligned} \delta_{\mathrm{T}} &= 0 \quad \text{if} \quad S_1(y,\,\lambda;\,\alpha,\,\delta_{\mathrm{VCi}}) > 0 \\ \delta_{\mathrm{T}} &= 1 \quad \text{if} \quad S_1(y,\,\lambda;\,\alpha,\,\delta_{\mathrm{VCi}}) < 0 \end{aligned} \tag{17}$$

where

$$S_1(y,\,\lambda;\,\alpha,\,\delta_{\mathrm{VCi}}) = \frac{\partial}{\partial \delta_{\mathrm{T}}} H(y,\,\lambda;\,\alpha,\,\delta_{\mathrm{VCi}},\,\delta_{\mathrm{T}}) = t_{\mathrm{cyc}} T_{\max}\left(\frac{\lambda_{\mathrm{V}}}{m} + \lambda_{\mathrm{f}}\sigma\right), \quad i = 1, \ldots n \tag{18}$$

is the switching function and $y = (V, \gamma, h, m_{\mathrm{f}})^{\mathrm{T}}$. Switching occurs when

$$S_1(y, \lambda; \alpha, \delta_{\mathrm{VCi}}) = 0$$

A singular arc means that the throttle is not on its boundary but takes on values interior to its admissible set. This is the case when $S_1(y,\,\lambda;\,\alpha,\,\delta_{\mathrm{VCi}}) = 0$ for a finite interval of τ. However, such a behavior was not observed.

The computer code applied in the numerical investigation is based on the method of multiple shooting and provides results with high accuracy[20,21].

REFERENCE SOLUTION

The results presented in this section are considered as a reference showing an engine model which can be controlled at an infinite control rate so that a bang-bang type switching is possible. All other characteristics are realistically modelled. This also applies to aerodynamics the main characteristics of which is the drag relationship shown in Fig. 2 as an example for two Mach numbers.

An optimal cycle of periodic maximum endurance cruise is presented in Figs. 3-5. As a main result also indicated, the use of variable camber provides a significant increase of endurance when compared to the best steady-state cruise.

From the history of state variables presented in Fig. 3 it follows that an optimal cycle is made up of two phases one of which is a climbing high-speed flight while the other is a descent at low speed. The speed attained during the high-speed phase stays below the region where the drag rise due to compressibility becomes significant.

Throttle setting and lift coefficient are shown in Fig. 4. As may be seen, a high throttle setting is combined with a small lift coefficient and vice versa. Thus, it is possible to achieve a climb at high speed and a descent at low speed. According to the engine model applied, the idling condition implies that thrust is reduzed to zero but fuel is still consumed.

Fig. 5 shows that a considerable usage of variable camber is made. For the phase at high speed and low lift coefficient, wing camber is kept at small values. The opposite holds for the phase where speed is low and lift coefficient is high. This means that the wing profile is more cambered when a high lift coefficient is demanded and it is decambered when the required lift coefficient becomes small.

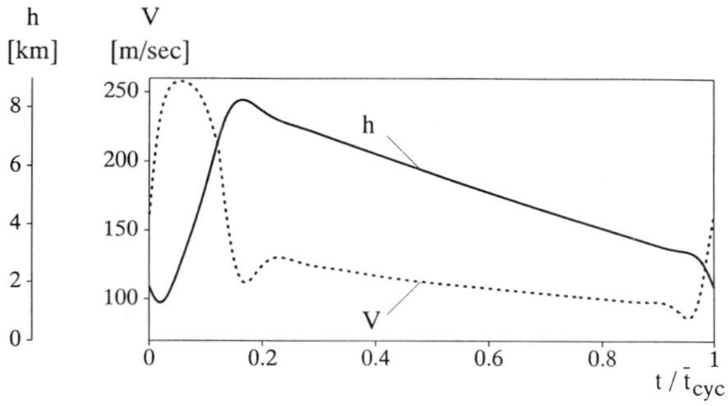

Figure 3. Periodic maximum endurance cruise with optimal variable camber control, altitude and speed (25.4 % endurance increase, $\bar{t}_{cyc} = 597.8$ sec)

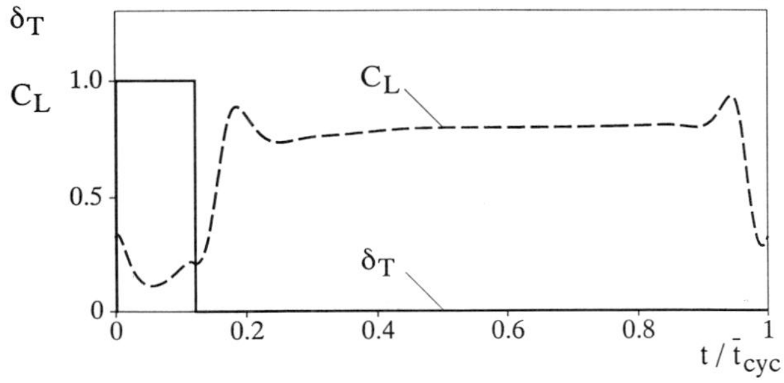

Figure 4. Periodic maximum endurance cruise with optimal variable camber control, lift coefficient and thrust setting (25.4 %endurance increase, $\bar{t}_{cyc} = 597.8$ sec)

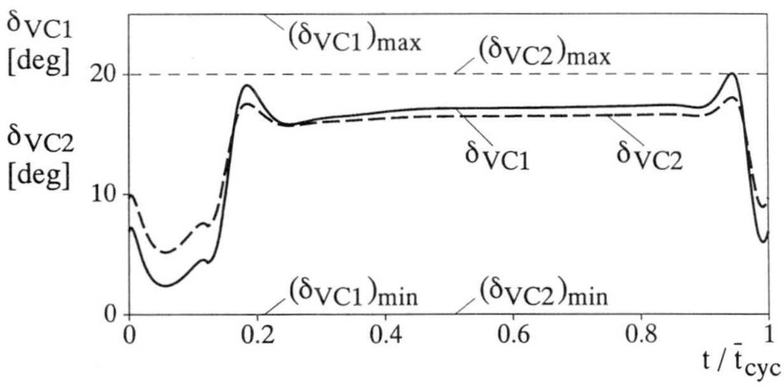

Figure 5. Periodic optimal endurance cruise with optimal variable camber control, optimal variable camber setting (25.4 %endurance increase, $\bar{t}_{cyc} = 597.8$ sec)

The control characteristics described provide an insight into the physical mechanism underlying the performance improvement due to variable camber control. They basically represent a correlation between camber setting and lift coefficient such that camber setting is continually adjusted in order to achieve the best lift/drag ratio at each lift coefficient. This means that drag which is a measure for the decrease of the energy state of the aircraft can be reduced by an optimally controlled camber setting when compared to a fixed camber wing.

EFFECT OF CONTROL RATE LIMITATIONS

The controls used for generating the optimal endurance trajectory are operated rapidly during some phases while they show little changes during others. For phases of rapid operation, the rate of change may be high and limitations which a practical design has may be exceeded.

A rate limitation may be of particular interest for the control of thrust because it shows a bang-bang characteristic which would mean an infinite rate of change. Therefore, the thrust model is expanded in order to account for a control rate limitation.

For this purpose, throttle setting δ_T is replaced as a control by its rate of change denoted by ζ_T. Accordingly, the mathematical model of the airplane of Eq. (2) is expanded by an additional differential equation

$$\frac{d\delta_T}{d\tau} = t_{cyc}\,\zeta_T \tag{19}$$

Boundary conditions are

$$\delta_T(0) = \delta_T(1) \tag{20}$$

The control ζ_T is subject to the following constraints

$$(\zeta_T)_{min} \leq \zeta_T \geq (\zeta_T)_{max} \tag{21}$$

The relation $0 \leq \delta_T \leq 1$ now represents a state variable constraint. This case can be treated by considering additional relations and conditions[22]. The relations

$$G_1 = \delta_T - 1 \leq 0\,, \quad G_2 = -\delta_T \leq 0 \tag{22}$$

are used for describing the constraint. Furthermore, an expanded Hamiltonian may be defined as

$$H^* = H + t_{cyc}(\lambda_\delta + \mu_1 - \mu_2)\zeta_T \tag{23}$$

The Lagrange multipliers $\mu_1 \geq 0$ and $\mu_2 \geq 0$ are introduced for treating the state variable constraint $0 \leq \delta_T \leq 1$ according to Eq. (22). The Lagrange multiplier λ_δ is determined by

$$\frac{d\lambda_\delta}{d\tau} = -t_{cyc}\left(\frac{\lambda_V}{m} + \lambda_f\sigma\right)T_{max} \tag{24}$$

where

$$\lambda_\delta(0) = \lambda_\delta(1) \tag{25}$$

The optimal controls relations Eqs. (17) and (18) are replaced by corresponding expressions for ζ_T. In case of δ_T not at its boundary, ζ_T takes on the values

$$\begin{aligned} \zeta_T &= (\zeta_T)_{min} \quad \text{if} \quad S_2(y^*, \lambda^*, \mu_1, \mu_2; \alpha, \delta_{VCi}) > 0 \\ \zeta_T &= (\zeta_T)_{max} \quad \text{if} \quad S_2(y^*, \lambda^*, \mu_1, \mu_2; \alpha, \delta_{VCi}) < 0 \end{aligned} \tag{26}$$

where S_2 is the switching function

$$\begin{aligned} S_2(y^*, \lambda^*, \mu_1, \mu_2; \alpha, \delta_{VCi}) &= \frac{\partial}{\partial \zeta_T} H^*(y^*, \lambda^*; \alpha, \delta_{VCi}, \zeta_T) \\ &= t_{cyc}(\lambda_\delta + \mu_1 - \mu_2), \quad i = 1, \ldots n \end{aligned} \tag{27}$$

and

$$y^* = (V, \gamma, h, m_f, \delta_T)^T, \lambda^* = (\lambda_V, \lambda_\gamma, \lambda_h, \lambda_f, \lambda_\delta)^T$$

When δ_T is at its boundary $(\delta_T)_{min} = 0$ or $(\delta_T)_{max} = 1$, the control takes on the value $\zeta_T = 0$. This result follows from Eq. (19) when taking into account that $(\delta_T)_{min}$ and $(\delta_T)_{max}$ are constant values. A characteristic such that $S_2 = 0$ for a finite interval of τ and $0 < \delta_T < 1$ was not observed in the numerical investigation.

The optimal endurance trajectory for the expanded thrust model is shown in Figs. 6-8. The main result is that endurance performance is practically not reduced when thrust control rate is limited.

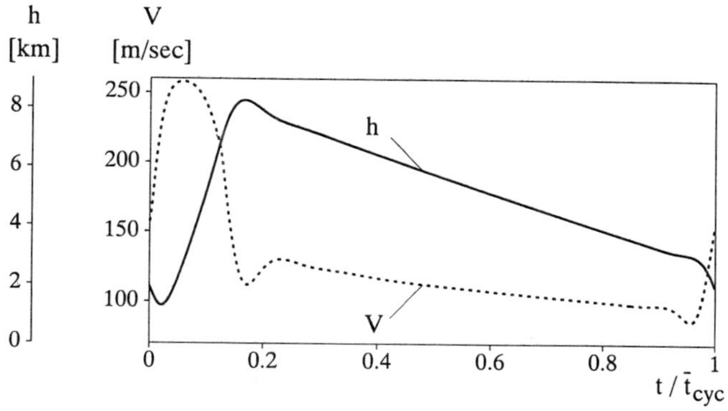

Figure 6. Periodic maximum endurance cruise with optimal variable camber control and a model for thrust dynamics, altitude and speed (25.3 % endurance increase, $\bar{t}_{cyc} = 600.3$ sec)

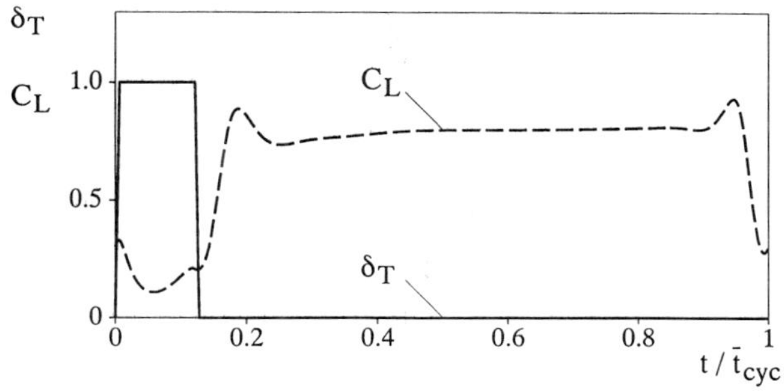

Figure 7. Periodic maximum endurance cruise with optimal variable camber control and a model for thrust dynamics, lift coefficient and thrust setting (25.3 % endurance increase, $\bar{t}_{cyc} = 600.3$ sec)

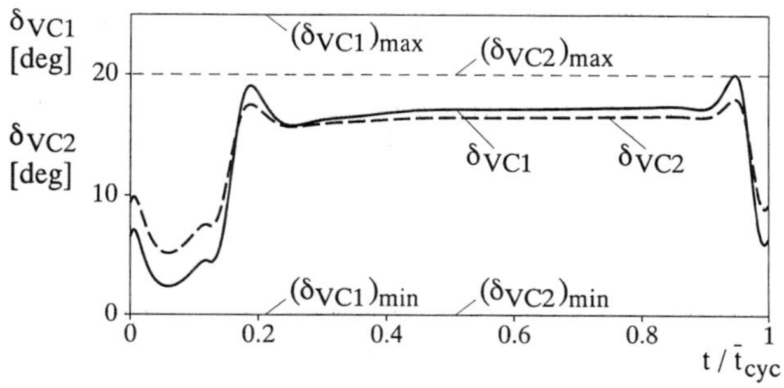

Figure 8. Periodic maximum endurance cruise with optimal variable camber control and a model for thrust dynamics, optimal variable camber setting (25.3 % endurance increase, $\bar{t}_{cyc} = 600.3$ sec)

The characteristics of state and control variables are quite similar to the previously considered case with bang-bang type thrust changes. There is some difference at the transitions from maximum thrust setting to idling and vice versa, where a continuous change now exists. This is illustrated in Fig. 9 which shows these thrust setting changes in more detail.

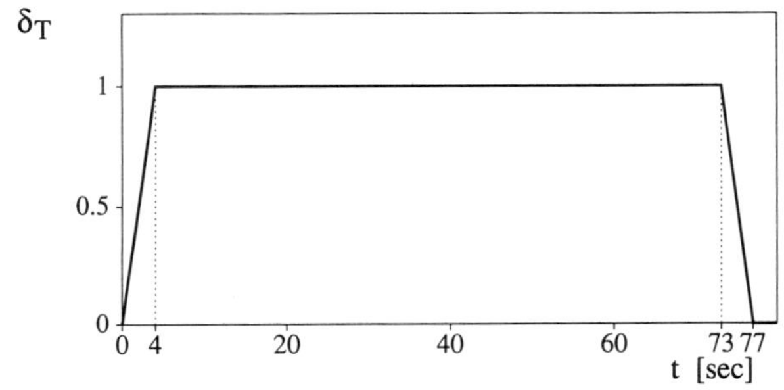

Figure 9. Details of thrust setting

Fuel consumption during an optimal period is shown in Fig. 10. As may be seen, a considerable portion of fuel is consumed when throttle setting is at an idling position and no thrust is produced. Nevertheless, fuel efficiency is significantly increased for the whole of the period when compared with the best steady-state cruise.

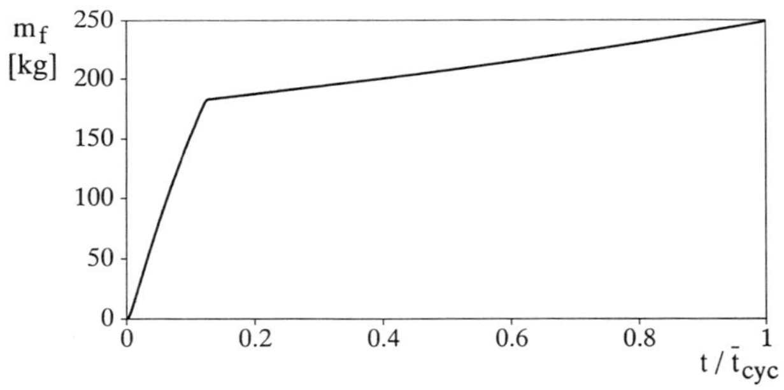

Figure 10. Fuel consumption during an optimal cycle (25.3 % endurance increase, $\bar{t}_{cyc} = 600.3$ sec)

Rate limitations may also be considered for the other controls which do not show a bang-bang type behavior but may have quite rapid changes in some phases of an optimal trajectory. This concerns the rates for deflecting the leading and trailing edge flaps as well as the rate of angle of attack. The numerical investigation shows that the maximum value of the deflection rates of the variable camber flaps stays below 1 deg/sec, Fig. 11. This is less than requirements for a practical design[24].

The maximum rate of angle of attack is of the order of 0.5 deg/sec, Fig. 12. This can be considered as a value within the performance capability of a high performance aircraft. As a consequence, a realistic limitation in angle of attack rate may not be exceeded so that no performance degradation needs to be expected.

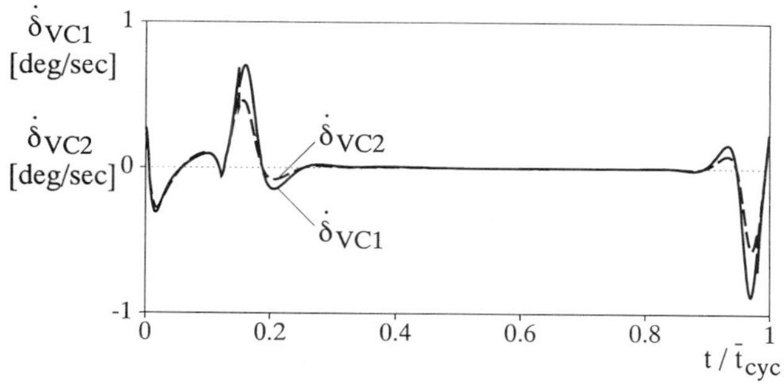

Figure 11. Control rate of variable camber (25.3 % endurance increase, $\bar{t}_{cyc} = 600.3$ sec)

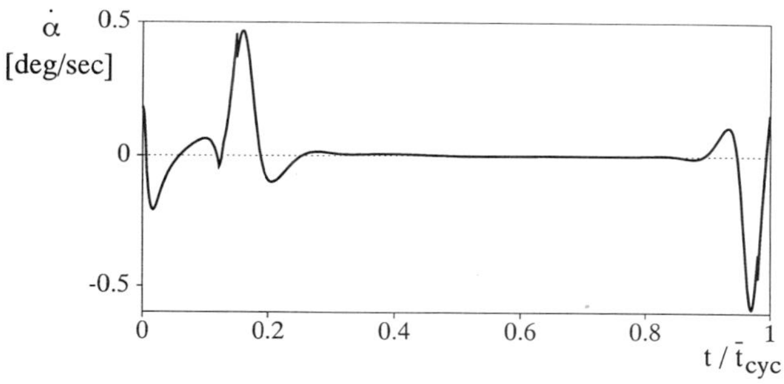

Figure 12. Control rate of angle of attack (25.3 % endurance increase, $\bar{t}_{cyc} = 600.3$ sec)

CONCLUSIONS

In addition to controls usually applied, variable camber is introduced for enhancing the endurance performance. The potential of variable camber can be fully utilized when the flight technique for maximum endurance is not a steady-state cruise but an unsteady process which consists of periodically repeated cycles.

It is shown that a significant increase in endurance performance can be achieved when compared with the best steady-state cruise. Variable camber provides a considerable contribution to this endurance increase because it improves the lift/drag ratios both in the high and low angle of attack range.

The control of thrust for a reference engine model shows a bang-bang type behavior switching between its boundary values. Since this is not realizable because of an infinite control rate required, an expanded thrust model accounting for a limitation of control rate is also considered. The results show that the endurance performance is practically not degraded in case of the expanded thrust model.

Further control rate limitations may concern the deflections for variable camber and angle of attack. The numerical investigation shows that the control rates during an optimal period are below limitations which a realistic design may have.

The optimal control technique applied is based on the minimum principle for deriving optimality conditions and on the method of multiple shooting for the numerical investigation. A realistic aircraft model is used for describing aerodynamics and powerplant characteristics which can be considered as representative of a high performance vehicle.

REFERENCES

1. Miele, A., "Flight Mechanics. Vol. I, Theory of Flight Paths," Reading, Massachusetts, Palo Alto, London: Addison-Wesley Publ. Corp., Pergamon Press 1962.
2. Speyer, J.L., "Nonoptimality of the Steady-State Cruise of Aircraft," AIAA Journal, Vol. 14, 1976, pp. 1604-1610.
3. Speyer, J.L., Dannemiller, D., Walker, D.,"Periodic Optimal Cruise of an Atmospheric Vehicle," Journal of Guidance, Control, and Dynamics, Vol. 8, 1985, pp. 31-38.
4. Gilbert, E.G., "Vehicle Cruise: Improved Fuel Economy by Periodic Control," Automatica, Vol. 12, 1976, pp. 159-166.
5. Gilbert, E.G., Lyons, D.T., "Improved Aircraft Cruise by Periodic Control: The Computation of Optimal Specific Range Trajectories," Proc. of 1980 Conf. Sci. Syst., Princeton Unversity, 1980.
6. Breakwell, J.V., Shoee, H., "Minimum Fuel Flight Paths for Griven Range," AIAA Paper No. 80-1660, 1980.
7. Breakwell, J.V., "Oscillatory Cruise – A Perspective," in: Lecture Notes in Control and Information Sciences, Vol. 95, Optimal Control, 1987, pp. 157-168.
8. Houlihan, S.C., Cliff, E.M., Kelley, H.J., "Study of Chattering Cruise," Journal of Aircraft, Vol. 19, 1982, pp. 119-124.
9. Bhardwaj, P., Kelley, H.J., Cliff, E.M., "Aircraft Cruise Performance Optimization Using Chattering Controls," Proceeding of the 10th IFAC Congress, Munich, 1987, pp. 200-205.
10. Grimm, W., Well, K.-H., Oberle, H.J., "Periodic Control for Minimum Fuel Aircraft Trajectories," Journal of Guidance, Control, and Dynamics, Vol. 9, 1986, pp. 169-174.
11. Menon, P.K.A., "Study of Aircraft Cruise," Journal of Guidance, Control, and Dynamics, Vol. 12, No. 5, pp. 631-639, 1989.
12. Sachs, G., Christodoulou, T., "Endurance Increase by Cyclic Control," Journal of Guidance, Control, and Dynamics, Vol. 9, 1986, pp. 58-63
13. Sachs, G., Christodoulou, T., "Reducing Fuel Consumption of Subsonic Aircraft by Optimal Cyclic Cruise," Journal of Aircraft, Vol. 24, 1987, pp. 616-622.
14. Sachs, G., Mehlhorn, R., "Endurance Increase by Periodic Optimal Camber Control," AIAA Guidance, Navigation and Control Conference Proceedings, pp. 635-641, 1991.
15. Siewert, R.F., Whitehead, R.E., " Analysis of Advanced Variable Camber Concepts," AGARD-CP-241, pp. 14-1 - 14-21, 1978.
16. "AFTI/F-111 Mission Adaptive Wing Briefing to Industry," AFWAL-TR-88-3082, Flight Dynamics Laboratory, Air Force Wright Aeronautical Laboratories, Wright-Pattersom Air Force Base, Ohio, 1988.
17. Szodruch, J., Hilbig, R., "Variable Wing Camber for Transport Aircraft," Progress in Aerospace Sciences, Vol. 25, pp. 297-328, 1988.
18. Smith, S.B., Nelson, D.W., "Determination of the Aerodynamic Characteristics of the Mission Adaptive Wing," Journal of Aircraft, Vol. 27, pp. 950-958, 1990.
19. ICAO Standard Atmosphere, International Civil Aviation Organization, Montreal, 1964.
20. Bulirsch, R., "Die Mehrzielmethode zur numerischen Lösung von nichtlinearen Randwertproblemen und Aufgaben der optimalen Steuerung," Report of the Carl-Cranz-Gesellschaft, Oberpfaffenhofen, Germany, 1971.
21. Oberle, H.J., "Numerische Berechnung optimaler Steuerungen von Heizung und Kühlung für ein realistisches Sonnenhausmodell," Institut für Mathematik der Technischen Universität München, TUM-M8310, 1983.
22. Bock, H.G., "Numerische Behandlung von zustandsbeschränkten und Chebychef-Steuerungsproblemen," Course R1.06 der Carl-Cranz-Gesellschaft, Oberpfaffenhofen, Germany, 1983.
23. Marchal, C., "Chattering Arcs and Chattering Controls," Journal of Optimization Theory and Applications, Vol. 11, No. 5, pp. 441-468, 1973
24. Gilbert, W.W., "Development of a Mission Adaptive Wing System for a Tactical Aircraft," AIAA Paper No. 80-1886, 1980.

OPTIMAL THRUST AND AERODYNAMIC CONTROLS IN HYPERSONIC FLIGHT

N.X.Vinh[1] and Sudhakar Medepalli [2]

[1]Professor, Department of Aerospace Engineering
The University of Michigan
Ann Arbor, Michigan 48109-2140
[2]Mechanical Dynamics Inc.
2301 Commonwealth Blvd.
Ann Arbor, MI 48105

ABSTRACT

The general equations of motion for a thrusting, lifting vehicle flying inside a non rotating planetary atmosphere with a central gravitational force field are presented in vector form. It is shown that the optimal thrust and aerodynamic controls are governed by the Lawden's primer vector $\bar{p}_v$, the adjoint vector associated with the velocity vector. The optimal control laws are presented and the integrals of the motion derived in vector form. For the case where the thrust is used at an intermediate level such as the case of prolonged thrusting flight inside the atmosphere, additional relations are derived and the explicit expression for the modulated thrust magnitude on this sustained arc is obtained. Finally, using canonical transformation, the two sets of state and adjoint equations together with the resulting optimal control laws and the integrals of the motion are obtained in terms of the usual trajectory variables (r, θ, ϕ, V, γ, ψ) employed in hypersonic flight.

INTRODUCTION

With the advent of the space shuttle and with projects such as the National Aerospace Plane and Mission to Mars planned for the near future, hypersonic flight of spacecraft through planetary atmospheres is becoming a reality. To take advantage of the atmospheric flight segment, future spacecraft are being designed with lifting capability. To compensate for the energy loss due to the atmospheric drag, they also need

Mechanics and Control, Edited by R.S. Guttalu
Plenum Press, New York, 1994

to produce thrust. With energy conservation becoming a vital issue, one needs to find an optimal way to utilize the thrust and the lift forces to maximize the performance of a mission for a given amount of energy consumption. Stated differently, one needs to minimize the cost to achieve certain performance. These considerations come under the general category of optimal control problems.

Hohmann was one of the first to formulate an optimal orbital transfer problem which led to the famous Hohmann transfer between two coplanar circular orbits. Lawden [1] developed a general theory of optimal orbital transfers. Both of the above studies, however, are concerned with pure gravitational fields with no atmosphere. It was London [2] who, with his pioneering work in 1962 established that by utilizing a spacecraft's lifting capability, significant savings can be obtained in the propellant consumption for certain types of orbital transfers near planetary atmospheres. Since then, a number of researchers studied various optimal aeroassisted orbital transfer problems [3]-[10]. Most of these early studies have two features in common. They all express the equations of motion in terms of the orbital elements and they are all posed as parametric optimization problems.

For a spacecraft modeled as a point mass and capable of lifting and thrusting, there are five controls in general. The thrust magnitude and two angles to specify the thrust direction constitute the thrust controls while the angle of attack and the bank angle comprise the aerodynamic controls. It is quite common in the literature on atmospheric flight to assume that the spacecraft has a plane of symmetry made up of the velocity vector and the lift vector. It is also quite common to assume that the thrust vector always lies in this plane of symmetry. This is the so called symmetric flight condition [11, 12], a term borrowed from aircraft flight. This assumption reduces the controls by one, leaving the so called thrust angle of attack as the sole thrust direction control. If we further assume that the thrust line is fixed with respect to the body of the spacecraft, we are left with only three controls. Most of the early researchers studied the effect of each of these controls, while keeping the rest of them constant. It was also shown by Mease et. al. that using orbital elements as state variables leads to singularity problems for certain orbits which can be avoided by using the trajectory variables instead [9]. The survey article by Walberg [10] lists a number of these early references.

Meanwhile, classical calculus of variations had already been applied extensively to rocket powered flight in vacuum by Lawden [13], Breakwell [14] and others as well as to atmospheric flight problems over a flat earth with constant gravity by Leitmann [15], Miele [16], Bryson [17] and others. These results, along with the newly announced Pontryagin's Maximum Principle (PMP) [18] were then applied to atmospheric flight problems over a spherical earth with a central gravitational field [19]. The optimal time histories of the controls were obtained as functions of the state and the adjoint variables. The assumptions regarding the nature of the atmospheric density and the spacecraft's aerodynamic characteristics were borrowed from aircraft flight. Even though these assumptions were not very realistic, they served as a good starting point and provided valuable insights. Numerous problems were studied by a number of authors such as Miele, Vinh, Breakwell, Speyer etc .... An extensive survey of the literature on aeroassisted orbital transfers is beyond the scope of this paper. The survey article by Mease [20] is very useful in this regard.

It has been recognized a long time ago that certain relations among the state and the adjoint variables, called the integrals of motion, play an important role in dynamic systems. If the number of these integrals is the same as the order of the dynamic system, we can obtain, at least in theory, closed form solutions for the optimal controls. It has been observed in the above orbital transfer problems that a sufficient number of integrals of motion do not exist. Different approaches such as using reduced order models, singular perturbation methods and numerical solution of the resulting Two Point Boundary Value Problem (TPBVP) have been followed by various authors to overcome this difficulty.

From the above discussion, it seems as though the theory of optimization of trajectories in atmospheric flight is complete. However, there is a class of controls, termed the "singular controls" for which neither the PMP nor the classical calculus of variations gives us any information and new necessary conditions are needed to check for their optimality. These occur, for example, when the control appears linearly in the equations of motions. These are quite common in optimal orbital transfer problems both in vacuum and in the atmosphere whenever we assume a constant ejection velocity type propulsion system with a bounded thrust magnitude. The part of the trajectory over which these singular controls occur is called the "singular arc." These arcs are important because they might form a part of the optimal trajectory. Two singular arcs were studied quite extensively in the literature, the first one being the Lawden's spiral [13] and the second one the cruise arc in aircraft flight [23] - [26]. One of the earliest successful methods that deals with problems involving singular controls is the Green's theorem approach of Miele [27]. Although this method provides both necessary and sufficient conditions for optimality, it is severely restricted in its applicability. Kelley [28] was the first to develop new necessary conditions for the optimality of the singular arcs and the method of obtaining the singular control. The method was generalized by Kelley, Kopp and Moyer [29] and the new necessary conditions were called the Generalized Legendre-Clebsch condition (Kelley-Contensou test in Europe).

This paper presents a summary of all the important results in the theory of optimal aeroassisted orbital maneuvers. We follow a vectorial approach to obtain all the results for generality and conciseness starting with the equations of motion expressed in an inertial reference frame as given in [12]. We state a very general optimal control problem of the Mayer type and follow a geometric approach in the hodograph plane [12] to obtain the resulting optimal aerodynamic controls. Here we assume that the thrust direction is constrained along the velocity vector, a fairly common assumption in atmospheric flight problems. Thus the thrust magnitude is the sole propulsive control. We show that, due to our assumption of a constant ejection velocity type propulsion system, we have three kinds of possible sub-arcs for the thrust magnitude control along the optimal trajectory. Using the elegant geometric ideas mentioned above, we obtain the explicit expression for the magnitude of the thrust along a singular sub-arc. All the existing integrals of motion are also stated. Through an example canonical transformation to the flight path variables, we illustrate how one can translate all the above results, expressed in vector form, to any set of state variables one chooses.

Note that throughout this paper we denote vector quantities by a bar overhead and unit vectors by a hat overhead. Further if $\bar{a}$ is any vector, then $\hat{e}_a$ corresponds to a unit vector along $\bar{a}$.

EQUATIONS OF MOTION

We make the following general assumptions regarding the spacecraft.

- The spacecraft is modeled as a point mass.
- The only external forces acting on the spacecraft are the thrust generated by the propulsion system, the aerodynamic forces due to the atmosphere and the gravitational force of the planet.

- The spacecraft propulsion system is of the Constant Ejection Velocity (CEV) type.

With these assumptions, the dynamics of the spacecraft are given by the following first order ordinary differential equations according to the Newton's laws of motion.

$$\begin{aligned}\frac{d\bar{r}}{dt} &= \bar{v},\\ \frac{d\bar{v}}{dt} &= \frac{\bar{T}}{m}+\frac{\bar{A}}{m}+\bar{g},\\ \frac{dm}{dt} &= -\frac{T}{c}.\end{aligned} \tag{1}$$

Here, $\bar{r}$ is the position vector measured from the center of the planet, $\bar{v}$ is the velocity of the spacecraft, m is the mass of the spacecraft and t refers to time. All the derivatives are evaluated in an inertial reference frame fixed at the center of the planet. The external forces are the thrust $\bar{T}$, the aerodynamic force $\bar{A}$ and the gravitational force $\bar{g}$. T is the magnitude of the thrust and c is the constant ejection velocity of the propulsion system.

We further make the following assumptions regarding the nature of the various forces given above.

- The planet is spherical, non-rotating with a stationary atmosphere. Thus we neglect the coriolis and the centrifugal force terms that arise due to the rotation of the planet.

- The gravitational force is central and attractive, i.e., it is directed along the radius vector towards the center of the planet. Its magnitude is assumed to be a function of the spacecraft radial distance alone. Thus, we can write

$$\bar{g} = -g(r)\hat{e}_r, \tag{2}$$

where, $\hat{e}_r$ is the unit vector in the direction of the radius vector.

- The aerodynamic force is composed of the lift and the drag forces which are assumed to have the following form [12].

$$\begin{aligned}\bar{A} &= \bar{L}+\bar{D},\\ \bar{L} &= \frac{1}{2}\rho(r)Sv^2C_L\hat{e}_l,\\ \bar{D} &= -\frac{1}{2}\rho(r)Sv^2C_D\hat{e}_v.\end{aligned} \tag{3}$$

Here, $\rho(r)$ is the density of the atmosphere, assumed to be a function of the altitude (or the radial distance) alone, S is the effective surface area of the vehicle, v is the magnitude of the velocity, C_L and C_D are the lift and the drag coefficients respectively, $\hat{e}_l$ is the unit vector in the direction of the lift force and $\hat{e}_v$ is the unit vector in the direction of the velocity. Thus we can write,

$$\bar{A} = \frac{1}{2}\rho(r)Sv^2\bar{C}_a \tag{4}$$

where,

$$\bar{C}_a = C_L\hat{e}_l - C_D\hat{e}_v. \tag{5}$$

Since the lift force is in a plane orthogonal to the velocity vector, we have the following relation between $\hat{e}_l$ and $\hat{e}_v$.

$$\hat{e}_l \cdot \hat{e}_v = 0. \tag{6}$$

- We assume that the spacecraft has a plane of symmetry and that both the velocity vector and the lift vector remain in this plane throughout the flight. This is the so called symmetric flight condition where the side slip angle is zero. Thus the attitude of the spacecraft can be described by the angle of attack, which is the angle between the velocity vector and an axis fixed in the body of the spacecraft in the symmetric plane [12].

- There exists a general drag polar relating the lift and the drag coefficients. Thus,

$$C_D = C_D(C_L). \tag{7}$$

  In general, both the lift and the drag coefficients are functions of the angle of attack and the Mach number. But we ignore the dependency on the Mach number and use the above relation instead of explicit equations for C_L and C_D in terms of the angle of attack.

- Since we are only concerned with the atmospheric flight segment of the spacecraft throughout this paper, we assume that the thrust direction is fixed with respect to the body of the spacecraft. Further, assuming that the angle between this fixed direction and the velocity vector (the so called thrust angle of attack) is small, we can write the thrust force explicitly as

$$\bar{T} = T\hat{e}_v. \tag{8}$$

- The thrust magnitude is bounded as follows:

$$0 \leq T \leq T_{max}. \tag{9}$$

  Even though T_{max} is in general a function of the altitude as well as the Mach number, we assume that it is a constant for the propulsion system used.

With these assumptions, the equations of motion given in (1) can be written more explicitly as

$$\begin{aligned} \frac{d\bar{r}}{dt} &= \bar{v}, \\ \frac{d\bar{v}}{dt} &= \frac{T}{m}\hat{e}_v + \frac{\rho S}{2m}v^2[C_L\hat{e}_l - C_D\hat{e}_v] - g\hat{e}_r, \\ \frac{dm}{dt} &= -\frac{T}{c}. \end{aligned} \tag{10}$$

THE OPTIMAL CONTROL PROBLEM

The state variables in our problem are the position $\bar{r}$, the velocity $\bar{v}$ and the mass m. They are governed by the first order ordinary differential equations given by (1). The initial conditions at time $t = t_0$ are

$$\begin{aligned} \bar{r}(t_0) &= \bar{r}_0 \\ \bar{v}(t_0) &= \bar{v}_0 \\ m(t_0) &= m_0 \end{aligned} \tag{11}$$

and the final conditions at $t = t_f$ are given by

$$\Psi(t_f, \bar{r}_f, \bar{v}_f, m_f) = 0, \quad \Psi \in \Re^r \tag{12}$$

where, $\bar{r}_f = \bar{r}(t_f)$, $\bar{v}_f = \bar{v}(t_f)$ and $m_f = m(t_f)$.

The control variables are the thrust magnitude T, the lift vector direction $\hat{e}_l$ and the angle of attack α. Since $\hat{e}_l$ always lies in the symmetric plane and is always orthogonal to the velocity vector in that plane, we need just one angle to specify the orientation of the symmetric plane with respect to some reference plane. We do this by defining the bank angle σ as the angle between the orbital plane (the plane containing $\bar{r}$ and $\bar{v}$) and the symmetric plane (see Fig.7). As mentioned earlier, we use the lift coefficient instead of the angle of attack as the other control variable. The bounds on the control variables are

$$\begin{gathered} 0 \le T \le T_{max} \\ 0 \le C_L \le C_{L max} \\ |\sigma| \le \sigma_{max} \end{gathered} \tag{13}$$

Finally, the performance index is a scalar function of the final state variables and final time and can be written as

$$J = J(t_f, \bar{r}_f, \bar{v}_f, m_f) \,. \tag{14}$$

Statement of the Optimal Control Problem

It is desired to obtain the time histories of the admissible controls $T(\cdot)$, $C_L(\cdot)$ and $\sigma(\cdot)$ such that the performance index (14) is maximized while satisfying the differential constraints (1) and the end conditions (11) and (12). By admissible controls we mean piece-wise continuous controls which satisfy the constraints (13).

Clearly, the problem stated above is of the Mayer type. Following Pontryagin's maximum principle, we form the Hamiltonian by introducing the adjoint variables $\bar{p}_r$, $\bar{p}_v$ and p_m as given below.

$$H = \bar{p}_r \cdot \bar{v} + \bar{p}_v \cdot \left(\frac{\bar{A}}{m} + \bar{g}\right) + \frac{T}{m}\left(\bar{p}_v \cdot \hat{e}_v - p_m \frac{m}{c}\right) \tag{15}$$

Optimal Aerodynamic Control

In order to obtain the optimal aerodynamic control, we must maximize the Hamiltonian with respect to the controls σ and C_L. We do this by introducing the concept of the domain of maneuverability [12, 30].

Definition 1 *Given the equations of state,*

$$\dot{\mathbf{x}} = \mathbf{f}(\mathbf{x}, \mathbf{u}, t) \tag{16}$$

where $\mathbf{x}$ *is the state vector,* $\mathbf{u}$ *is the control vector and* t *is the time,* the natural domain of maneuverability $D(\mathbf{x}, t)$ *is the instantaneously usable (reachable) domain of the hodograph space (velocity space).*

Thus the control space Σ is transformed into the maneuverability domain D through the transformation, $\mathbf{v} = \dot{\mathbf{x}} = \mathbf{f}(\mathbf{x}, \mathbf{u}, t)$ (see Fig.1(a)). Maximizing $H = \bar{p}^T \bar{f}$

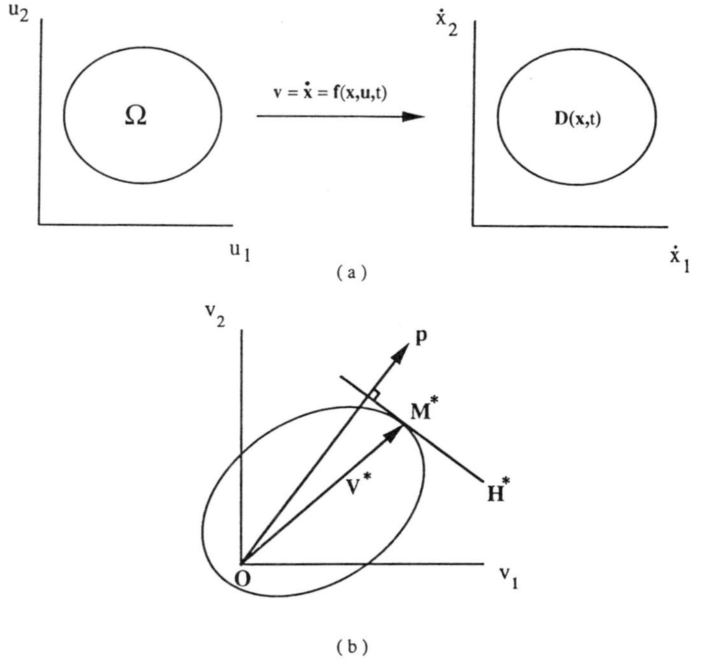

Figure 1. (a) Domain of Maneuverability, (b) Optimal Condition.

with respect to the controls is equivalent to finding a point M^* on the convex portion of the boundary of D such that the projection of the vector $V^* = OM^*$ on the adjoint vector $\bar{p}$ is maximized as shown in Fig.1(b).

Optimal Bank Control : In Fig.7, let us consider the $PLMN$ right-hand triad such that PL and PM are in the orbital plane and PN is orthogonal to it. In this system, the domain of maneuverability described by the terminus of the vector $\bar{A}$ is the surface of revolution $\sum$ about the v-axis. This surface is bounded by $C_L \leq C_{Lmax}$ and $\sigma \leq \sigma_{max}$, as shown in Fig.2(a).

In order to maximize the Hamiltonian with respect to the aerodynamic controls, we must select $\bar{A}$ such that at each instant the dot product $(\bar{p}_v \cdot \bar{A})$ is maximized. This can be achieved by selecting the terminus of $\bar{A}$ at the contact point of the plane tangent to $\sum$ while orthogonal to $\bar{p}_v$ as shown in Fig.2(b). Thus, $\bar{A}$ or, equivalently, the vehicle plane of symmetry, must be rotated out of the orbital plane about the velocity vector by an angle σ such that the vectors $\bar{A}$, $\bar{v}$ and $\bar{p}_v$ are coplanar [12]. This geometric construction can be expressed analytically as

$$(\bar{p}_v \times \bar{v}) \cdot \bar{A} = 0 \tag{17}$$

from which, the bank control can be obtained explicitly as:

$$\cos\sigma = \frac{[(\bar{p}_v \times \bar{v}) \cdot (\bar{r} \times \bar{v})]}{|(\bar{p}_v \times \bar{v})||(\bar{r} \times \bar{v})|} \tag{18}$$

If the resulting bank angle is larger than σ_{max}, we set $\sigma = \sigma_{max}$.

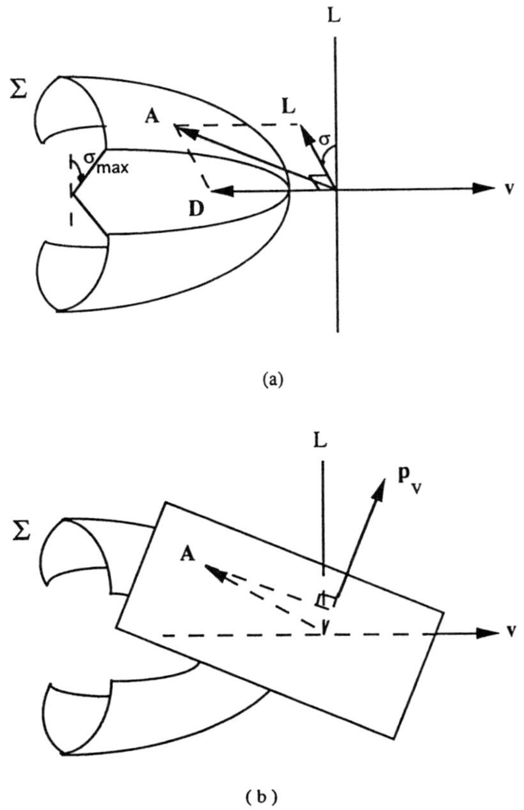

Figure 2. (a) The Drag Polar, (b) Optimal Bank Control.

Optimal Lift Control : Consider the ($\bar{v}$, $\bar{p}_v$) plane which now contains the aerodynamic force vector $\bar{A}$ also, as shown in Fig.3(a). This plane intersects the surface of revolution $\sum$ along a meridian which is precisely the drag polar given by (7). The terminus of $\bar{A}$ is chosen at a point on this drag polar such that the tangent to the polar at that point is orthogonal to $\bar{p}_v$ as shown in Fig.3(a). Let ϵ be the angle between $\bar{v}$ and $\bar{p}_v$. This is also the angle between the rotated lift direction and the tangent. Thus,

$$\tan \epsilon = \frac{\partial C_D}{\partial C_L} \tag{19}$$

gives the governing equation for the optimal lift control [11, 12]. If the resulting $C_L > C_{Lmax}$, we set $C_L = C_{Lmax}$.

Remark 1 *The above discussion for the lift control is valid only if the optimal bank control is interior, i.e., $\sigma \leq \sigma_{max}$ and the vectors $\bar{A}$, $\bar{v}$ and $\bar{p}_v$ are coplanar. Otherwise, we must replace $\bar{p}_v$ by its projection onto the ($\bar{v}$, $\bar{A}$) plane and let ϵ be the angle between $\bar{v}$ and this projection.*

Remark 2 *Note that beyond certain value of ϵ, say, ϵ_{max}, the lift control is always*

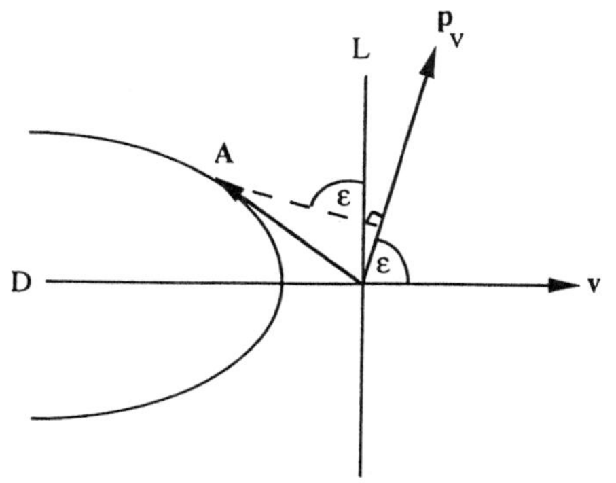

(a)

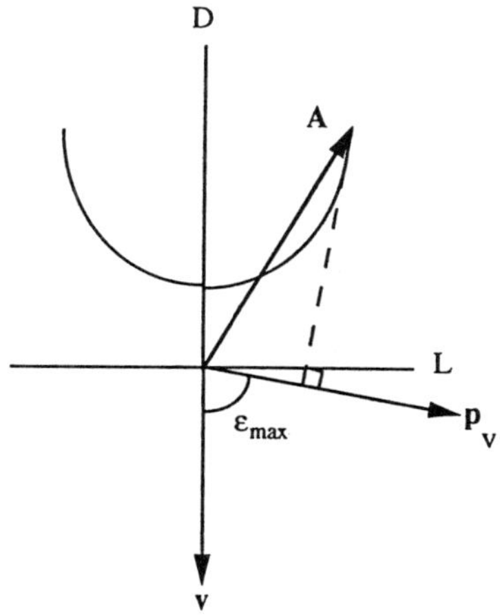

(b)

Figure 3. (a) Optimal Lift Control, (b) Maximum Lift Condition.

given by $C_L = C_{Lmax}$. This condition occurs when the line orthogonal to $\bar{p}_v$ is tangent to the drag polar at $C_L = C_{Lmax}$ as shown in Fig.3(b).

Optimal Thrust Control

Since we assumed that the thrust direction is fixed along the velocity vector, the only thrust control is the magnitude T, which appears linearly in the Hamiltonian. We define the coefficient of the thrust magnitude in the Hamiltonian as the *switching function* Φ. Thus,

$$\Phi = (\bar{p}_v \cdot \hat{e}_v) - p_m \frac{m}{c} \tag{20}$$

and we have the following optimal thrust magnitude control.

$$If \begin{cases} \Phi > 0, & T = T_{max} \\ \Phi < 0, & T = 0 \\ \Phi \equiv 0, \forall t \in [t_1, t_2] & T \ \ is \ \ variable \end{cases} \tag{21}$$

where, $[t_1, t_2] \in [t_0, t_f]$.

The first case corresponds to the boost arc, the second one the coast arc and the last the sustained arc. In the case of the sustained arc, we must differentiate the switching function an even number of times before the thrust magnitude appears linearly. We postpone the discussion of the sustained arc case to a later stage in this paper.

Remark 3 *The adjoint vector $\bar{p}_v$ associated with the velocity vector plays a crucial role in determining the optimal controls as was evident from the above discussion. This vector has been named the* primer vector *by Lawden* [1].

The Adjoint Equations

According to the maximum principle, the differential equations governing the adjoint variables are given by

$$\begin{aligned} \dot{\bar{p}}_r &= -\frac{\partial H}{\partial \bar{r}} \\ \dot{\bar{p}}_v &= -\frac{\partial H}{\partial \bar{v}} \\ \dot{p}_m &= -\frac{\partial H}{\partial m} \end{aligned} \tag{22}$$

where we use the optimal control values in the partial derivatives on the right hand side.

At $t = t_f$, the adjoint variables satisfy the transversality conditions

$$\bar{p}(t_f) = \frac{\partial J}{\partial \bar{x}_f} + \left(\frac{\partial \Psi}{\partial \bar{x}_f}\right)^T \bar{\Lambda} \tag{23}$$

where, $\bar{\Lambda} \in \Re^r$ is a constant vector and $\bar{x} = [\bar{r} \;\; \bar{v} \;\; m]^T$ is the vector of state variables. Thus we have a two point boundary value problem with the state variables specified at t_0 and the adjoint variables specified at t_f.

Integrals of the Motion

The complexity of the optimal control problem can be reduced if there exist certain relationships among the state, the adjoint and the control variables, called the first integrals of motion.

In our case, since the equations of motion are independent of time, we have,

$$\frac{dH}{dt} = \frac{\partial H}{\partial t} = 0 \tag{24}$$

Thus,

$$H = c_0 \tag{25}$$

is a first integral where c_0 is the constant of integration. If the final time is free, we have, $c_0 = 0$.

Due to the assumed spherical symmetry about the center of the planet, we have the following first integral in vector form [12, 30].

$$(\bar{r} \times \bar{p}_r) + (\bar{v} \times \bar{p}_v) = \bar{c} \tag{26}$$

The above four integrals are valid throughout the optimal trajectory. Furthermore, on a coast arc, since the thrust magnitude is zero, we have, $m =$ *constant*. On a sustained arc, in the case of atmospheric flight, we must differentiate the switching function twice, so that we have the two additional relations, $\Phi = 0$ and $\dot{\Phi} = 0$. In any case, we have seven adjoint variables and at most six relationships among them. Thus, the problem is not completely integrable and we must use some kind of numerical scheme to solve the TPBVP.

THE THRUST MAGNITUDE ON A SINGULAR ARC

In this section we derive the expression for the thrust magnitude on a singular arc. We first express the switching function solely as a function of the state and the adjoint variables. This implies that we must express all the non-singular controls in terms of the state and the adjoint variables. We do this with the help of the geometric ideas developed earlier. Then we differentiate the switching function twice and substitute the corresponding state and adjoint equations to obtain the expression for the thrust magnitude.

Aerodynamic Controls as Functions of the State and the Adjoint Variables

To simplify the computations, we introduce the transformations

$$\begin{aligned} q &= -c \log \frac{m}{m_0} \\ \tau &= \frac{T}{m} \end{aligned} \tag{27}$$

Since the equation relating q and m represents a Legrange's point transformation, the canonicity condition from the theory of canonical transformations implies that [21]

$$p_m dm = p_q dq \tag{28}$$

Since $dq = -\frac{c}{m} dm$, we get

$$p_m = -\frac{c}{m} p_q \tag{29}$$

Substituting (27) in the equations of motion given earlier in (1) we get

$$\begin{aligned} \dot{\bar{r}} &= \bar{v} \\ \dot{\bar{v}} &= \tau \hat{e}_v + \left(\frac{e^{q/c}}{m_0} \right) \bar{A} - \bar{g} \\ \dot{q} &= \tau \end{aligned} \tag{30}$$

Here $\bar{A}$ is written as

$$\bar{A} = \frac{1}{2} \rho S v^2 \bar{C}_a \tag{31}$$

where,

$$\bar{C}_a = -C_D \hat{e}_v + C_L \hat{e}_l \tag{32}$$

We assume that the lift and the drag coefficients are related by a parabolic drag polar given by

$$C_D = C_{D0} + K {C_L}^2 \tag{33}$$

In terms of τ, the switching function has the form

$$\Phi = (\bar{p}_v^T \hat{e}_v) + p_q \tag{34}$$

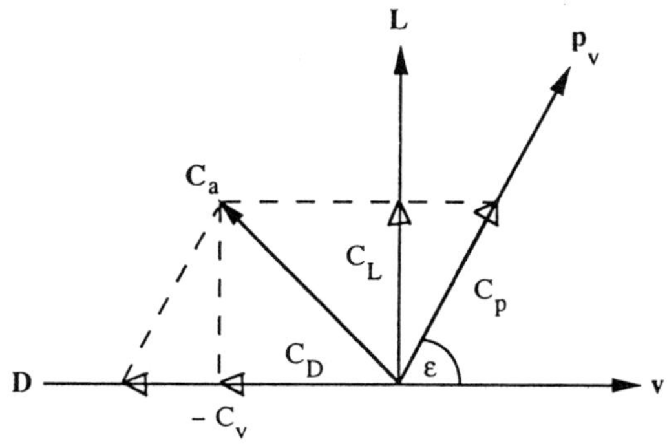

Figure 4. Resolving $\bar{C}_a$ along $\bar{v}$ and $\bar{p}_v$.

which, upon using (29) is seen to be identical to (20). Hence the singular arcs corresponding to T and τ are the same. We derive the expression for τ along the singular arc and then find the corresponding T using Eq.(27) instead of directly deriving the expression for T.

The non-singular controls in our problem are the lift coefficient C_L and the bank angle σ, which specifies the direction of the lift force with respect to the orbital plane. Earlier we showed geometrically that the bank angle is optimal when the velocity vector, the aerodynamic force vector and the primer vector all lie in the same plane. In other words, $\hat{e}_v$, $\bar{C}_a$ and $\bar{p}_v$ are coplanar. Thus we can also resolve the vector $\bar{C}_a$ along the non-orthogonal directions $\hat{e}_v$ and $\hat{e}_{p_v}$, where,

$$\hat{e}_{p_v} = \frac{\bar{p}_v}{|\bar{p}_v|} \tag{35}$$

This is shown in Fig.4. Thus we can write

$$\bar{C}_a = C_v\hat{e}_v + C_p\hat{e}_{p_v} \tag{36}$$

Now the optimal lift coefficient is chosen such that (see Fig.4)

$$\frac{\partial C_D}{\partial C_L} = 2KC_L = \tan\epsilon \tag{37}$$

From Fig.4, we can write

$$\hat{e}_{p_v} = (\cos\epsilon)\hat{e}_v + (\sin\epsilon)\hat{e}_l \tag{38}$$

and substituting into (36), we get

$$\bar{C}_a = (C_v + C_p\cos\epsilon)\hat{e}_v + (C_p\sin\epsilon)\hat{e}_l \tag{39}$$

Equating the two expressions for $\bar{C}_a$ in (32) and (39), we get

$$\begin{aligned} C_v + C_p\cos\epsilon &= -C_D \\ C_p\sin\epsilon &= C_L \end{aligned} \tag{40}$$

which, by using (37), can be written as

$$\begin{aligned} C_p &= \frac{C_L}{\sin\epsilon} \\ &= \frac{C_L}{\tan\epsilon}\frac{1}{\cos\epsilon} \\ &= \frac{1}{2K}\frac{1}{\cos\epsilon} \end{aligned} \tag{41}$$

Using (33), (37), (40) and the first relation in (41), we have

$$\begin{aligned} C_v &= -C_D - \frac{C_L}{\tan\epsilon} \\ &= -C_{D0} - K{C_L}^2 - \frac{1}{2K} \\ &= -C_{D0} - K\left(\frac{\tan\epsilon}{2K}\right)^2 - \frac{1}{2K} \\ &= -C_{D0} - \frac{1}{4K} - \frac{1}{4K}\frac{1}{\cos^2\epsilon} \end{aligned} \tag{42}$$

By defining

$$\begin{aligned} E^* &= \frac{{C_L}^*}{{C_D}^*} \\ {C_D}^* &= 2C_{D0} \\ {C_L}^* &= \sqrt{\frac{C_{D0}}{K}} \end{aligned} \tag{43}$$

we get

$$\frac{1}{4K} = E^{*2}C_{D0} \tag{44}$$

From Fig.4, we notice that

$$\cos\epsilon = (\hat{e}_{p_v}^T\hat{e}_v) \tag{45}$$

Substituting (44) and (45) in (41) and (42), we get

$$\begin{aligned} C_p &= \frac{2E^{*2}C_{D0}}{(\hat{e}_{p_v}^T\hat{e}_v)} \\ C_v &= -C_{D0}\left[1 + E^{*2} + \frac{E^{*2}}{(\hat{e}_{p_v}^T\hat{e}_v)^2}\right] \end{aligned} \tag{46}$$

From (31), (36) and (46) we note that

$$\bar{A} = \frac{\rho S C_{D0}}{2}v^2\left[-\left\{1 + E^{*2} + \frac{E^{*2}v^2|\bar{p}_v|^2}{(\bar{p}_v^T\bar{v})^2}\right\}\hat{e}_v + \frac{2E^{*2}v}{(\bar{p}_v^T\bar{v})}\bar{p}_v\right] \tag{47}$$

Thus we expressed the aerodynamic force as a function of the state and the adjoint variables only. From the above expression, we note the following two relations.

$$(\bar{p}_v^T\bar{A}) = \frac{\rho S C_{D0}}{2}\left[-(1 + E^{*2})v(\bar{p}_v^T\bar{v}) + \frac{E^{*2}v^3|\bar{p}_v|^2}{(\bar{p}_v^T\bar{v})}\right] \tag{48}$$

and

$$(\bar{v}^T\bar{A}) = \frac{\rho S C_{D0}}{2}v^2\left[(E^{*2} - 1)v - \frac{E^{*2}v^3|\bar{p}_v|^2}{(\bar{p}_v^T\bar{v})^2}\right] \tag{49}$$

The Hamiltonian and the adjoint equations can be written as

$$H = (\bar{p}_r^T\bar{v}) + \left(\frac{e^{q/c}}{m_0}\right)(\bar{p}_v^T\bar{A}) + (\bar{p}_v^T\bar{g}) + \tau\left[(\bar{p}_v^T\hat{e}_v) + p_q\right] \tag{50}$$

and

$$
\begin{aligned}
\dot{\bar{p}}_r = -\frac{\partial H}{\partial \bar{r}} &= -\left(\frac{\rho_r}{\rho}\right)\left(\frac{e^{q/c}}{m_0}\right)\frac{(\bar{p}_v^T\bar{A})}{r}\bar{r} + \frac{g}{r}\bar{p}_v + \frac{(\bar{p}_v^T\bar{r})}{r^2}\left(g_r - \frac{g}{r}\right)\bar{r} \\
\dot{\bar{p}}_v = -\frac{\partial H}{\partial \bar{v}} &= -\bar{p}_r - \frac{\tau}{v}\left[\bar{p}_v - \frac{(\bar{p}_v^T\bar{v})}{v^2}\bar{v}\right] \\
&\quad -\frac{\rho S}{2}\left(\frac{e^{q/c}}{m_0}\right)\left[C_v v\bar{p}_v + C_v(\bar{p}_v^T\bar{v})\hat{e}_v + 2C_p|\bar{p}_v|\bar{v}\right] \\
\dot{p}_q = -\frac{\partial H}{\partial q} &= -\frac{1}{c}\left(\frac{e^{q/c}}{m_0}\right)(\bar{p}_v^T\bar{A})
\end{aligned}
\tag{51}
$$

where ρ_r and g_r are, respectively, the partial derivatives of ρ and g with respect to r.

Note that the derivatives of various scalar quantities can be written as

$$
\frac{dr}{dt} = \frac{(\dot{\bar{r}}^T\bar{r})}{r} = \frac{(\bar{v}^T\bar{r})}{r} \tag{52}
$$

$$
\frac{dv}{dt} = \frac{(\dot{\bar{v}}^T\bar{v})}{v} = \tau + \left(\frac{e^{q/c}}{m_0}\right)\frac{(\bar{v}^T\bar{A})}{v} + \frac{(\bar{v}^T\bar{g})}{v} \tag{53}
$$

and

$$
\frac{d|\bar{p}_v|}{dt} = \frac{(\dot{\bar{p}}_v^T\bar{p}_v)}{|\bar{p}_v|} \tag{54}
$$

We can easily show the following two important relations which are used later.

$$
\begin{aligned}
(\dot{\bar{p}}_v^T\bar{v}) &= -(\bar{p}_r^T\bar{v}) - 2\left(\frac{e^{q/c}}{m_0}\right)(\bar{p}_v^T\bar{A}) \\
(\dot{\bar{p}}_v^T\bar{p}_v) &= -(\bar{p}_r^T\bar{p}_v) - \frac{\tau}{v}\left[|\bar{p}_v|^2 - \frac{(\bar{p}_v^T\bar{v})^2}{v^2}\right] \\
&\quad -\left(\frac{e^{q/c}}{m_0}\right)\frac{1}{v^2}\left[(\bar{v}^T\bar{A})|\bar{p}_v|^2 + (\bar{p}_v^T\bar{A})(\bar{p}_v^T\bar{v})\right]
\end{aligned}
\tag{55}
$$

At this stage we are in a position to obtain the successive derivatives of the switching function, which is given as

$$
\Phi = (\bar{p}_v^T\hat{e}_v) + p_q = \frac{(\bar{p}_v^T\bar{v})}{v} + p_q = 0 \tag{56}
$$

Differentiating with respect to time, we get

$$
\dot{\Phi} = \frac{(\dot{\bar{p}}_v^T\bar{v})}{v} + \frac{(\bar{p}_v^T\dot{\bar{v}})}{v} - \frac{(\bar{p}_v^T\bar{v})}{v^2}\dot{v} = 0 \tag{57}
$$

Substituting (53) for $\dot{v}$ and using the state and the adjoint equations given by (30) and (51), we get

$$
\begin{aligned}
\dot{\Phi} &= \frac{(\bar{p}_v^T\bar{g})}{v} - \frac{(\bar{p}_v^T\bar{v})(\bar{v}^T\bar{g})}{v^3} - \frac{(\bar{p}_r^T\bar{v})}{v} \\
&\quad -\left(\frac{e^{q/c}}{m_0}\right)\left[\left(1+\frac{v}{c}\right)\frac{(\bar{p}_v^T\bar{A})}{v} + \frac{(\bar{p}_v^T\bar{v})(\bar{v}^T\bar{A})}{v^3}\right] = 0
\end{aligned}
\tag{58}
$$

We can get an alternate expression to be satisfied along a singular arc by using relations (50) and (58) and the Hamiltonian integral given by (25) where $c_0 = 0$ if the end conditions are also independent of time. Thus we get

$$(v\dot{\Phi} + H^*) = \bar{p}_v^T \left[2\bar{g} - \frac{(\bar{v}^T\bar{g})}{v^2}\bar{v} - \left(\frac{e^{q/c}}{m_0}\right)\frac{v}{c}\bar{A} - \left(\frac{e^{q/c}}{m_0}\right)\frac{(\bar{v}^T\bar{A})}{v^2}\bar{v}\right] = c_0 \tag{59}$$

Geometrically, when $c_0 = 0$, this equation gives the locus of points on a plane orthogonal to the primer vector. This plane is the switching surface in the configuration space.

To obtain the second time derivative of Φ, we can use either of the two expressions given by (58) or (59). We will use the former expression for this purpose. From the expression for $\dot{\Phi}$ in (58), we notice that it is a function of the variables $\bar{r}$, $\bar{v}$, q, $\bar{p}_v$ and their magnitudes. Thus we can use the chain rule for differentiating each term in that expression.

$$\frac{d}{dt}\left[\frac{(\bar{p}_v^T\bar{g})}{v}\right] = \frac{(\dot{\bar{p}}_v^T\bar{g})}{v} + \frac{(\bar{p}_v^T\dot{\bar{g}})}{v} - \frac{(\bar{p}_v^T\bar{g})}{v^2}\dot{v} \tag{60}$$

where, using the explicit expression for $\bar{g}$ given in (2), we get

$$\dot{\bar{g}} = -\frac{g}{r}\dot{\bar{r}} + \left(-\frac{g_r}{r} + \frac{g}{r^2}\right)\dot{r}\bar{r} \tag{61}$$

Similarly,

$$\begin{aligned}\frac{d}{dt}\left[-\frac{(\bar{p}_v^T\bar{v})(\bar{v}^T\bar{g})}{v^3}\right] &= -\frac{(\dot{\bar{p}}_v^T\bar{v})(\bar{v}^T\bar{g})}{v^3} - \frac{(\bar{p}_v^T\dot{\bar{v}})(\bar{v}^T\bar{g})}{v^3} \\ &\quad -\frac{(\bar{p}_v^T\bar{v})(\dot{\bar{v}}^T\bar{g})}{v^3} - \frac{(\bar{p}_v^T\bar{v})(\bar{v}^T\dot{\bar{g}})}{v^3} + 3\frac{(\bar{p}_v^T\bar{v})(\bar{v}^T\bar{g})}{v^4}\dot{v}\end{aligned} \tag{62}$$

where, once again, $\dot{\bar{g}}$ is given by (61). Differentiating the third term in (58), we get

$$\frac{d}{dt}\left[-\frac{(\bar{p}_r^T\bar{v})}{v}\right] = -\frac{(\dot{\bar{p}}_r^T\bar{v})}{v} - \frac{(\bar{p}_r^T\dot{\bar{v}})}{v} + \frac{(\bar{p}_r^T\bar{v})}{v^2}\dot{v} \tag{63}$$

Finally the last term, upon differentiation, gives

$$\frac{d}{dt}\left[-\left(1+\frac{v}{c}\right)\left(\frac{e^{q/c}}{m_0}\right)\frac{(\bar{p}_v^T\bar{A})}{v} - \left(\frac{e^{q/c}}{m_0}\right)\frac{(\bar{p}_v^T\bar{v})(\bar{v}^T\bar{A})}{v^3}\right] \tag{64}$$

$$\begin{aligned}&= \frac{d}{dt}\left[\frac{\rho S C_{D0}}{2}\left(\frac{e^{q/c}}{m_0}\right)\left\{\left(2+\frac{v}{c}+E^{*2}\frac{v}{c}\right)(\bar{p}_v^T\bar{v}) - \frac{E^{*2}v^3|\bar{p}_v|^2}{(\bar{p}_v^T\bar{v})c}\right\}\right] \\ &= -\left(\frac{e^{q/c}}{m_0}\right)\left[\left(1+\frac{v}{c}\right)\frac{(\bar{p}_v^T\bar{A})}{v} + \frac{(\bar{p}_v^T\bar{v})(\bar{v}^T\bar{A})}{v^3}\right]\left[\frac{\dot{q}}{c} + \left(\frac{\rho_r}{\rho}\right)\dot{r}\right] \\ &\quad +\frac{\rho S C_{D0}}{2}\left(\frac{e^{q/c}}{m_0}\right)\left[(1+E^{*2})\frac{(\bar{p}_v^T\bar{v})}{c} - \frac{3E^{*2}v^2|\bar{p}_v|^2}{(\bar{p}_v^T\bar{v})c}\right]\dot{v} \\ &\quad +\frac{\rho S C_{D0}}{2}\left(\frac{e^{q/c}}{m_0}\right)\left[\left(2+\frac{v}{c}+E^{*2}\frac{v}{c}\right) + \frac{E^{*2}v^3|\bar{p}_v|^2}{(\bar{p}_v^T\bar{v})^2c}\right]\left[(\dot{\bar{p}}_v^T\bar{v}) + (\bar{p}_v^T\dot{\bar{v}})\right] \\ &\quad -\rho S C_{D0}\left(\frac{e^{q/c}}{m_0}\right)\frac{E^{*2}v^3}{(\bar{p}_v^T\bar{v})c}(\bar{p}_v^T\dot{\bar{p}}_v)\end{aligned} \tag{65}$$

Upon substituting for all the time derivatives in the above expression, collecting terms and simplifying, we get

$$\ddot{\Phi} = \mathcal{A} + \tau\mathcal{B} = 0 \tag{66}$$

where,

$$\mathcal{B} = -2\frac{(\bar{p}_v^T\bar{g})}{v^2} + 2\frac{(\bar{p}_v^T\bar{v})(\bar{v}^T\bar{g})}{v^4} + \frac{\rho S C_{D0}}{2}\left(\frac{e^{q/c}}{m_0}\right)\left(\frac{v^2}{c^2} + 4\frac{v}{c} + 2\right)\frac{(\bar{p}_v^T\bar{v})}{v} + \frac{\rho S C_{D0}}{2}\left(\frac{e^{q/c}}{m_0}\right)E^{*2}\frac{v}{c^2}\left[(\bar{p}_v^T\bar{v}) - \frac{v^2|\bar{p}_v|^2}{(\bar{p}_v^T\bar{v})}\right] \tag{67}$$

and

$$\begin{aligned}
\mathcal{A} = & -\left(\frac{\rho_r}{\rho}\right)\left(\frac{e^{q/c}}{m_0}\right)\frac{(\bar{v}^T\bar{r})}{r}\left[\frac{(\bar{p}_v^T\bar{A})}{c} + \frac{(\bar{p}_v^T\bar{v})(\bar{v}^T\bar{A})}{v^3}\right] \\
& + \frac{g}{vr}\left[2\frac{(\bar{p}_v^T\bar{r})(\bar{v}^T\bar{r})}{r^2} - \frac{(\bar{p}_v^T\bar{v})(\bar{v}^T\bar{r})^2}{v^2r^2} - (\bar{p}_v^T\bar{v})\right] \\
& + \frac{g_r}{vr^2}\left[-2(\bar{p}_v^T\bar{r})(\bar{v}^T\bar{r}) + \frac{(\bar{p}_v^T\bar{v})(\bar{v}^T\bar{r})^2}{v^2}\right] \\
& + 3\frac{(\bar{p}_v^T\bar{v})(\bar{v}^T\bar{g})^2}{v^5} - g^2\frac{(\bar{p}_v^T\bar{v})}{v^3} - 2\frac{(\bar{p}_v^T\bar{g})(\bar{v}^T\bar{g})}{v^3} \\
& - 2\frac{(\bar{p}_r^T\bar{g})}{v} + 2\frac{(\bar{p}_r^T\bar{v})(\bar{v}^T\bar{g})}{v^3} \\
& + \left(\frac{e^{q/c}}{m_0}\right)\left[\frac{(\bar{p}_r^T\bar{v})(\bar{v}^T\bar{A})}{v^3} - \left(1 - \frac{v}{c}\right)\frac{(\bar{p}_r^T\bar{A})}{v} - \rho S C_{D0}(\bar{p}_r^T\bar{v})\right] \\
& - \rho S C_{D0}\left(\frac{e^{q/c}}{m_0}\right)^2(\bar{p}_v^T\bar{A}) \\
& + \frac{\rho S C_{D0}}{2}\left(\frac{e^{q/c}}{m_0}\right)(\bar{p}_v^T\bar{g})\left[\left(4 + \frac{v}{c}\right) + E^{*2}\left(-2 + \frac{v}{c}\right) + \frac{E^{*2}v^2|\bar{p}_v|^2}{(\bar{p}_v^T\bar{v})^2}\left(2 + \frac{v}{c}\right)\right] \\
& + \frac{\rho S C_{D0}}{2}\left(\frac{e^{q/c}}{m_0}\right)\frac{(\bar{p}_v^T\bar{v})(\bar{v}^T\bar{g})}{v^2} \\
& \quad \left[\left(-2 + \frac{v}{c}\right) + E^{*2}\left(4 + \frac{v}{c}\right) - \frac{E^{*2}v^2|\bar{p}_v|^2}{(\bar{p}_v^T\bar{v})^2}\left(4 + 3\frac{v}{c}\right)\right]
\end{aligned} \tag{68}$$

Remark 4 *Note that in the expression for $\mathcal{B}$, if we set $\rho = 0$, which corresponds to the case of flight in vacuum, we get*

$$\mathcal{B} = -2\frac{(\bar{p}_v^T\bar{g})}{v^2} + 2\frac{(\bar{p}_v^T\bar{v})(\bar{v}^T\bar{g})}{v^4} \tag{69}$$

which is not zero in general and hence the singular arc is still of order one. This result is contrary to the second order singular arc of the Lawden's spiral. The reason for this contradiction is the fact that here we constrain the thrust direction to be along the velocity vector where as in the Lawden's spiral, the thrust direction is free.

CANONICAL TRANSFORMATION TO FLIGHT PATH VARIABLES

So far, we have derived all our results in terms of general vector state variables $\bar{r}$ and $\bar{v}$ and the corresponding adjoint variables $\bar{p}_r$ and $\bar{p}_v$. Often, it is more convenient to express the results in a certain coordinate system, in terms of a suitable set of state variables. Thus, we need a transformation from $\bar{r}$, $\bar{v}$ to the new set of variables and it must have certain properties in order to generate the same trajectory. Such a transformation is called a *Canonical Transformation* [21, 22].

Definition 2 *Let* $\mathbf{x}$ *and* $\mathbf{p}$ *be the state and the adjoint variables with Hamiltonian* H *such that they satisfy the canonical equations*

$$\dot{\mathbf{x}} = \frac{\partial H}{\partial \mathbf{p}} \quad and \quad \dot{\mathbf{p}} = -\frac{\partial H}{\partial \mathbf{x}} \tag{70}$$

Then the transformation to a new set of variables $\mathbf{X} = \mathbf{X}(\mathbf{x}, \mathbf{p})$ *and* $\mathbf{P} = \mathbf{P}(\mathbf{x}, \mathbf{p})$ *is called a* Canonical Transformation *if it preserves the canonical structure, i.e., if*

$$\dot{\mathbf{X}} = \frac{\partial H}{\partial \mathbf{P}} \quad and \quad \dot{\mathbf{P}} = -\frac{\partial H}{\partial \mathbf{X}} \tag{71}$$

where the Hamiltonian is now expressed in terms of $\mathbf{X}$ *and* $\mathbf{P}$.

Here, we are concerned only with a special type of transformation called the *Legrange's point transformation* [21]. We already noted that the equations of motion are autonomous and the Hamiltonian is independent of time. Under these conditions, the sufficient condition for the transformation to be canonical is that the following equation is satisfied [21].

$$\mathbf{p} \cdot d\mathbf{x} = \mathbf{P} \cdot d\mathbf{X} \tag{72}$$

Flight Path Variables

The most commonly used set of variables for atmospheric flight over a spherical planet are the flight path variables given by the radial distance to the spacecraft r, the longitude θ, the latitude ϕ, the velocity V, the flight path angle γ and the heading angle ψ. See Fig.5 for the definition of these variables. For illustration, we derive a canonical transformation between ($\bar{r}$, $\bar{v}$) and (r, θ, ϕ, V, γ and ψ) and obtain the equations of motion using this transformation. We first derive the expressions for (p_r, p_θ, $\cdots$, p_ψ) in terms of $\bar{p}_r$ and $\bar{p}_v$ and then derive the expressions for the latter in terms of the former. Of course, both the sets of expressions will be functions of the two sets of state variables.

First of all, let us express $d\bar{r}$ and $d\bar{v}$ as functions of dr, $d\theta$, ..., $d\psi$. Refer to Fig.6 for various reference frames we use to represent different vector quantities. OXYZ is an inertial reference frame fixed at the center of the planet, which is assumed to be non-rotating with a stationary atmosphere. The unit vectors $\hat{\imath}$, $\hat{\jmath}$ and $\hat{k}$ are along the X, Y and Z axes respectively. The position vector is represented in terms of the spherical coordinates in the rotating reference frame Oxyz, with unit vectors $\hat{e}_r$, $\hat{e}_\theta$ and $\hat{e}_\phi$. Finally, the velocity vector is represented in the flight path reference frame PLMN with unit vectors $\hat{e}_\gamma$, $\hat{e}_V$ and $\hat{e}_\psi$. The transformation matrices relating the various unit vector triads as well as the angular velocities of the rotating frames are given in the Appendix.

Now using equations (99) and (101), we can write,

$$\begin{aligned} \bar{r} &= r\hat{e}_r \\ \bar{v} &= V\hat{e}_v \\ \bar{\omega} &= \dot{\theta}\hat{k} - \dot{\phi}\hat{e}_\theta \\ \bar{\Omega} &= \dot{\psi}\hat{e}_r - \dot{\gamma}\hat{e}_\psi \end{aligned} \tag{73}$$

where $\bar{\omega}$ is the angular velocity of Oxyz in OXYZ and $\bar{\Omega}$ is the angular velocity of PLMN in Oxyz. Note that the total angular velocity of PLMN in OXYZ is $(\bar{\omega} + \bar{\Omega})$. Thus,

$$d\bar{\omega} = (d\theta)\hat{k} - (d\phi)\hat{e}_\theta \tag{74}$$

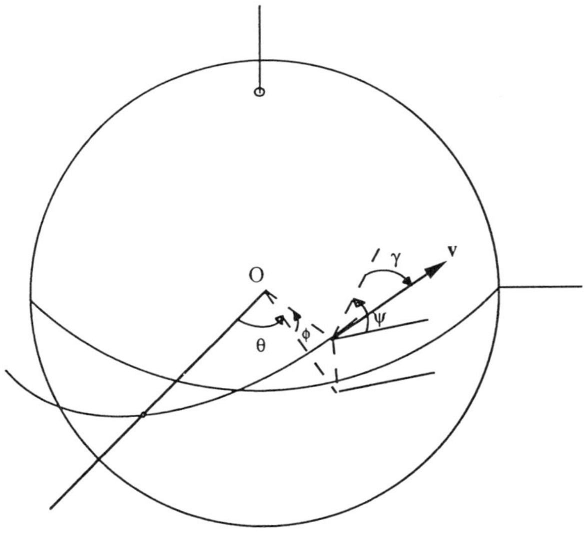

Figure 5. Definition of the Flight Path Variables.

and

$$d\bar{\Omega} = (d\psi)\hat{e}_r - (d\gamma)\hat{e}_\psi \tag{75}$$

Using the above two equations, we can write [31]

$$\begin{aligned} d\bar{r} &= (dr)\hat{e}_r + r(d\hat{e}_r) \\ &= (dr)\hat{e}_r + r(d\bar{\omega} \times \hat{e}_r) \\ &= (dr)\hat{e}_r + (d\theta)(\hat{k} \times \bar{r}) - (d\phi)(\hat{e}_\theta \times \bar{r}) \end{aligned} \tag{76}$$

Similarly,

$$\begin{aligned} d\bar{v} &= (dV)\hat{e}_v + V(d\hat{e}_v) \\ &= (dV)\hat{e}_v + V[(d\bar{\omega} + d\bar{\Omega}) \times \hat{e}_v] \\ &= (dV)\hat{e}_v + (d\theta)(\hat{k} \times \bar{v}) - (d\phi)(\hat{e}_\theta \times \bar{v}) + (d\psi)(\hat{e}_r \times \bar{v}) - (d\gamma)(\hat{e}_\psi \times \bar{v}) \end{aligned} \tag{77}$$

Writing the canonicity condition given in (72) for our problem, we get,

$$\bar{p}_r \cdot d\bar{r} + \bar{p}_v \cdot d\bar{v} = p_r dr + p_\theta d\theta + p_\phi d\phi + p_V dV + p_\gamma d\gamma + p_\psi d\psi \tag{78}$$

Substituting the expressions for $d\bar{r}$ and $d\bar{v}$ from equations (76) and (77) and equating the coefficients of dr, $d\theta$, ..., $d\psi$, we have the following expressions for p_r, p_θ, ..., p_ψ.

$$\begin{aligned} p_r &= \bar{p}_r \cdot \hat{e}_r \\ p_\theta &= \bar{p}_r \cdot (\hat{k} \times \bar{r}) + \bar{p}_v \cdot (\hat{k} \times \bar{v}) \\ p_\phi &= -\bar{p}_r \cdot (\hat{e}_\theta \times \bar{r}) - \bar{p}_v \cdot (\hat{e}_\theta \times \bar{v}) \\ p_V &= \bar{p}_v \cdot \hat{e}_v \\ p_\gamma &= -\bar{p}_v \cdot (\hat{e}_\psi \times \bar{v}) \\ p_\psi &= \bar{p}_v \cdot (\hat{e}_r \times \bar{v}) \end{aligned} \tag{79}$$

Using the property of the triple scalar product that

$$\bar{a} \cdot (\bar{b} \times \bar{c}) = \bar{b} \cdot (\bar{c} \times \bar{a}) \ , \tag{80}$$

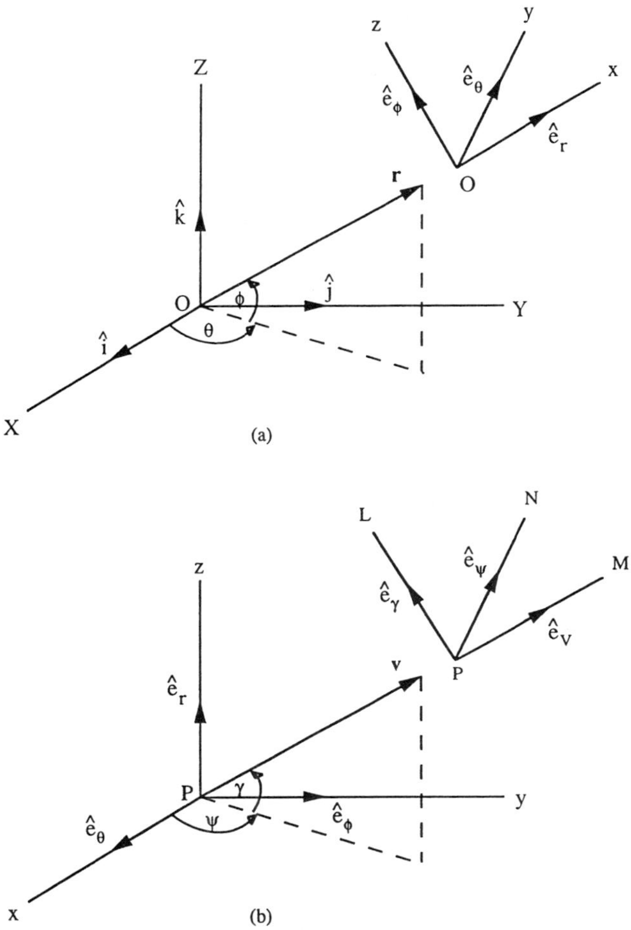

Figure 6. The Reference Frames for: (a) Radius Vector, (b) Velocity Vector.

where $\bar{a}$, $\bar{b}$ and $\bar{c}$ are any three vectors, we can re-write

$$
\begin{aligned}
p_\theta &= \hat{k}\cdot[(\bar{r}\times\bar{p}_r)+(\bar{v}\times\bar{p}_v)] \\
p_\phi &= -\hat{e}_\theta\cdot[(\bar{r}\times\bar{p}_r)+(\bar{v}\times\bar{p}_v)] \\
p_\psi &= \hat{e}_r\cdot[(\bar{r}\times\bar{p}_r)+(\bar{v}\times\bar{p}_v)]
\end{aligned}
\tag{81}
$$

Note that in the final equation we used the fact that

$$\hat{e}_r\cdot(\bar{r}\times\bar{p}_r)=0 \tag{82}$$

Finally, noting from Fig.7 that

$$
\begin{aligned}
\hat{e}_\gamma &= \frac{1}{\cos\gamma}\left[\hat{e}_r-\frac{(\bar{r}\cdot\bar{v})}{rV}\hat{e}_v\right] \\
\hat{e}_v &= \frac{\bar{v}}{V} \\
\hat{e}_\psi &= \frac{(\bar{r}\times\bar{v})}{rV\cos\gamma}
\end{aligned}
\tag{83}
$$

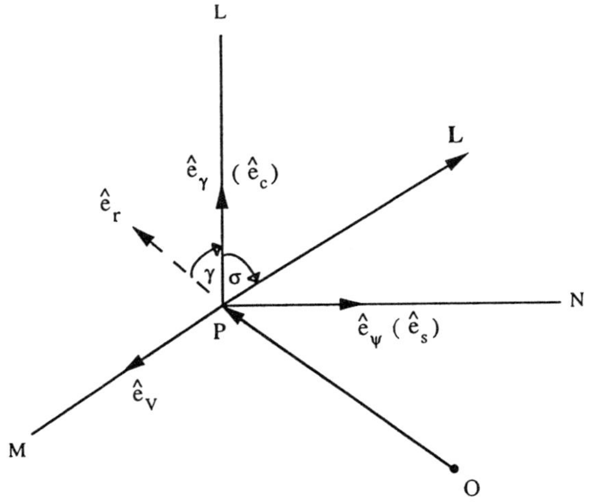

Figure 7. The Lift Vector and the Symmetry Plane.

we can write p_γ as

$$\begin{aligned} p_\gamma &= \frac{(\bar{r} \times \bar{v})}{rV \cos\gamma} \cdot (\bar{p}_v \times \bar{v}) \\ &= \frac{\bar{r} \cdot [\bar{v} \times (\bar{p}_v \times \bar{v})]}{rV \cos\gamma} \end{aligned} \tag{84}$$

In summary, using the first integral given in (26), we have

$$\begin{aligned} p_r &= \bar{p}_r \cdot \hat{e}_r \\ p_V &= \bar{p}_v \cdot \hat{e}_v \\ p_\theta &= \hat{k} \cdot \bar{c} \\ p_\phi &= -\hat{e}_\theta \cdot \bar{c} \\ p_\psi &= \hat{e}_r \cdot \bar{c} \\ p_\gamma &= \frac{\bar{r} \cdot [\bar{v} \times (\bar{p}_v \times \bar{v})]}{rV \cos\gamma} \end{aligned} \tag{85}$$

Thus we have expressed the adjoint variables associated with the flight path variables in terms of their vector counterparts. Now we will obtain the inverse transformations. Instead of expressing dr, $d\theta$, ..., $d\psi$ in terms of $d\bar{r}$ and $d\bar{v}$ and then equating the coefficients, we make use of Eq. (85) to obtain the result in a more straightforward manner. Let us seek expressions for $\bar{p}_r$ and $\bar{p}_v$ in the following form.

$$\bar{p}_r = p_{r_1}\hat{e}_r + p_{r_2}\hat{e}_\theta + p_{r_3}\hat{e}_\phi \tag{86}$$

and

$$\bar{p}_v = p_{v_1}\hat{e}_\gamma + p_{v_2}\hat{e}_v + p_{v_3}\hat{e}_\psi \tag{87}$$

It is easier to derive the expressions for p_{v_1}, p_{v_2} and p_{v_3} first. Using (79) and noting that (see Fig.7)

$$(\hat{e}_v \times \hat{e}_\psi) = \hat{e}_\gamma \quad and \quad (\hat{e}_r \times \hat{e}_v) = \cos\gamma \hat{e}_\psi \ , \tag{88}$$

we get

$$\begin{aligned} p_V &= (\bar{p}_v \cdot \hat{e}_v) \\ p_\gamma &= V\bar{p}_v \cdot (\hat{e}_v \times \hat{e}_\psi) = V(\bar{p}_v \cdot \hat{e}_\gamma) \\ p_\psi &= V\bar{p}_v \cdot (\hat{e}_r \times \hat{e}_v) = V\cos\gamma(\bar{p}_v \cdot \hat{e}_\psi) \end{aligned} \tag{89}$$

Thus,

$$\begin{aligned} p_{v_1} &= (\bar{p}_v \cdot \hat{e}_\gamma) = \frac{p_\gamma}{V} \\ p_{v_2} &= (\bar{p}_v \cdot \hat{e}_v) = p_V \\ p_{v_3} &= (\bar{p}_v \cdot \hat{e}_\psi) = \frac{p_\psi}{V \cos\gamma} \end{aligned} \tag{90}$$

and we can write

$$\bar{p}_v = \left(\frac{p_\gamma}{V}\right)\hat{e}_\gamma + (p_V)\hat{e}_v + \left(\frac{p_\psi}{V\cos\gamma}\right)\hat{e}_\psi \tag{91}$$

Similarly, from equation(79), we can write

$$p_{r_1} = (\bar{p}_r \cdot \hat{e}_r) = p_r \tag{92}$$

Using (91) for $\bar{p}_v$, and noting that

$$\hat{e}_v \cdot (\hat{k} \times \bar{v}) = 0\ , \tag{93}$$

we get

$$p_\theta = (\bar{p}_r \cdot \hat{e}_\theta) r \cos\phi + \frac{p_\gamma}{V}[\hat{e}_\gamma \cdot (\hat{k} \times \bar{v})] + \frac{p_\psi}{V\cos\gamma}[\hat{e}_\psi \cdot (\hat{k} \times \bar{v})] \tag{94}$$

Using the property of the triple scalar product given in (80), we obtain

$$p_{r_2} = (\bar{p}_r \cdot \hat{e}_\theta) = \frac{p_\theta}{r\cos\phi} + \frac{p_\gamma}{r\cos\phi}(\hat{k} \cdot \hat{e}_\psi) - \frac{p_\psi}{r\cos\phi}\frac{(\hat{k} \cdot \hat{e}_\gamma)}{\cos\gamma} \tag{95}$$

In a similar way we have

$$p_\phi = (\bar{p}_r \cdot \hat{e}_\phi) r - \frac{p_\gamma}{V}[\hat{e}_\gamma \cdot (\hat{e}_\theta \times \bar{v})] - \frac{p_\psi}{V\cos\gamma}[\hat{e}_\psi \cdot (\hat{e}_\theta \times \bar{v})] \tag{96}$$

and

$$p_{r_3} = (\bar{p}_r \cdot \hat{e}_\phi) = \frac{p_\phi}{r} - \frac{p_\gamma}{r}(\hat{e}_\theta \cdot \hat{e}_\psi) + \frac{p_\psi}{r\cos\gamma}(\hat{e}_\theta \cdot \hat{e}_\gamma) \tag{97}$$

In summary, we can write

$$\begin{aligned} \bar{p}_r &= p_r \hat{e}_r \\ &+ \left[\frac{p_\theta}{r\cos\phi} + \frac{p_\gamma}{r\cos\phi}(\hat{k} \cdot \hat{e}_\psi) - \frac{p_\psi}{r\cos\phi}\frac{(\hat{k} \cdot \hat{e}_\gamma)}{\cos\gamma}\right]\hat{e}_\theta \\ &+ \left[\frac{p_\phi}{r} - \frac{p_\gamma}{r}(\hat{e}_\theta \cdot \hat{e}_\psi) + \frac{p_\psi}{r\cos\gamma}(\hat{e}_\theta \cdot \hat{e}_\gamma)\right]\hat{e}_\phi \end{aligned} \tag{98}$$

Using Eqs.(85), (91) and (98), we can translate all the results that we obtained in vector form in this paper in terms of the flight path variables.

CONCLUSIONS

We have presented the general equations of motion of a thrusting and lifting spacecrat in atmospheric flight. We stated a Mayer type optimal control problem and outlined the nature of the optimal aerodynamic controls using the elegant geometric concept of the domain of maneuverability in the hodograph space. We also showed that the thrust magnitude control leads to three sub-arcs. Using the results for the optimal aerodynamic controls, we derived an explicit expression for the optimal thrust magnitude on the intermediate thrust arc. All the results are derived in vector form and we illustrated how one can translate the results in terms of a desired set of state variables with the help of canonical transformations. Thus, one has a choice of working with either vector or scalar variables. Note that the same procedure can be followed to obtain the results for the case of free thrust direction control.

APPENDIX : TRANSFORMATION MATRICES

While obtaining the canonical transformation we have introduced several reference frames to express various vector quantities. Here we derive the transformation matrices relating the unit vector triads of any two of those frames. Refer to Fig.6 for the definitions of the different reference frames.

Transformation Matrix From OXYZ to Oxyz

From Fig. 6(a), we note that the Oxyz reference frame is obtained from the OXYZ frame by a $+\theta$ rotation about the $\hat{k}$ axis followed by a $-\phi$ rotation about the new $\hat{\jmath}$ axis, which is the same as the $\hat{e}_\theta$ axis. Thus, the angular velocity of Oxyz with respect to OXYZ can be written as

$$\bar{\omega} = \dot{\theta}\hat{k} - \dot{\phi}\hat{e}_\theta \tag{99}$$

and

$$\begin{bmatrix} \hat{e}_r \\ \hat{e}_\theta \\ \hat{e}_\phi \end{bmatrix} = R_2(-\phi)R_3(\theta)\begin{bmatrix} \hat{\imath} \\ \hat{\jmath} \\ \hat{k} \end{bmatrix} \tag{100}$$

Transformation Matrix From Pxyz to PLMN

From Fig. 6(b), we note that the PLMN reference frame is obtained from the Pxyz frame by a $+\psi$ rotation about the $\hat{e}_r$-axis followed by a $-\gamma$ rotation about the new $\hat{e}_\phi$-axis, which is the same as the $\hat{e}_\psi$ axis. Thus, the angular velocity of PLMN with respect to Pxyz (or Oxyz) is

$$\bar{\Omega} = \dot{\psi}\hat{e}_r - \dot{\gamma}\hat{e}_\psi \tag{101}$$

and

$$\begin{bmatrix} \hat{e}_\gamma \\ \hat{e}_V \\ \hat{e}_\psi \end{bmatrix} = R_3(-\gamma)R_1(\psi)\begin{bmatrix} \hat{e}_r \\ \hat{e}_\theta \\ \hat{e}_\phi \end{bmatrix} \tag{102}$$

The definition of the matrices R_1, R_2 and R_3 used in Eqs.(100) and (102) is given as follows. For any $\eta \in \Re^1$, we have

$$R_1(\eta) = \begin{bmatrix} 1 & 0 & 0 \\ 0 & \cos\eta & \sin\eta \\ 0 & -\sin\eta & \cos\eta \end{bmatrix} \tag{103}$$

$$R_2(\eta) = \begin{bmatrix} \cos\eta & 0 & -\sin\eta \\ 0 & 1 & 0 \\ \sin\eta & 0 & \cos\eta \end{bmatrix} \tag{104}$$

and

$$R_3(\eta) = \begin{bmatrix} \cos\eta & \sin\eta & 0 \\ -\sin\eta & \cos\eta & 0 \\ 0 & 0 & 1 \end{bmatrix} \tag{105}$$

REFERENCES

[1] Lawden, D. F., "Optimal Rocket Trajectories," *ARS Journal*, Vol. 27, No. 12, 1957.

[2] London, H. S., "Change of Satellite Orbit Plane by Aerodynamic Maneuvering," *Journal of the Aerospace Sciences*, March 1962, pp. 323-332.

[3] Nyland, F. S., "The Synergetic Plane Change for Orbiting Spacecraft," *Proceedings of the 5th International Symposium on Space Technology and Science*, Tokyo, Sept 1963, AGNE Publishing Inc., Tokyo.

[4] Cuadra, E. and Arthur, P. D., "Orbit Plane Change by External Burning Aerocruise," *Journal of Spacecraft*, Vol. 3, No. 3, Mar 1966, pp. 347-352.

[5] Parsons, W. D., "Analytic Solution of the Synergetic Turn," *Journal of Spacecraft and Rockets*, Vol. 3, No. 11, November 1966, pp. 1675-1678.

[6] Dickmanns, E.D., "The Effect of Finite Thrust and Heading Constraints on the Synergetic Plane Change Maneuver for a Space Shuttle Orbiter-Class Vehicle," NASA TN D-7211, October 1973.

[7] Ikawa, H. and Rudiger, T. F., "Synergetic Maneuvering of Winged Spacecraft for Orbital Plane Change," *Journal of Spacecraft*, Vol. 19, No. 6, Nov-Dec 1982, pp. 513-520

[8] Cervisi, R. T., "Analytical Solution for a Cruising Plane Change Maneuver," *Journal of Spacecraft*, Vol. 22, No. 2, Mar-Apr 1985, pp. 134-140.

[9] Mease, K. D., Lee, J. Y. and Vinh, N. X., "Orbital Changes During Hypersonic Aerocruise," *The Journal of the Astronautical Sciences*, Vol. 36, Nos. 1/2, Jan-June 1988, pp. 103-137.

[10] Walberg, G. D., "A Survey of Aeroassisted Orbit Transfer," *Journal of Spacecraft*, Vol. 22, No. 1, Jan-Feb 1985, pp. 3-18.

[11] Miele, A., *Flight Mechanics, Theory of Flight Paths*, Addison-Wesley, Reading 1963.

[12] Vinh, N. X., *Optimal Trajectories in Atmospheric Flight*, Elsevier, New York 1981.

[13] Lawden, D. F., *Optimal Trajectories for Space Navigation*, Butterworths, London, 1963.

[14] Breakwell, J. V., "The Optimization of Trajectories," *SIAM Journal on Applied Mathematics*, Vol. 7, No. 2, 1959.

[15] Leitmann, G., "A Calculus of Variations Solution of Goddard's Problem," *Astronautica Acta*, Vol. 2, No. 2, 1956.

[16] Miele, A., "The Calculus of Variations in Applied Aerodynamics and Flight Mechanics," in *Optimization Techniques*, George Leitmann, ed., Academic Press, New York, 1962.

[17] Bryson, A. E., Jr. and Lele, M. M., "Minimum Fuel Lateral Turns at Constant Altitude," *AIAA Journal*, Vol. 7, No. 3, Mar 1969, pp. 559-560.

[18] Pontryagin, L. S., et al., *The Mathematical Theory of Optimal Processes*, Wiley Interscience, New York, 1962.

[19] Vinh, N. X., "General Theory of Optimal Trajectory for Rocket Flight in a Resisting Medium," *Journal of Optimization Theory and Applications*, Vol. 11, No. 2, 1973, pp. 189-202.

[20] Mease, K. D., "Optimization of Aeroassisted Orbital Transfer : Current Status," *Journal of the Astronautical Sciences*, Vol. 36, No. 1/2, 1988, pp. 7-33.

[21] Lanczos, C., *The Variational Principles of Mechanics*, Dover Publications, New York, 1986.

[22] Greenwood, D. T., *Classical Dynamics*, Prentice Hall, Englewood Cliffs, NJ, 1977.

[23] Schultz, R. L. and Zagalasky, N. R., "Aircraft Performance Optimization," *Journal of Aircraft*, Vol. 9, No. 2, Feb 1972, pp. 108-114.

[24] Schultz, R. L., "Fuel Optimality of Cruise," *Journal of Aircraft*, Vol. 11, No. 9, Sept 1974, pp. 586-587.

[25] Speyer, J. L., "On the Fuel Optimality of Cruise," *Journal of Aircraft*, Vol. 10, No. 12, Dec 1973, pp. 763-765.

[26] Speyer, J. L., "Nonoptimality of the Steady-State Cruise for Aircraft," *AIAA Journal*, Vol. 14, No. 11, Nov 1976, pp. 1604-1610.

[27] Miele, A., "Extremization of Linear Integrals by Green's Theorem," in *Optimization Techniques*, George Leitmann, ed., Academic Press, New York, 1962.

[28] Kelley, H. J., "A Second Variation Test for Singular Extremals," *AIAA Journal*, Vol. 2, 1964, pp. 1380-1382.

[29] Kelley, H. J., Kopp, R. E. and Moyer, H. G., "Singular Extremals," in *Topics in Optimization*, George Leitmann, ed., Academic Press, New York, 1967.

[30] Marec, J. P., *Optimal Space Trajectories*, Elsevier, New York 1979.

[31] Greenwood, D. T., *Principles of Dynamics*, Prentice Hall, Englewood Cliffs, NJ, 1965.

CONTROL OF MECHANICAL SYSTEMS

SEQUENTIAL DESIGN OF A LINEAR QUADRATIC CONTROLLER FOR THE DEEP SPACE NETWORK ANTENNAS

Wodek Gawronski
Jet Propulsion Laboratory
California Institute of Technology
Pasadena, CA 91109

ABSTRACT

A linear quadratic controller design procedure is proposed for the NASA Deep Space Network antennas. The procedure is developed for an antenna model divided into tracking and flexible subsystems. Controllers for the flexible and tracking parts are designed separately by adjusting the performance index weights. Ad hoc weights are chosen for the tracking part of the controller and the weights for the flexible part are adjusted. Next, the weights for the tracking part are determined, followed by the flexible controller final tune-up. The controller for the flexible part is designed separately for each mode; thus the design procedure consists of weight adjustment for small-size subsystems. The effect of weights on system performance is a crucial property, and is analytically described in this paper. The procedure is illustrated with the control system design for the DSS-13 antenna.

INTRODUCTION

A linear quadratic (LQ) controller design procedure for the Deep Space Network (DSN) antennas is presented. Alvarez and Nickerson [1] have used the LQ approach for controller design of the DSS-14 antenna. In the Alvarez and Nickerson approach the gearbox flexible mode was included in the rigid body model of the antenna. In recently designed antenna structures (such as the DSS-13 antenna) significant flexible deformations are observed during tracking operations. The antenna rate-loop model described in [4,5] consists of 21 flexible modes up to 10 Hz. Controllers for these antennas should suppress flexible motion while following the tracking command. The method presented in this paper allows the design of a controller with a flexible motion suppression capability through sequential adjustment of the weights of the LQ performance index.

Procedures presented Ref. [1] and [2], separate controller design for the elevation and azimuth drives. This approach, effective for slow and/or rigid antennas, cannot be justified for fast and/or flexible antennas. In the latter

Mechanics and Control, Edited by R.S. Guttalu
Plenum Press, New York, 1994

case the flexible properties of the full antenna significantly differ from the properties of the elevation-only, or azimuth-only model of the antenna; thus the separate design of controllers for elevation and azimuth drives would result in system instability.

For flexible antennas there is a quasi-separation of the flexible motion and tracking motions. This property simplifies the controller design procedure. A controller for the flexible part is designed first, followed by a controller for the tracking part, with final corrections of the controller for the flexible part. The design of the controller for the flexible part is of sequential nature as well: a controller for each mode is designed separately.

The tracking performance requirements are included in the definition of the performance index. Although an LQ controller is an optimal one in the sense of minimization of the performance index, it may not satisfy the tracking error requirements. Indeed the closed loop system dynamics depends heavily on the choice of the weighting matrix, as illustrated with the DSS-13 antenna in Fig.1. In case 1, the weight 10 for the integral of the antenna position, the weight 1 for the position itself, and the weight 0 for the flexible modes have been chosen. The antenna performance, characterized in this case by its step response in Fig.1 (solid line), shows excessive flexible motion. In case 2, the weights are the same as those in the previous case, but the weights of the flexible modes are now set equal to 0.001. The closed-loop antenna performance in Fig.1 (dashed line) shows a significant deterioration of the antenna tracking capabilities. Therefore, proper adjustment of weights is an important task, and it is crucial to determine the effect of weights on system performance. A method for determining this weight effect is presented in this paper. The procedure is illustrated with the control system design of the DSS-13 antenna.

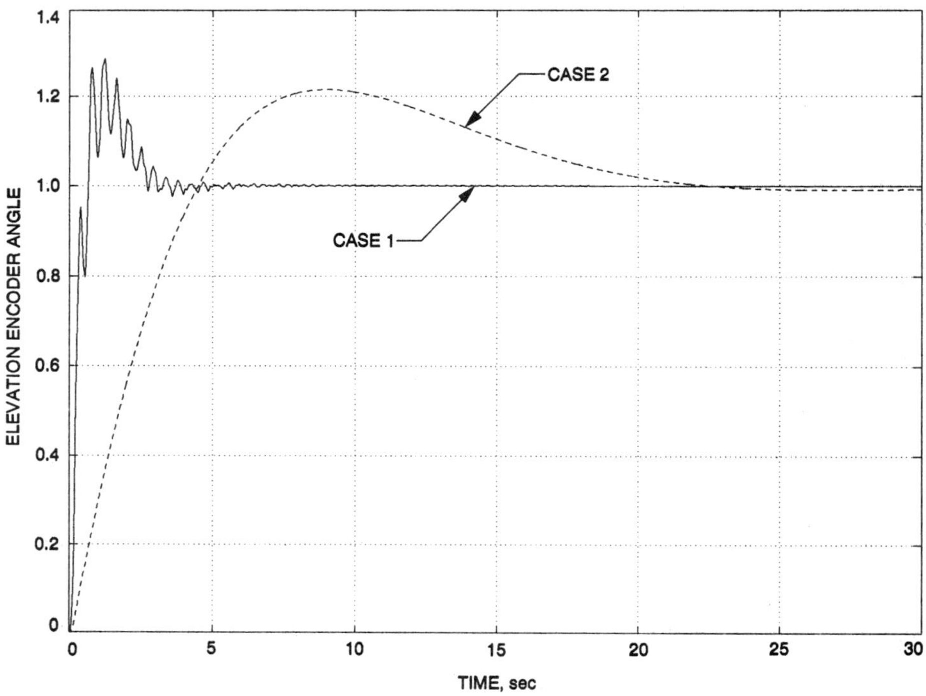

Figure 1. Step responses of the DSS-13 antenna with LQ controller for different weights.

QUASI-SEPARATION OF THE FLEXIBLE AND TRACKING SUBSYSTEMS

In this section the quasi-separation properties of an open-loop model (called also a rate-loop model) of a generic Deep Space Network antenna are studied. This study is based on the DSS-13 antenna model. This antenna represents the new generation of 34-meter-diameter antennas, which dynamics include a significant amount of non-negligible flexible motion [4,5].

The balanced state-space representation (A_p,B_p,C_p) of the DSS-13 antenna is derived in [4, 5]. Its rate command input in elevation (u_{pe}), and azimuth (u_{pa}) is denoted $u_p^T=[u_{pe}\ u_{pa}]$, and the elevation (y_{pe}) and azimuth (y_{pa}) angles are its outputs denoted $y_p^T=[y_{pe}\ y_{pa}]$. Its state vector x is divided into tracking x_t and flexible x_f states $x=[x_t^T\ x_f^T]^T$, where $x_t=[y_i^T,\ y_p^T]^T$, $y_i=[y_{ie},\ y_{ia}]^T$ - an integral of the elevation and azimuth position. The rate-loop representation (A,B,C) is divided respectively to the state vector x thus

$$A=\begin{bmatrix} A_t & A_{tf} \\ 0 & A_f \end{bmatrix},\quad B=\begin{bmatrix} B_t \\ B_f \end{bmatrix},\quad C=[C_t\ 0] \tag{1}$$

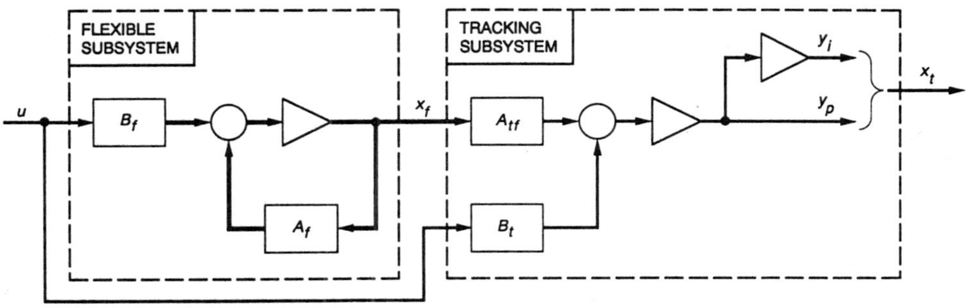

Figure 2. System configuration.

The block diagram of the rate-loop representation in Eq.(1) is shown in Fig.2, where the flexible and tracking parts are distinguished. In this representation B_t is small in comparison with B_f (typically $\|B_t\|<10^{-5}$, and $\|B_f\|>1$). Also A_{tf} is small in comparison with A_t and A_f (typically $\|A_{tf}\|<10^{-3}$, $\|A_f\|>10$, and $\|A_t\|=1$). Both properties appear as a result of small $\|C_p\|$, shown earlier. Thus, the states of the tracking part are much weaker than the states of the flexible part, see Fig.2. The strong states of the flexible subsystem and the weak states of the tracking subsystem are also visible in Fig.3, which presents the transfer function plots of the rate-loop systems due to elevation rate command. This property is a foundation of the control design strategy described below.

In the LQ design the performance index J

$$J=\int_0^\infty (x^TQx+u^TRu)dt \tag{2}$$

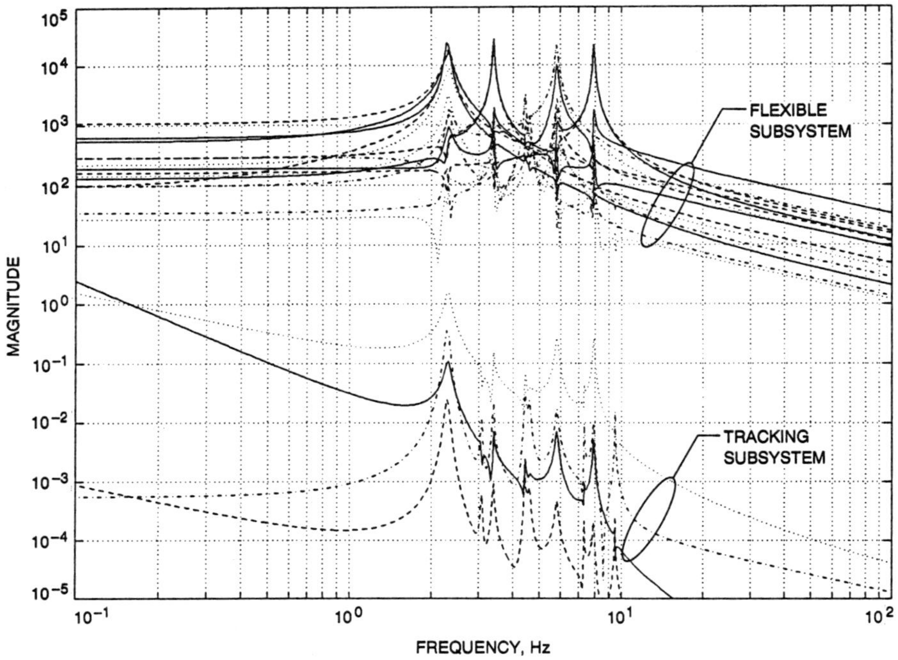

Figure 3. Magnitude of transfer function of tracking and flexible subsystems for the elevation rate input.

is minimized with respect to the controller gain matrix K. The minimum is obtained for $K=R^{-1}B^TS$, and S is a solution of the Riccati equation [8]

$$A^TS+SA-SBR^{-1}B^TS+Q=0 \tag{3}$$

In the above equations R is a positive definite input weight matrix, while Q is positive semi-definite state weight matrix. It is assumed that $R=\rho I$, since both inputs (elevation and azimuth commands) are of equal importance, and the choice of $\rho=1$ is made without loss of generality. Divide S and K into parts related to the triple (A,B,C) in Eq.(1)

$$S=\begin{bmatrix} S_t & S_{tf} \\ S_{tf}^T & S_f \end{bmatrix}, \quad K=[K_t \ \ K_f] \tag{4}$$

so that Eq.(3) can be written as follows

$$A_t^TS_t+S_tA_t-S_tB_tB_t^TS_t+Q_t-\Delta_{tf}=0 \tag{5a}$$

$$A_t^TS_{tf}+S_{tf}A_f+S_tA_{tf}-K_t^TK_f=0 \tag{5b}$$

$$A_f^TS_f+S_fA_f+S_fB_fB_f^TS_f+Q_f-\Delta_{ft}=0 \tag{5c}$$

where

$$K_t=B_t^TS_t+B_f^TS_{tf}^T, \quad K_f=B_t^TS_{tf}+B_f^TS_f, \tag{6a}$$

$$\Delta_{tf}=S_tB_tB_f^TS_{tf}^T+S_{tf}B_fK_t, \quad \Delta_{ft}=A_{tf}^TS_{tf}+S_{tf}^TA_{tf}-S_{tf}^TB_tK_f+S_fB_fB_t^TS_{tf} \tag{6b}$$

Taking a closer look at Eqs.(6) one can notice that there exist weights Q_t and Q_f such that the gain K_f depends on the flexible subsystem only. Namely, for large enough Q_f, such that $\|Q_f\| \gg \|\Delta_{ft}\|$, the solution S_f of (5c) is independent of tracking system, and for small Q_t one obtains $\|B_t^T S_{tf}\| \ll \|B_f^T S_f\|$. In terms of Eq.(6a) the latter inequality means that the gain K_f depends only on the flexible subsystem. However, due to the weak-strong relationship between flexible and tracking subsystems, the situation is not quite symmetric: there are no such Q_t and Q_f that the gain K_t depends only on tracking subsystem. To understand this, note that the term "small" has different meaning for Q_f and Q_t. Magnitudes of small Q_f and small Q_t are of different order, namely small $Q_f < 10^{-7}$ and small $Q_t < 1$. Therefore, increasing Q_t in order to obtain $\|Q_t\| \gg \|\Delta_{tf}\|$ one obtains $\|B_f^T S_{ft}\|$ and $\|B_t^T S_t\|$ of the same magnitude. According to Eq.(6a) the latter fact means that the gain K_t depends on the flexible as well as on the tracking subsystem, and the solution S_t of Eq.(5a) is dependent of flexible subsystem. This property can be validated by observation of the closed loop transfer functions for different weights, as in Fig.4, where the magnitude of closed loop transfer function (elevation angle-to-elevation command) for $Q_t=I_4$, and $Q_f=0\times I_{26}$ is shown as a solid line, for $Q_t=I_4$, and $Q_f=10^{-7}\times I_{26}$ is shown as a dashed line, and for $Q_t=10\times I_4$, and $Q_f=0\times I_{36}$ is shown as a dot-dashed line. It follows from the plots that the variations of Q_f changed the properties of the flexible subsystem only, while the variations of Q_t changed the properties of both subsystems.

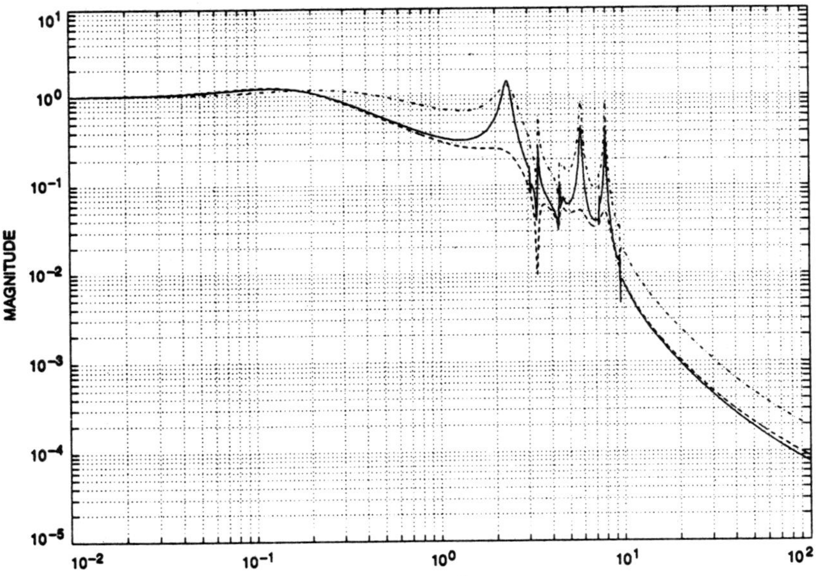

Figure 4. Magnitude of transfer function of the closed loop system for different weights: $Q_t=I_4$, $Q_f=0\times I_{26}$,(solid line), $Q_t=I_4$, $Q_f=10^{-7}\times I_{26}$ (dashed line), and $Q_t=10\times I_4$, $Q_f=0\times I_{26}$ (dot-dashed line).

The independence of flexible system gains from tracking system properties is a consequence of small B_t, A_{tf}, and Q_t. However, the weight Q_t should be large enough to achieve the pointing performance requirements, and the increase of Q_t causes the increasing dependency of the gains on the tracking system. For this reason the above independence is in fact a quasi-independence, which leads to a quasi-separation in the first stage of controller design. The design consists of the initial choice of relatively small weights for the tracking subsystem, and determination of the controller gains of the flexible subsystem. It is followed by adjustment of weights of the tracking subsystem, and a final tuning of the flexible weights.

PROPERTIES OF THE LQ CONTROLLER FOR FLEXIBLE STRUCTURES

The properties of an LQ controller for a flexible subsystem are discussed in this section. In this application a linear system with distinct complex conjugate pairs of poles, and with small real parts of the poles, is considered a flexible structure. In the following a balanced state-space representation of a flexible structure is discussed. The balanced representation of flexible structures is close (but not identical) to a modal one [3,6,7]. For LQ synthesis purposes a balanced rather than modal representation is recommended, since the balanced reduction (necessary in controller design) yields more accurate results than the modal reduction, especially for closely spaced poles [9].

Since the LQ controller for the flexible subsystem is determined separately from the tracking subsystem, in this section the flexible subsystem is considered only. Its state-space representation (A,B,C) is controllable and observable (subscript "f" is dropped in this section for simplicity of notation), and its controllability (W_c) and observability (W_o) grammians are equal and diagonal, $W_c=W_o=\Gamma$, where Γ is a positive definite diagonal matrix which satisfies the following Lyapunov equations

$$A\Gamma^2+\Gamma^2A^T+BB^T=0, \qquad A^T\Gamma^2+\Gamma^2A+C^TC=0 \tag{7}$$

For a balanced flexible system with n components (or $2n$ states), the balanced grammian has the following form

$$\Gamma\cong diag(\gamma_1,\gamma_1,\gamma_2,\gamma_2,\dots,\gamma_n,\gamma_n), \tag{8}$$

and the matrix A is almost block-diagonal [3,6], with dominant 2×2 blocks on the main diagonal

$$A\cong diag(A_i), \qquad A_i=\begin{bmatrix} -\zeta_i\omega_i & -\omega_i \\ \omega_i & -\zeta_i\omega_i \end{bmatrix}, \quad i=1,\dots,n \tag{9}$$

where ω_i is the i-th natural frequency of the structure, ζ_i is the i-th modal damping. Combination of Eqs.(7) and (9) gives

$$\gamma_i^2(A_i+A_i^T)\cong -B_iB_i^T\cong -C_i^TC_i \tag{10}$$

For the LQ controller it is assumed a diagonal weight matrix

$$Q=diag(q_iI_2), \qquad \text{with } 0<q_i\ll 1,\ i=1,\dots,n \tag{11}$$

In this case the following is true.

Proposition 1. $S\cong diag(s_iI_2)$ is the solution of (3), where

$$s_i=-0.5\gamma_i^{2}(1-\beta_i), \quad i=1,\dots,n \tag{12}$$

where

$$\beta_i=\sqrt{1+2q_i\gamma_i^2/\zeta_i\omega_i} \tag{13}$$

Proof is presented in the Appendix.

Next it will be shown that the weighting Q as in Eq.(11) shifts the i-th pair of complex poles of flexible structure, and leaves the remaining pairs of poles almost unchanged. Only the real part of the pair of poles is changed (moving the pole apart from the imaginary axis, thus stabilizing the system), and the imaginary part of the poles remains unchanged.

Proposition 2. For the weight Q as in Eq.(11) the closed-loop pair of flexible poles $(\lambda_{cri}, \pm j\lambda_{cii})$ is obtained from the open-loop poles $(\lambda_{ori}, \pm j\lambda_{oii})$

$$(\lambda_{cri}, \pm j\lambda_{cii}) = (\beta_i\lambda_{ori}, \pm j\lambda_{oii}), \qquad i=1,\ldots,n \tag{14}$$

where $\beta_i$is defined in Eq.(13). For proof see the Appendix.

The imaginary part of the poles is shifted by β_i, while the real part of the poles remains unchanged. The plots of β_i with respect to $\zeta_i\omega_i$ and γ_i are shown in Fig.5. They show relatively large β_i even for small q_i, i.e., significant pole shift to the left. Also, since β_i increases with γ_i, and decreases with $\zeta_i\omega_i$, there is a significant pole shift for highly observable and controllable modes with small damping. In terms of the transfer function profile, the weight q_i suppresses the resonant peak at frequency ω_i while leaving the natural frequency unchanged (see Fig.6). Due to weak coupling between modes, the assignment of one mode insignificantly influences other modes, thus the weight assignment can be done for each mode separately.

CONTROLLER DESIGN ALGORITHM

The procedure for the antenna LQ controller design is a sequential one. First, for the *ad hoc* (but relatively small) chosen weights of the tracking subsystem, the weights of flexible subsystem are determined. Second, the adjustment of the weights of the tracking system is performed, followed by final adjustment of weights of flexible system. The weights of the flexible subsystem are determined sequentially, which gives more insight into the system performance, and simplifies the procedure.

The controller order is determined as a part of the weight tuning process. Only the modes which influence the plant performance are considered. If the number of flexible modes is n_f, the number of disregarded modes is n_o, the size of the tracking system is n_t, then the controller order n_c is $n_c=n_t+2(n_f-n_o)$.

The following LQ controller design algorithm is proposed:

1. Determine the plant state-space representation in the form of Eq.(1), consisting of flexible and tracking parts.
2. Choose an *ad hoc* but reasonably small weights for the tracking part $Q_t=Q_{tah}$.
3. For each balanced coordinate of the flexible part choose the weight q_i, $(i=1,\ldots,n_f)$ and define the weight matrix $Q_{fi}=diag(0,0,\ldots,q_i,q_i,0,0,\ldots 0)$, so that the closed-loop system performance for the the weight $Q_i=diag(Q_{tah}, Q_{fi})$ is maximized. For example, determine the weights q_i to impose the required pole shift; or to suppress the *i-th* resonant peak to the required level without depreciating other properties of the closed-loop transfer function. Note that for small q_i, only *i-th* pair of poles is shifted (to the left), and the remaining poles are almost unaffected. Disregard the modes for which the weighting does not improve the closed-loop system performance.
4. For the already determined weight Q_f, tune weight Q_t to obtain improvements in tracking properties of the antenna.
5. Adjust weights of flexible subsystem, if necessary.

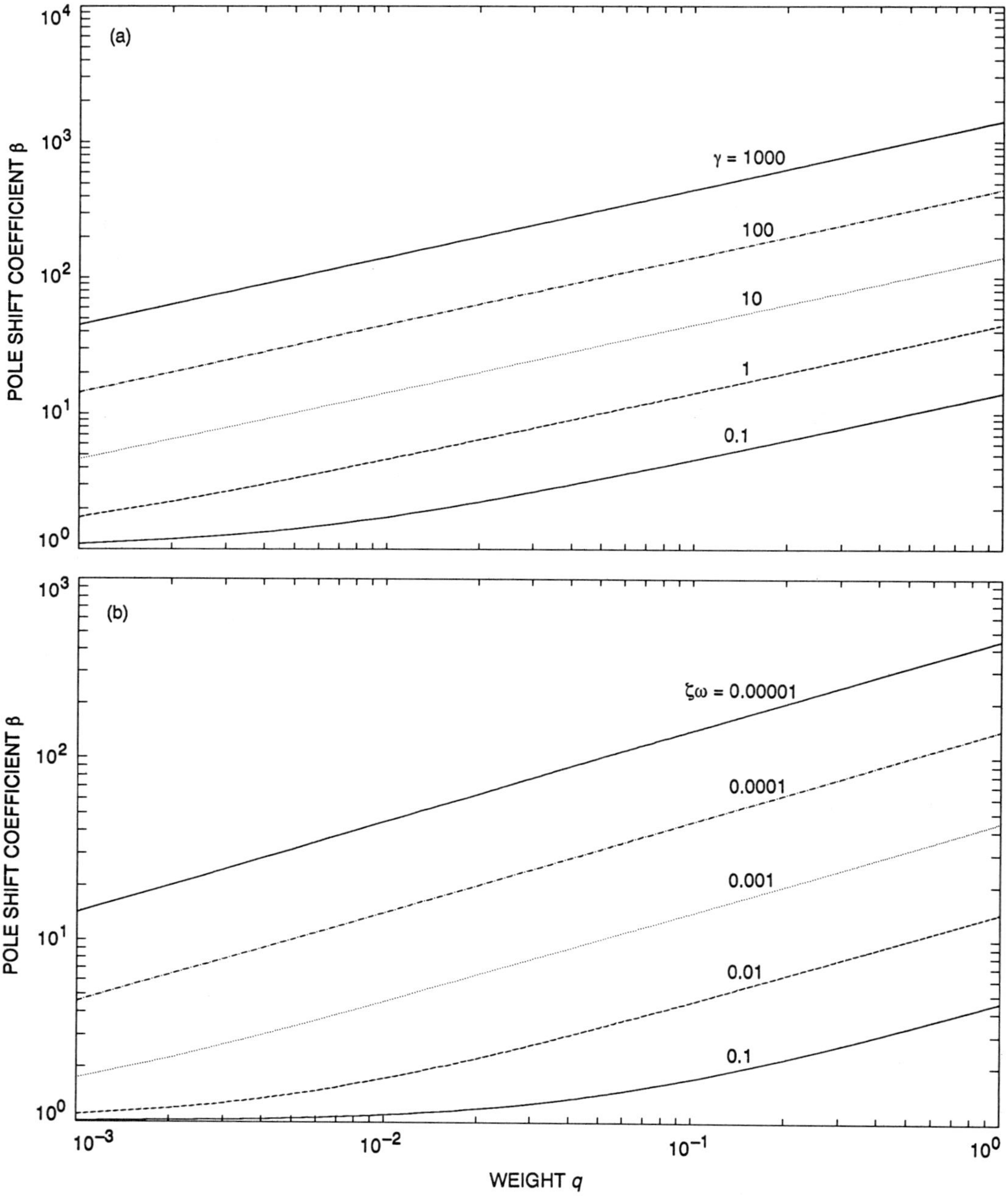

Figure 5. Pole shift coefficient β versus: (a) weight q and Hankel singular value γ^2, (b) weight q and modal damping $\zeta\omega$.

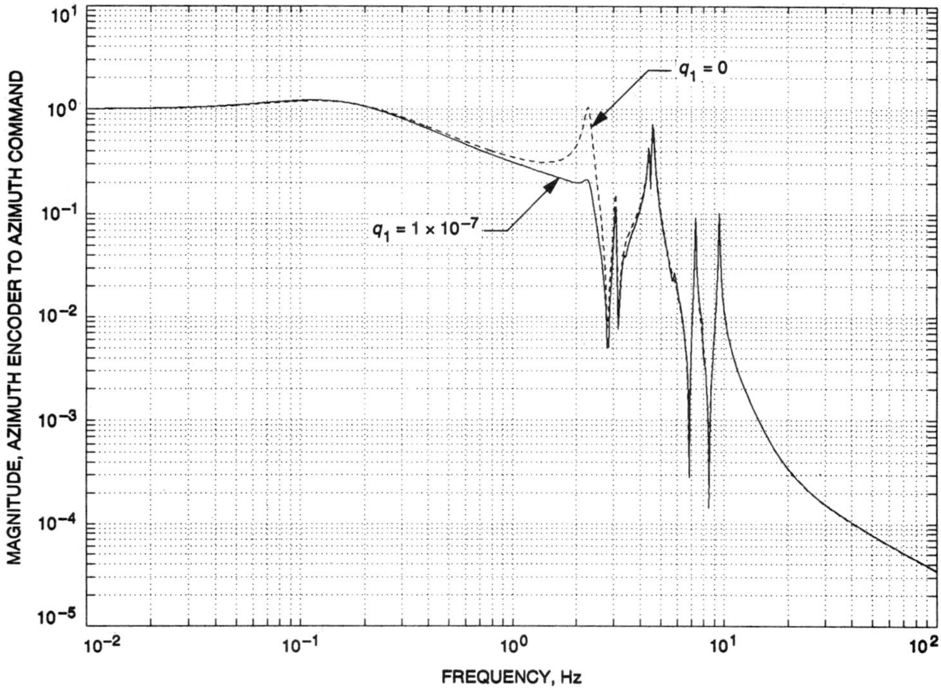

Figure 6. Vibration suppression with a single weight.

CLOSED-LOOP SYSTEM

The LQ controller configuration for the DSN antenna model is shown in Fig.7a. The tracking command y_c is compared with the antenna position y_p, and the error $\varepsilon=y_p-y_c$ and the integral of the error are the controller inputs. The equations for the plant triple (A,B,C), given by (1) are as follows

$$\dot{x}=Ax+Bu_f, \quad y_p=C_px, \quad x_f=C_fx \tag{15}$$

Denoting the output error ε, and the integral of the error z, one obtains

$$\varepsilon=y_p-u=C_px-u, \quad \dot{z}=\varepsilon \tag{16}$$

thus

$$u_f=-k_fC_fx-k_oC_px-k_iz+k_ou. \tag{17}$$

For the closed-loop state $x_{cl}=[z\ x^T]^T$, one obtains the closed-loop triple

$$A_{cl}=\begin{bmatrix} 0 & C_p \\ -Bk_i & A-Bk_fC_f-Bk_oC_p \end{bmatrix}, \quad B_{cl}=\begin{bmatrix} -I \\ Bk_o \end{bmatrix}, \quad C_{cl}=[0 \quad C_p] \tag{18}$$

The antenna states are not directly measured, thus the state observer is included in the closed-loop system in Fig.7b. Based on plant equation (15) the estimator equations are obtained

$$\dot{\hat{x}}=A\hat{x}+Bu_f+K_e(y_p-\hat{y}_p), \quad \hat{y}_p=C_p\hat{x}, \quad \hat{x}_f=C_f\hat{x} \tag{19}$$

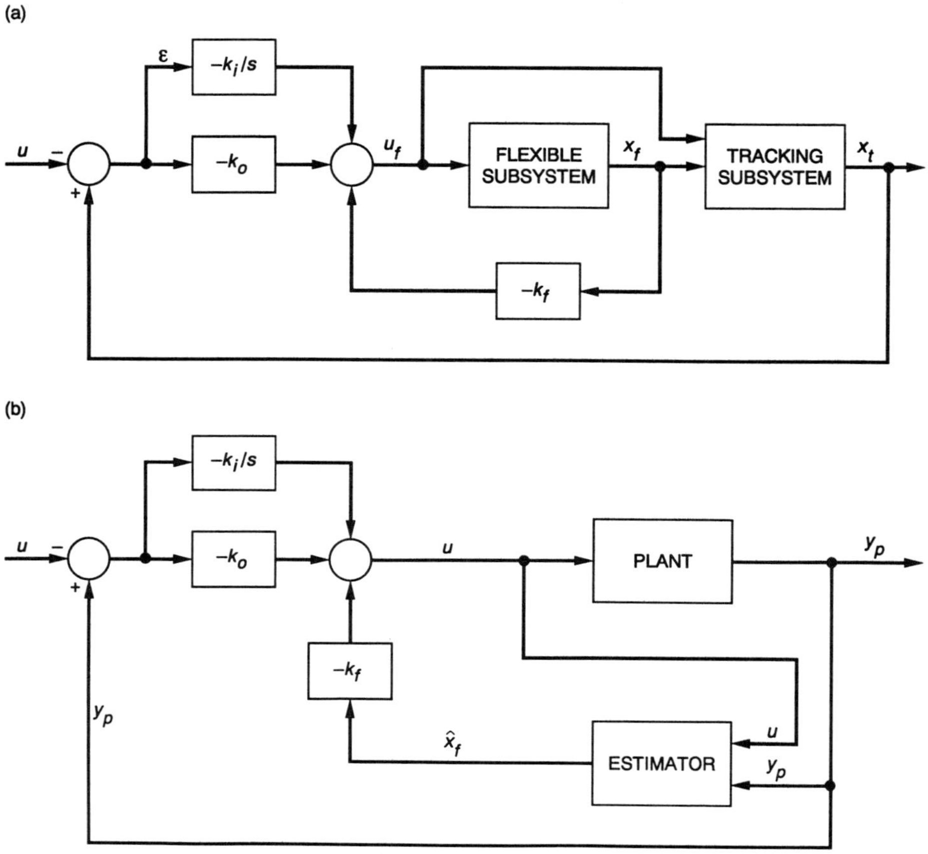

Figure 7. Block diagrams: (a) antenna closed-loop system, (b) antenna closed-loop system with an observer.

The equations (15), (16), (17) give the triple for the closed-loop system with observer

$$A_{co}=\begin{bmatrix} 0 & C_p & 0 \\ -Bk_i & A-Bk_oC_p & -Bk_fC_f \\ -Bk_i & K_eC_p-Bk_oC_p & A-K_eC_p-Bk_fC_f \end{bmatrix},\quad B_{co}=\begin{bmatrix} -I \\ Bk_o \\ Bk_o \end{bmatrix},\quad C_{co}=[0\ C_p\ 0] \tag{20}$$

CONTROLLER DESIGN FOR THE DSS-13 ANTENNA

The DSS-13 antenna is a recently designed antenna with a 34-m dish (Fig.8), which has significant flexible deformations observed during tracking operations. Its model development is presented in [4, 5]. The model consists of two tracking states (azimuth and elevation angle), and 13 flexible modes (or 26 balanced states). The preliminary weights $q_{ie}=q_{pe}=q_{ia}=q_{pa}=1$ for the tracking subsystem (for y_i and y_p) and zero weights for the flexible subsystem ($q_1=q_2=\ldots=q_{13}=0$ for all 13 modes) have been chosen for the the LQ controller design. The magnitudes of the closed-loop transfer function in Fig.9 show that flexible motion of the antenna is excessive, and should be damped out. This is achieved by adjusting weights for the flexible subsystem. For the tracking weights as before, the weight for the first mode (2.32 Hz) is chosen to be $q_1=10^{-7}$, and the remaining weights are zero, obtaining the closed-loop system responses as in Fig.10. One can see that the 2.32-Hz resonance peak in the azimuth command response has disappeared. The elevation motion is unaffected however, since the AZ mode is almost non-existent in the elevation motion.

Next, assuming q_{ie}, q_{ia}, q_{pe}, q_{pa}, q_1, as given above, and setting a weight for the second mode (2.64 Hz) $q_2=10^{-7}$ while the remaining weights are zero, one obtains the LQ control system responses as in Fig.11. As a result of a non-zero weight, the 2.64-Hz resonant peak has disappeared in the elevation command motion.

The weight should be chosen carefully. Too small weight (e.g. 3×10^{-9} in the case considered) will not suppress the resonant peak (Fig.12a). Too large weight (e.g. 10^{-5}) will deteriorate the tracking performance: for the overweighted mode the transfer function is pressed down within wide frequency range (Fig.12b). The proper weight suppresses the resonant peak, leaving the other peaks unchanged (Fig.6).

Similar procedures have been applied for the third (4.26-Hz), fourth (3.77-Hz), fifth (7.88-Hz), sixth (4.47-Hz), seventh (3.38-Hz), eighth (5.98-Hz), ninth (7.32-Hz), and tenth (9.48-Hz) modes, with weight 10^{-7} for each mode. As a result, the suppression of the remaining flexible motion and resonant peaks is observed in Fig.13. Weights for the remaining modes (eleventh through thirteenth) have been set to zero. The controller order is 24 for the plant order of 30.

The root locus of the closed-loop system due to weight variations of the 7.92-Hz mode is shown in Fig.14. The figure shows the horizontal departure of poles into the left-hand side direction (stabilizing property). It confirms the properties of the weighted LQ design described previously.

In the next step, the tracking properties of the system are improved by proper weight setting of the tracking subsystem. Namely, setting the integral weight to $q_{ie}=q_{ia}=70$ and the proportional weight to $q_{pe}=q_{pa}=100$ improves the system tracking properties, as shown in Fig.15 (extended bandwidth - up to 2 Hz). However, by improving the tracking properties, the transfer function has been raised dramatically in the frequency region 1 - 3 Hz, which forces the first two modes located in this region to appear again in the step response. By sacrificing a bit of the tracking properties, the flexible motion in the step response is

Figure 8. DSS-13 antenna

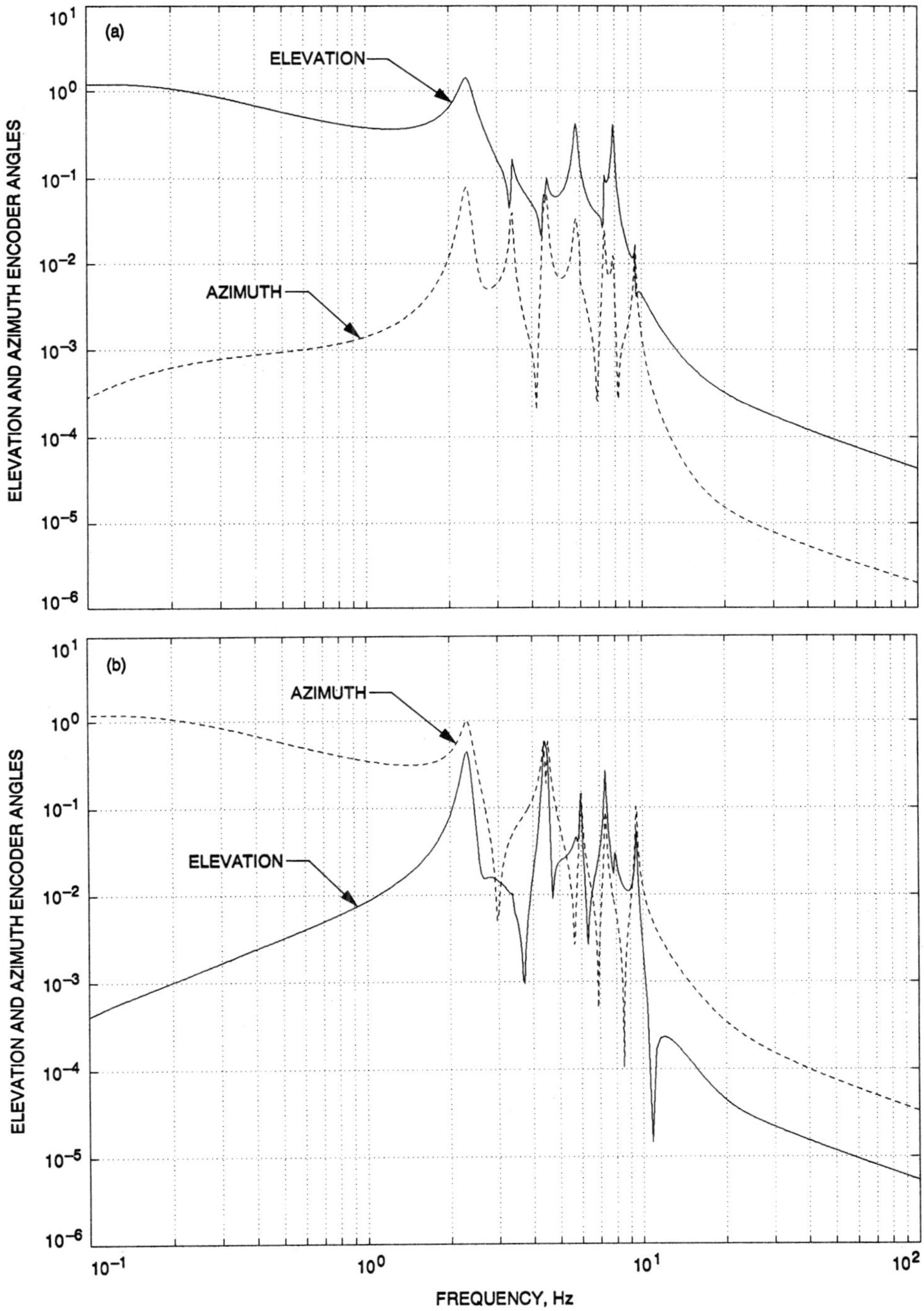

Figure 9. Closed-loop transfer function for unit proportional and unit integral weight in azimuth and elevation, and zero weight for flexible subsystem: (a) elevation command, and (b) azimuth command.

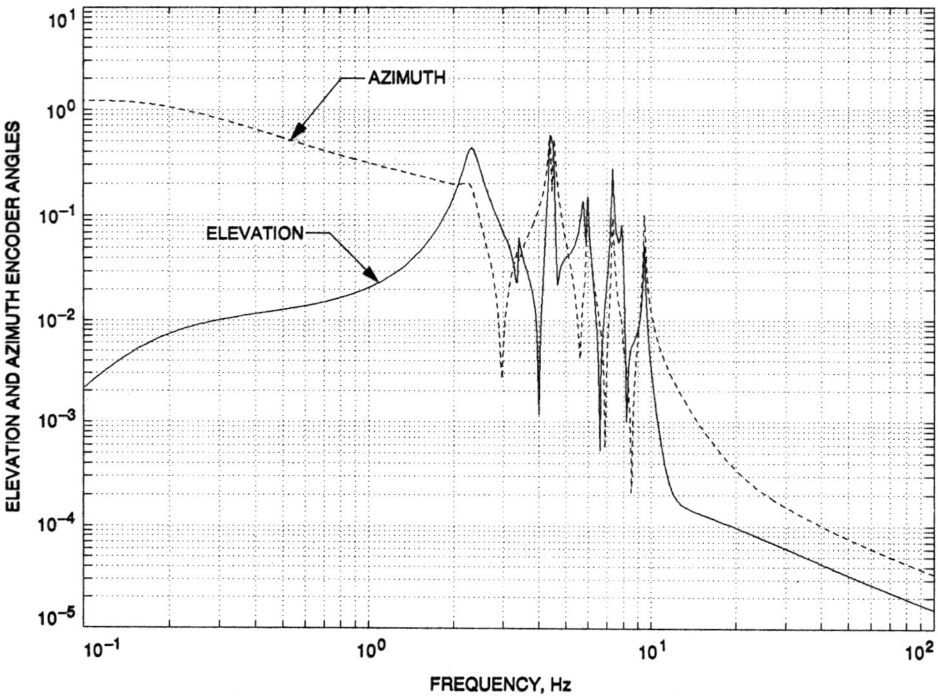

Figure 10. Transfer function (elevation encoder to elevation command and azimuth encoder to azimuth command), for weights the same as in Fig.9 but $q_{f1}=10^{-7}$.

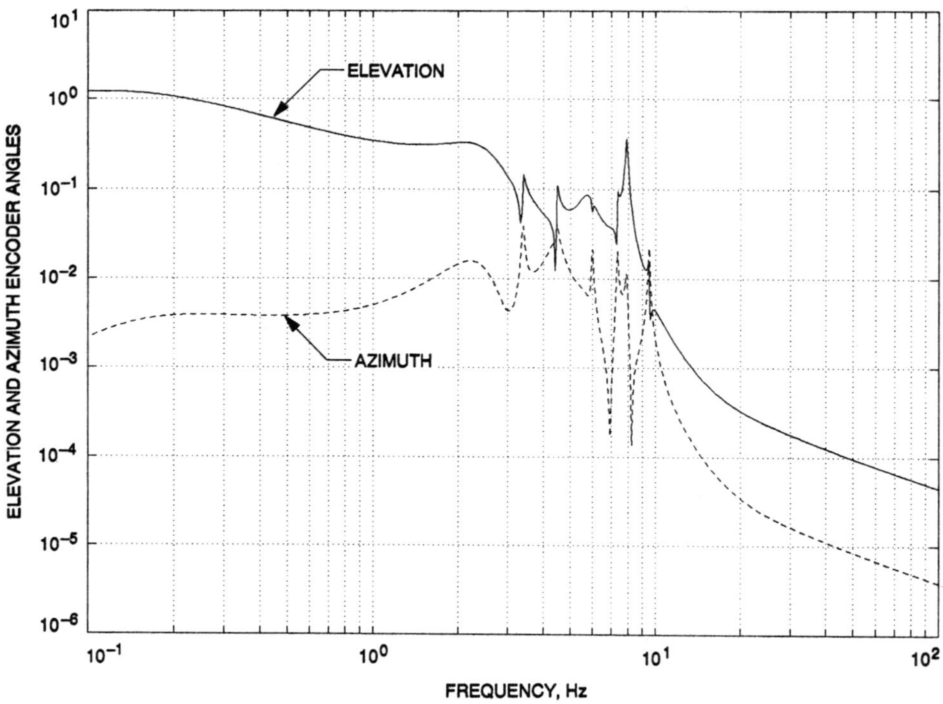

Figure 11. Transfer function (elevation encoder to elevation command and azimuth encoder to azimuth command) for weights the same as in Fig.9 but $q_{f1}=q_{f2}=10^{-7}$.

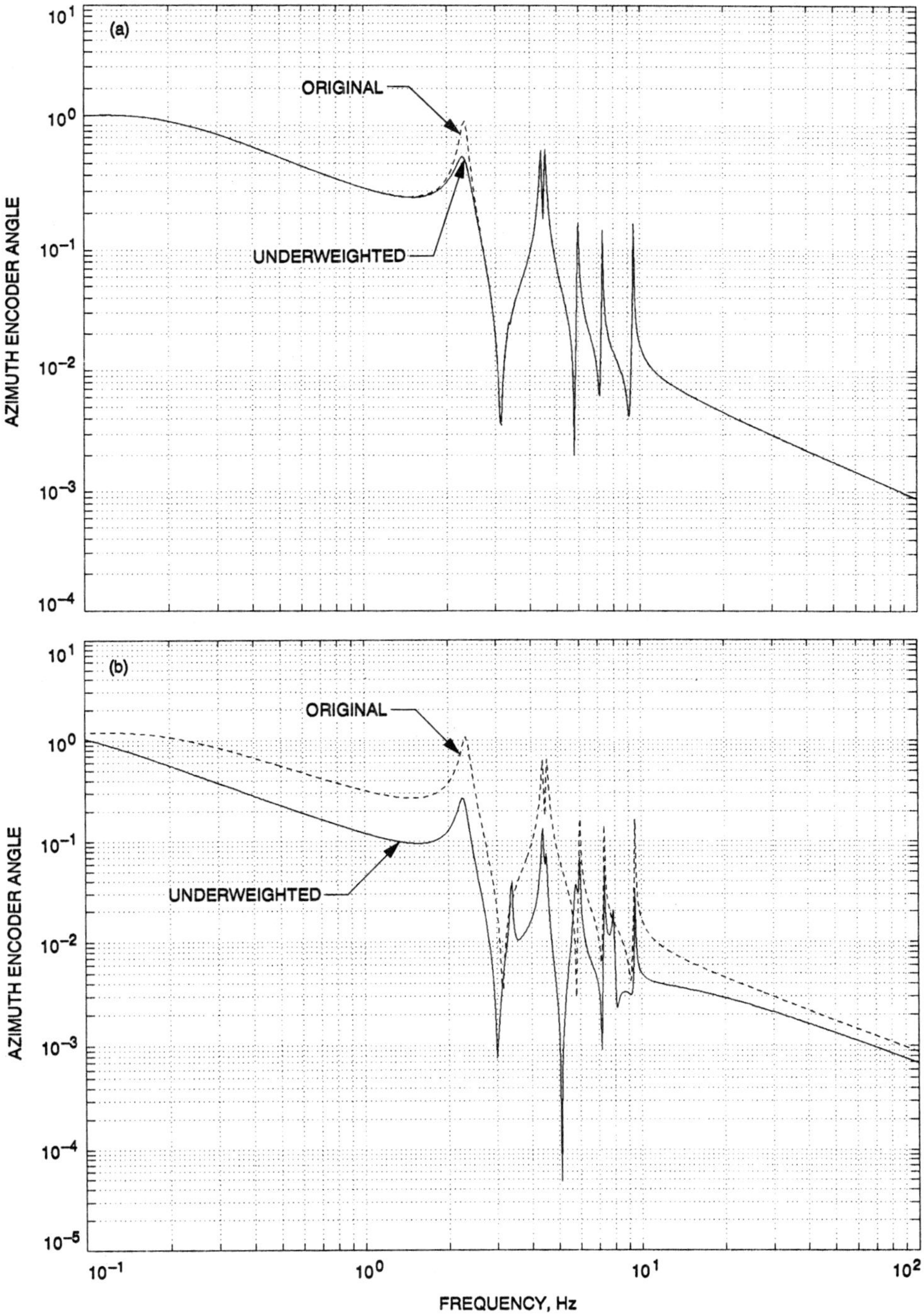

Figure 12. The first component: (a) underweighted, (b) overweighted.

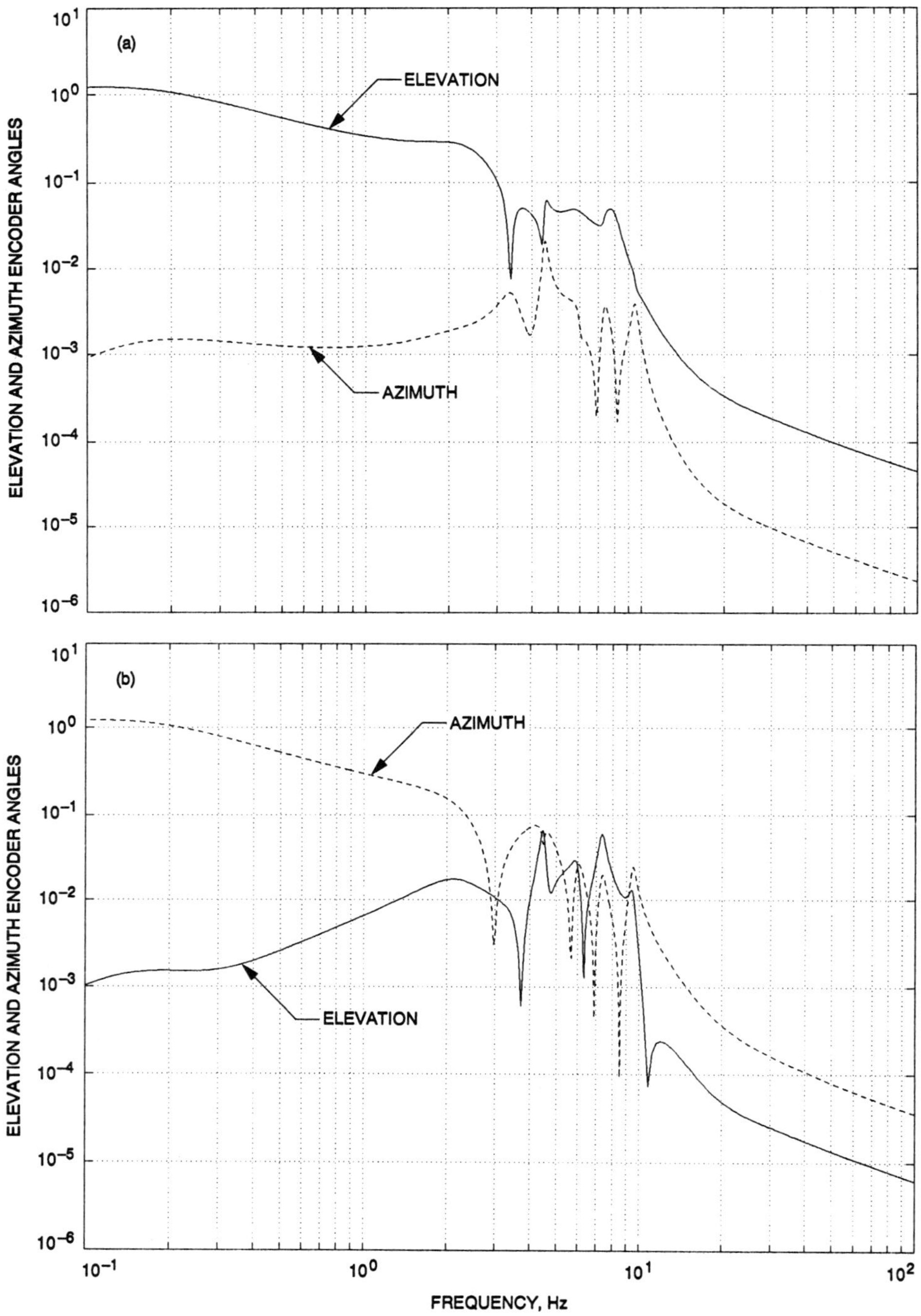

Figure 13. Transfer function (elevation encoder to elevation command and azimuth encoder to azimuth command) for weights the same as in Fig.9 but $q_{f1}=\ldots=q_{f10}=10^{-7}$.

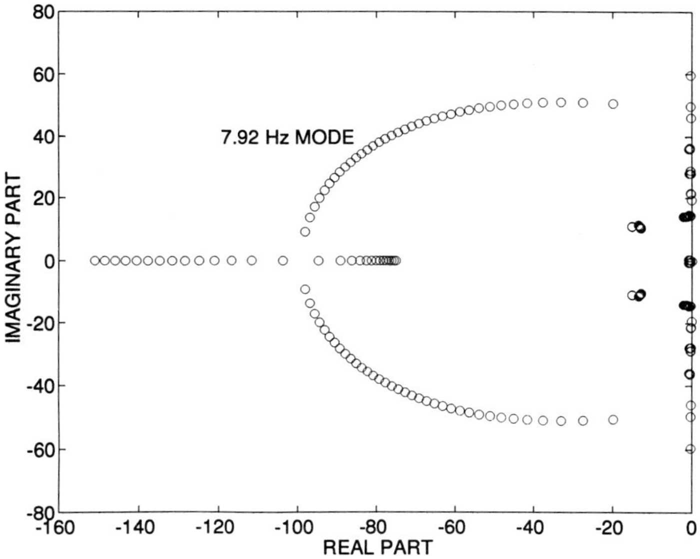

Figure 14. Root locus for 7.92 Hz mode versus weight.

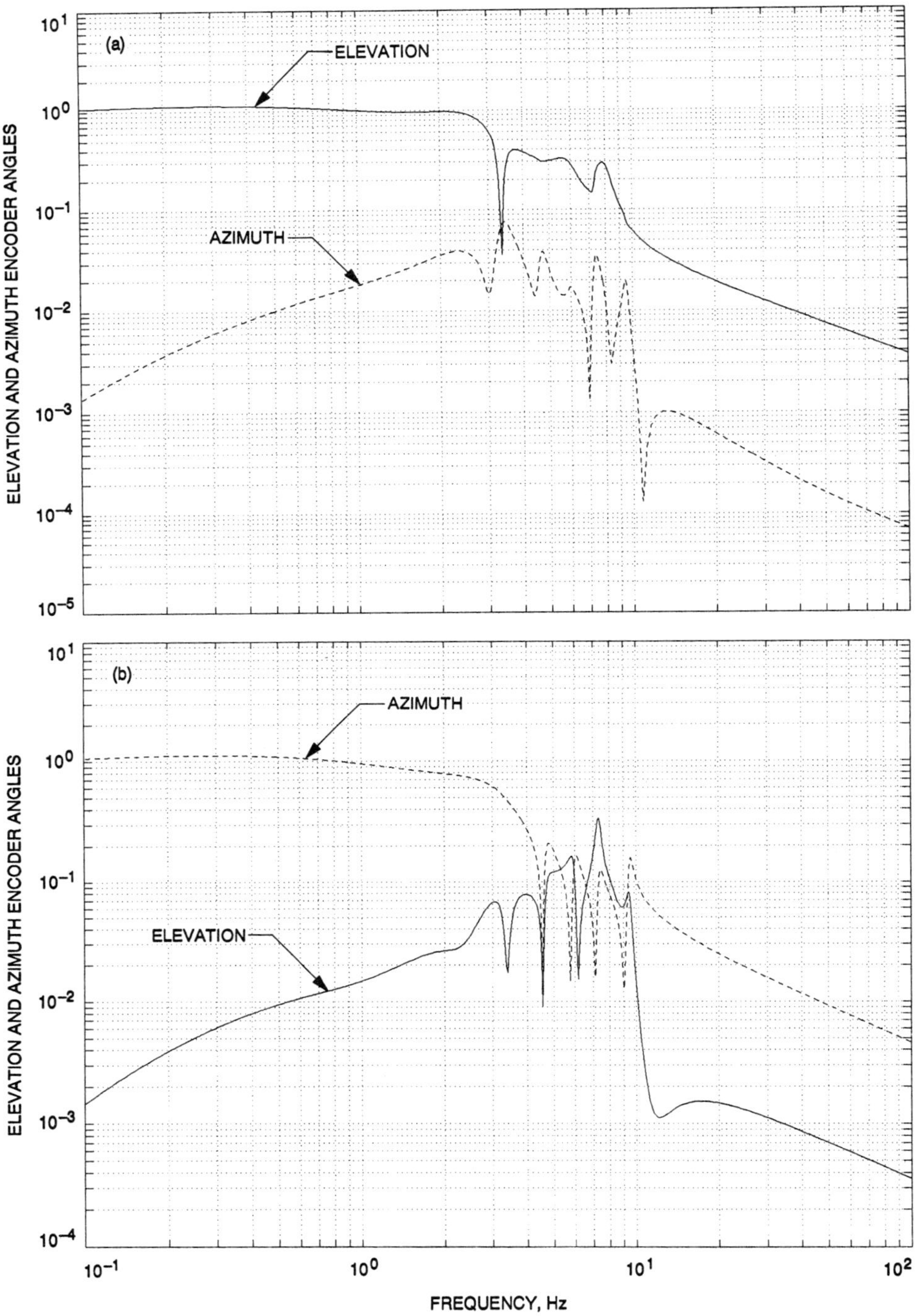

Figure 15. Closed-loop transfer function for proportional weight 100 and integral weight 70 (both in azimuth and elevation), and weights for flexible subsystem equal to 10^{-7}: elevation encoder to elevation command, and azimuth encoder to azimuth step command.

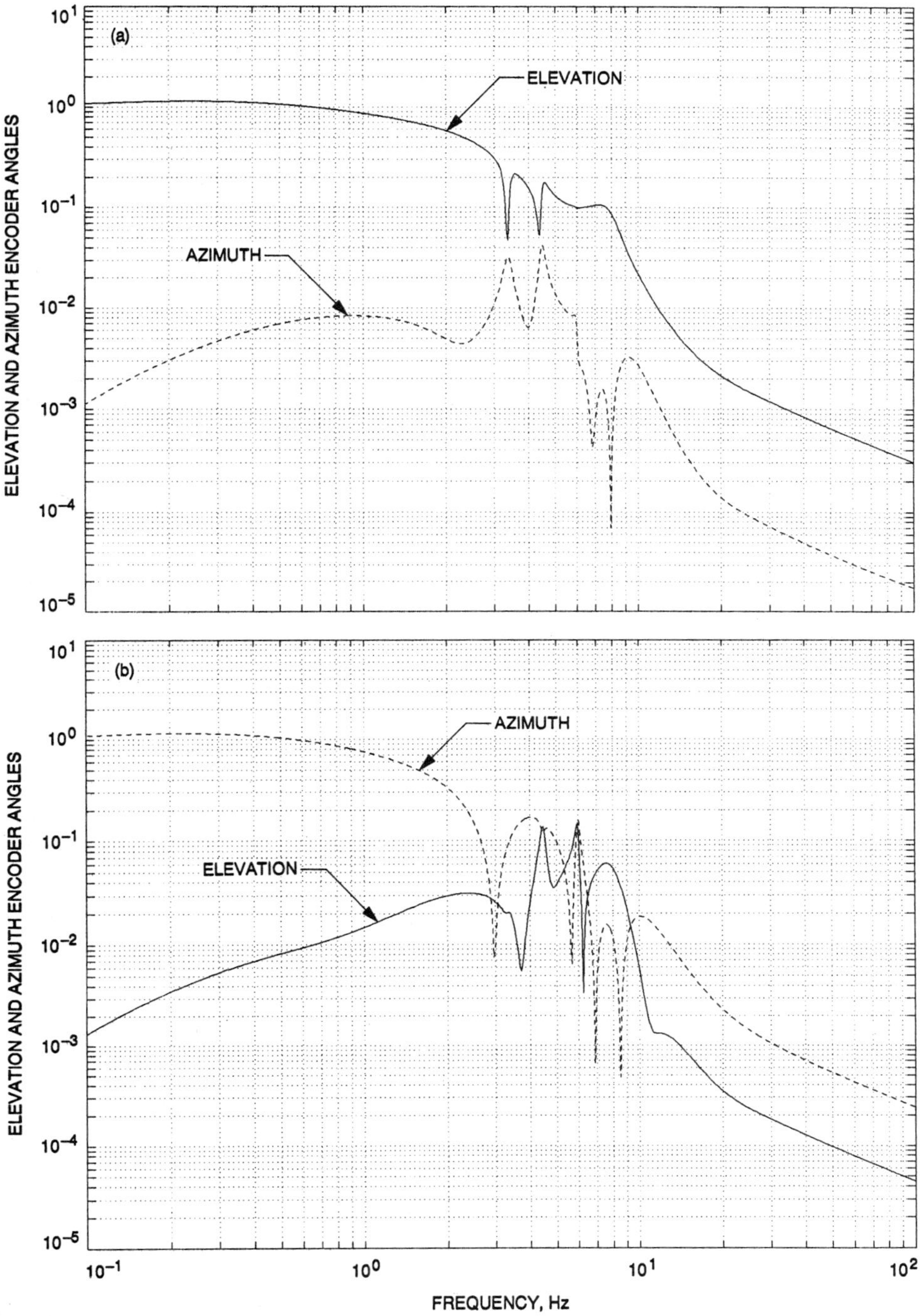

Figure 16. Closed-loop transfer function for proportional weight 100 and integral weight 70 (in azimuth and elevation), and for flexible weights $q_{f1}=\ldots=q_{f6}=10^{-6}$, $q_{f7}=q_{f8}=10^{-7}$, $q_{f9}=q_{f10}=10^{-5}$: elevation encoder to elevation command (solid line), and azimuth encoder to azimuth command (dashed line).

removed. This is done by increasing slightly the weights of the flexible subsystem, setting them as follows: $q_1=q_2=q_3=q_4=q_5=q_6=10^{-6}$, $q_7=q_8=10^{-7}$, $q_9=q_{10}=10^{-5}$. The closed-loop system response with the satisfactory tracking performance is shown now in Fig.16 (1-Hz bandwidth).

CONCLUSIONS

A new procedure for the DSN antenna controller design has been proposed. The antenna model is divided into flexible and tracking parts rather then into elevation and azimuth parts. In a sequential design strategy a controller for the flexible subsystem is designed first, followed by a controller design for the tracking subsystem with the final tune-up of the flexible part. This approach results in a significant improvement of the performance of the antenna closed-loop system through a sequential weight adjustment of the state vector. The analytical relationships between weight and either gains or root locus have been derived in this paper. The order of the controller is determined through the monitoring the closed-loop performance for each flexible mode. The DSS-13 antenna tracking controller design has been used to illustrate the procedure.

ACKNOWLEDGMENTS

The author would like to thank B. Parvin for helpful comments. This research was performed at the Jet Propulsion Laboratory, California Institute of Technology, under a contract with the National Aeronautics and Space Administration.

REFERENCES

1. L.S. Alvarez, and J. Nickerson, "Application of Optimal Control Theory to the Design of the NASA/JPL 70-Meter Antenna Axis Servos." *TDA Progress Report*, vol.42-97, Jet Propulsion Laboratory, California Institute of Technology, May 1989.
2. G. Biernson, *Optimal Radar Tracking Systems*. Wiley, New York, 1990.
3. W. Gawronski, and J.-N. Juang, "Model Reduction for Flexible Structures," in: *Control and Dynamic Systems,* ed. C.T. Leondes, vol.36, pp.143-222, Academic Press: New York, 1990.
4. W. Gawronski, and J.A. Mellstrom, "Modeling and Simulations of the DSS-13 Antenna Control System". *TDA Progress Report,* 42-106, Jet Propulsion Laboratory, California Institute of Technology, August 1991.
5. W. Gawronski, and J.A. Mellstrom, "Control and Dynamics of the Deep Space Network Antennas," in *Control and Dynamic Sysytems,* ed. C.T. Leondes, vol.57, Academic Press: New York, 1993.
6. W. Gawronski, and T. Williams, "Model Reduction for Flexible Space Structures," *Journal of Guidance, Control, and Dynamics,* vol.14, No.1, Jan. 1991, pp.68-76.
7. E.A. Jonckheere, "Principal Component Analysis of Flexible Systems - Open Loop Case," *IEEE Trans. on Automat. Control,* vol.27, 1984.
8. H. Kwakernaak, and R. Sivan, *Linear Optimal Control Systems,* Wiley: New York, 1972.
9. T. Williams, and W. Gawronski, "Model Reduction for Flexible Spacecraft with Clustered Natural Frequencies", *Proc. 3rd Annual Conference on Aerospace Computational Control*, Oxnard, CA 1989.

APPENDIX

Proof of Proposition 1. For a flexible structure in balanced representation the state matrix A is diagonally dominant (with 2×2 blocks on the main diagonal), and for $R=I$ and Q as in Eq.(11), the solution S of the Riccati equation (3) is also diagonally dominant with 2×2 blocks S_i on the main diagonal

$$S_i=s_iI_2, \quad s_i>0, \quad i=1,\ldots,n \tag{A1}$$

Thus, the equation (3) turns into a set of the following equations

$$s_i(A_i+A_i^T)-s_iB_iB_i^T+q_iI_2=0, \quad i=1,\ldots,n \tag{A2}$$

For a balanced system $B_iB_i^T\cong-\gamma_i^2(A_i+A_i^T)$, see Eq.(10), and $A_i+A_i^T=-2\zeta_i\omega_iI_2$, see Eq.(9). Therefore Eq.(A2) is now

$$s_i^2+s_i/\gamma_i-0.5q_i/\zeta_i\omega_i\gamma_i^2=0, \quad i=1,\ldots,n \tag{A3}$$

There are two solutions of Eq.(A3), but for $q_i=0$ it is required that $s_i=0$. Therefore Eq.(12) represents the unique solution of Eq.(A3).

Proof of Proposition 2. For small q_i the matrix A of the closed loop system is diagonally dominant $A_o=diag(A_{oi})$, $i=1,\ldots,n$, and $A_{oi}=A_i-B_iB_i^Ts_i$. Introducing Eq.(10) one obtains

$$A_{oi}=A_i+2s_i\gamma_i^2(A_i+A_i^T), \tag{A5}$$

and introducing A_i as in Eq.(9) to Eq.(A5) one obtains

$$A_{oi}=\begin{bmatrix} -\beta_i\zeta_i\omega_i & -\omega_i \\ \omega_i & -\beta_i\zeta_i\omega_i \end{bmatrix}, \tag{A6}$$

with β_i as in Eq.(13).

MANEUVERING AND CONTROL OF SPACE STRUCTURES

Leonard Meirovitch

Department of Engineering Science and Mechanics
Virginia Polytechnic Institute and State University
Blacksburg, VA 24061

ABSTRACT

In certain spacecraft reorientation of the line of sight requires slewing of the whole spacecraft. If the spacecraft is flexible, then the slewing maneuver is likely to cause elastic vibration, as well as perturbations in the maneuvering angular motion. the latter are undesirable effects and must be controlled. Other spacecraft can be regarded as articulated flexible multibody systems and only certain smaller components must be reoriented relative to an inertial space. In this case, it is more convenient to stabilize the main spacecraft relative to the inertial space and reorient the elastic components relative to the main spacecraft. This paper reviews first the literature on flexible singlebody and multibody maneuvers. Then, it discusses the problem of maneuvering and control of articulated flexible space structures in detail.

INTRODUCTION

In many space applications, the mission consists of reorienting the line of sight in a flexible spacecraft, such as in the reorientation of the line of sight of a space telescope. Because in this case the line of sight can be regarded as being embedded in the undeformed structure, reorientation of the line of sight implies reorientation of the whole spacecraft. This type of maneuvering represents single body maneuvering. In other cases, the spacecraft represents an articulated flexible multibody system and the mission consists of maneuvering one or more of the bodies. Quite often, one of the bodies plays the role of a platform. In such cases, it is advisable to maneuver the various bodies relative to the platform while stabilizing the platform in an inertial space. We refer to this class of maneuvers as multibody maneuvers. In this paper, both types of maneuvers are reviewed, but only multibody maneuvers are discussed in detail.

A problem intimately related to the maneuvering of space structures is that of maneuvering of flexible robots. Hence, we include flexible robots in our discussion.

LITERATURE REVIEW

This section contains a review of papers in the general area of flexible single body and multibody maneuvering.

Mechanics and Control, Edited by R.S. Guttalu
Plenum Press, New York, 1994

Single Body Maneuvering

The problem of control of a flexible robot arm is discussed by Book and Majette [1]. The arm is modeled both in the frequency domain and in the time domain, the first by means of the transfer matrix approach and the second by means of a modal representation. Control gains in the latter are determined by the pole allocation method. As an alternative to representing the motion of a flexible robot arm by means of natural modes, Skaar [2] proposes to use transfer functions relating the response of one or more points within a distributed-parameter system to the control input(s). Turner and Junkins [3] consider the problem of optimal large-angle rotational maneuver of a flexible spacecraft while suppressing the elastic vibration at the same time. The mathematical model consists of a rigid hub rotating about a fixed axis and four flexible beams and control is achieved by means of a torque on the hub. Considering the linearized problem, an optimal controller is designed on the basis of the linear quadratic regulator (LQR) theory. The same basic problem was investigated by Breakwell [4], who describes a numerical procedure for generating a feedback control law permitting reorientation of the spacecraft, while leaving a number of bending modes unexcited at the end of the maneuver. The problem investigated in [3] is revisited by Turner and Chun [5], who consider torques on the elastic appendages, in addition to the torque on the hub. Chun, Turner and Juang [6] consider the problem of maneuvering a flexible spacecraft through a large angle using a disturbance-accommodating control tracking a desired output state in the form of the open-loop solution for the linear model. Juang, Horta and Robertshaw [7] present a slewing control experiment demonstrating a linear optimal control algorithm with terminal constrains.

The problem of simultaneous optimization of structural design and control design for maneuvering flexible spacecraft is investigated by Hale, Lisowski and Dahl [8]. As an example, they consider the same mathematical model as in [3], except that the beams are not uniform and their mass and stiffness distributions are subject to optimization. The work of [8] is extended by Hale and Lisowski [9] by comparing solutions of the parametric optimal control problem minimizing actual control cost with solutions that minimize the ratio of actual control cost to the cost of controlling optimally a rigid-body of the same inertial properties.

A different type of optimal control than the LQR is obtained in the case in which the interest lies in fast maneuvers. In this case, the cost is the maneuver time. In other cases, the object is to save fuel. The case in which the cost is a combination of fuel and time was investigated by Vander Velde and He [10]. The result is open-loop, on-off control. The maneuver and vibration suppression of a flexible spacecraft simulated by a free-free beam is investigated by Baruh and Silverberg [11]. They carry out a 90° slewing maneuver while controlling two rigid-body modes, one in translation and one in rotation, as well as the two lowest elastic modes. The control design is by the independent modal-space control method and the control law is bang-bang [12]. Skaar, Tang and Yalda-Mooshabad [13] demonstrate that open-loop, on-off maneuvering of a flexible spacecraft can be made attractive by selecting switching times so as to minimize the postmaneuver elastic energy. The problem of minimum-time slewing of flexible spacecraft is considered by Thompson, Junkins and Vadali [14]. Concerned that bang-bang control may cause spillover into the higher residual flexible modes, thus degrading the system response, they propose to alter the switching functions in a way that the control applied to the system is smooth and continuous throughout the entire maneuver. The same basic problem as in [14] is addressed by Singh, Kabamba and McClamroch [15], who consider planar, time-optimal, rest-to-rest slewing maneuvers of flexible spacecraft. The motion is described by a rigid-body rotation and n elastic degrees of freedom. The open-loop optimal control is shown to be bang-bang with multiple switches in each control variable. The problem in [14,15] has been considered more recently by Ben-Asher, Burns and Cliff [16].

A perturbation approach to the problem of flexible spacecraft maneuvering is proposed by Meirovitch and Quinn [17,18]. According to this approach, the rigid-body maneuver of the spacecraft is regarded as the unperturbed motion and the elastic motion and deviations from the rigid-body motions constitute the perturbations. The rigid-body maneuver is carried out open-loop and the perturbations are controlled closed-loop. The approach is used by Quinn and Meirovitch [19] to maneuver and control the model of the Spacecraft Control Laboratory Experiment (SCOLE). The same basic approach is used by Soga, Hirako, Ohkami, Kida and Yamaguchi [20] to carry out large-angle single-axis slewing of flexible spacecraft. They divide the problem into two parts, the first representing the open-loop slewing of the rigid-body mode and the second the closed-loop feedback control of the elastic modes. The approach is evaluated by means of a ground-based hardware experiment using a single-axis air bearing table and flexible cantilevered beam. The perturbation approach developed in [17-19] is used by Meirovitch and Sharony [21] and Sharony and Meirovitch [22] to maneuver and control a flexible spacecraft, but the feedback control of the perturbations is different. In particular, the feedback control involves a time-varying linear quadratic regulator capable of accommodating the disturbances caused by the rigid-body maneuver. The difference between [21] and [22] is that the first assumes perfect observation and the second uses a Luenberger observer.

Multibody Maneuvering

The problem of flexible multibody maneuvering has been attracting a great deal of interest due to the various applications in robotics and in spacecraft dynamics and control. Book, Maizza-Neto and Whitney [23] use a transfer matrix approach and a modal approach to model a two-beam manipulator. A variety of issue in the dynamics and control of flexible robots are considered by Baruh and Tadikonda [24]. The modeling is done by an approach akin to substructure synthesis [25] and flexibility is treated as a deterministic disturbance on the rigid-body motions. Still in the context of robotics, Quinn and Chang [26] present a dynamic formulation for multi-jointed open chains of flexible links.

The difference between earth-based robots and maneuvering articulated spacecraft lies in the motion of the base of the latter. Of course, in the case of flexible multibody maneuvering spacecraft, the derivation of the equations of motion becomes a very complex task. This task can be simplified to a significant extent by writing the equations of motion in terms of body axes, as they are the preferred axes for spacecraft control. In this regard, the state equations of motion for flexible bodies in terms of quasi-coordinates derived by Meirovitch [27] are particularly useful. The equations derived in [27] are used by Meirovitch and Kwak [28] to derive state equations capable of treating the problem of maneuvering and control of articulated flexible multibody spacecraft. The equations derived in [28] are used by Meirovitch and Kwak [29,30] to treat the problem in which flexible appendages are reoriented relative to a rigid platform, while the platform is stabilized relative to an inertial space and the vibration of the appendages is suppressed. Regarding the spacecraft as a collection of flexible substructures, France and Meirovitch [31] develop a substructure decentralized control approach to the maneuvering and control problem. Implementation of the maneuvering and control is carried out in [31] in discrete time. If the time-varying terms are relatively small, the control design can be carried out by means of a perturbation approach [32] requiring the solution of a matrix Riccati equation and a matrix Liapunov equation.

THE EQUATIONS OF MOTION FOR MANEUVERING AND CONTROL OF ARTICULATED FLEXIBLE MULTIBODY SYSTEMS

The structure under consideration consists of a central substructure o and a collection of maneuvering substructures a, as shown in Fig. 1. The motion is described

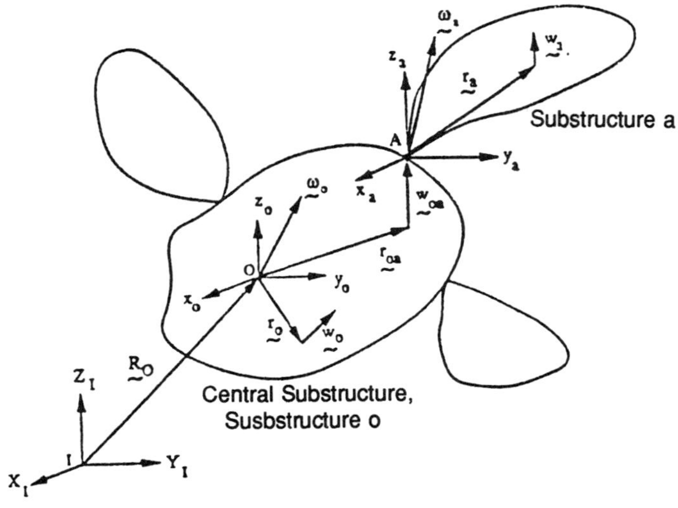

Figure 1. The Mathematical Model

conveniently by introducing the inertial axes $X_I Y_I Z_I$, and sets of body axes $x_o y_o z_o$ and $x_a y_a z_a$ attached to the various substructures in undeformed state. The position vector of typical points in o and a are respectively, $\mathbf{R}_o = \mathbf{R}_O + \mathbf{r}_o + \mathbf{w}_o$ and $\mathbf{R}_a = \mathbf{R}_O + \mathbf{r}_{oa} + \mathbf{r}_a + \mathbf{w}_a$ $(a = 1, 2, \ldots, N)$, where $\mathbf{R}_O$ is the radius vector from I to O, $\mathbf{r}_o$ the position vector of a typical point in $o, \mathbf{w}_o$ the elastic displacement vector of the same point, $\mathbf{r}_{oa}$ the vector from O to A, $\mathbf{w}_{oa}$ the elastic displacement vector of A, $\mathbf{r}_a$ the position vector of a typical point in a and $\mathbf{w}_a$ the elastic displacement vector of the same point. Note that all vectors in Eqs. (1) are expressed in terms of components along the local axes. Axes $x_o y_o z_o$ rotate with the angular velocity $\boldsymbol{\omega}$ relative to the inertial axes $X_I Y_I Z_I$, a set of intermediate axes attached to substructure o at A rotate with the angular velocity $\dot{\boldsymbol{\beta}}$ relative to $x_o y_o z_o$ because of elasticity, where the rotation angle $\boldsymbol{\beta} = \nabla \times \mathbf{w}_{oa}$ is small on the assumption that elastic deformations are small, and axes $x_a y_a z_a$ rotate with the angular velocity $\boldsymbol{\omega}_a$ relative to the intermediate axes at A. Then, using a tilde over a symbol to denote a skew symmetric matrix formed from the corresponding vector, the velocity vector of typical points o and a can be shown to be [33]

$$\dot{\mathbf{R}}_o = \mathbf{V}_O + \tilde{\omega}\left(\mathbf{r}_o + \mathbf{w}_o\right) + \dot{\mathbf{w}}_o \tag{1a}$$

$$\begin{aligned}\dot{\mathbf{R}}_a =& \mathbf{V}_O + \tilde{\omega}\left(\mathbf{r}_{oa} + \mathbf{w}_{oa}\right) + \dot{\mathbf{w}}_{oa} + \left[\left(\tilde{\omega} + \tilde{\dot{\beta}}\right) C_\beta^T C_a^T + C_\beta^T C_a^T \tilde{\omega}_a\right]\left(\mathbf{r}_a + \mathbf{w}_a\right) \\ &+ C_\beta^T C_a^T \dot{\mathbf{w}}_a, \quad a = 1, 2, \ldots, N\end{aligned} \tag{1b}$$

where $\mathbf{V}_o$ is the velocity vector of O, $C_\beta = C_\beta\left(\boldsymbol{\beta}\right)$ the matrix of direction cosines between the intermediate axes and $x_o y_o z_o$ and C_a is the matrix of direction cosines between $x_a y_a z_a$ and the intermediate axes. For future reference, we note that the relation between rigid-body motions about the body axes $x_o y_o z_o$ and those about the inertial axes $X_I Y_I Z_I$ is given by [27]

$$\mathbf{V}_o = C_o \dot{\mathbf{R}}_o, \quad \boldsymbol{\omega} = D\dot{\boldsymbol{\theta}} \tag{2a, b}$$

where C_o is the matrix of direction cosines between $x_o y_o z_o$ and $X_I Y_I Z_I$ and $D = D(\theta_1, \theta_2, \theta_3)$ is a transformation matrix.

The maneuver consists of reorienting substructures a relative to the inertial space. This can be achieved by stabilizing substructure o relative to the inertial space, while suppressing the elastic vibration of all substructures, and reorienting substructures a relative to substructure o. Hence, regarding the vectors $\boldsymbol{\omega}_a$ as known functions of time, the motion of the system is described by $\mathbf{V}_o$, $\boldsymbol{\omega}$, $\mathbf{w}_o$ and $\mathbf{w}_a$. The first two depend on time alone and the last two depend on corresponding spatial coordinates and time. As a result, the equations of motion consist of a set of hybrid (ordinary and partial) differential equations. Practical considerations dictate spatial discretization, which can be carried out by letting

$$\mathbf{w}_s(\mathbf{r}_s, t) = \Phi_s(\mathbf{r}_s)\,\mathbf{q}_s(t), \quad s = o, a; \quad a = 1, 2, \ldots, N \tag{3}$$

where Φ_s are matrices of admissible functions [25] and $\mathbf{q}_s$ are vectors of generalized coordinates.

We assume that the inertia of substructures a is relatively small, so that any rigid-body motions of substructure o caused by the maneuver are small, and that the elastic displacements are also small. As a result, the spatially discretized equations of motion can be linearized so as to read [33]

$$\begin{aligned}
&m_t \dot{\mathbf{V}}_O - \tilde{S}_t \dot{\omega} + 2\sum_{a=1}^{N} C_a^T \left(\widetilde{\tilde{S}_a \omega_a}\right) C_a \omega + \overline{\Phi}_t \ddot{\mathbf{q}}_o + 2\sum_{a=1}^{N} C_a^T \left(\widetilde{\tilde{S}_a \omega_a}\right) C_a \Upsilon_{oa} \dot{\mathbf{q}}_o \\
&+ \sum_{a=1}^{N} C_a^T \left(\tilde{\omega}_a \tilde{S}_a \omega_a + \tilde{S}_a \dot{\omega}_a\right) C_a \Upsilon_{oa} \mathbf{q}_o + \sum_{a=1}^{N} C_a^T \overline{\Phi}_a \ddot{\mathbf{q}}_a + 2\sum_{a=1}^{N} C_a^T \tilde{\omega}_a \overline{\Phi}_a \dot{\mathbf{q}}_a \\
&+ \sum_{a=1}^{N} C_a^T \left(\dot{\tilde{\omega}}_a + \tilde{\omega}_a^2\right) \overline{\Phi}_a \mathbf{q}_a = \mathbf{F} + \sum_{a=1}^{N} C_a^T \left(\tilde{\omega}_a \tilde{S}_a \omega_a + \tilde{S}_a \dot{\omega}_a\right)
\end{aligned} \tag{4a}$$

$$\begin{aligned}
&\tilde{S}_t \dot{\mathbf{V}}_O + I_t \dot{\omega} + \sum_{a=1}^{N} \left\{C_a^T \left[2\tilde{\omega}_a I_a - tr(I_a)\tilde{\omega}_a\right] C_a + 2\tilde{r}_{oa} C_a^T \left(\tilde{S}_a \omega_a\right) C_a\right\} \omega + \tilde{\Phi}_t \ddot{\mathbf{q}}_o \\
&+ \sum_{a=1}^{N} \left\{2\tilde{r}_{oa} C_a^T \left(\widetilde{\tilde{S}_a \omega_a}\right) + C_a^T \left[2\tilde{\omega}_a I_a - tr(I_a)\tilde{\omega}_a\right]\right\} C_a \Upsilon_{oa} \dot{\mathbf{q}}_o + \sum_{a=1}^{N} \dot{\Gamma}_{oa} \mathbf{q}_o + \sum_{a=1}^{N} \Pi_a \ddot{\mathbf{q}}_a \\
&+ 2\sum_{a=1}^{N} \left(\tilde{r}_{oa} C_a^T \tilde{\omega}_a \overline{\Phi}_a + C_a^T \Psi_a\right) \dot{\mathbf{q}}_a + \sum_{a=1}^{N} \dot{\hat{\Pi}}_a \mathbf{q}_a = \mathbf{M} + \sum_{a=1}^{N} \left[\tilde{r}_{oa} C_a^T \left(\tilde{\omega}_a \tilde{S}_a \omega_a + \tilde{S}_a \dot{\omega}_a\right)\right. \\
&\left. - C_a^T \left(\tilde{\omega}_a I_a \omega_a + I_a \dot{\omega}_a\right)\right]
\end{aligned} \tag{4b}$$

$$\begin{aligned}
&\overline{\Phi}_t^T \dot{\mathbf{V}}_O + \tilde{\Phi}_t^T \dot{\omega} + \sum_{a=1}^{N} \left\{2\Phi_{oa}^T C_a^T \left(\widetilde{\tilde{S}_a \omega_a}\right) + \Upsilon_{oa}^T C_a^T \left[2\tilde{\omega}_a I_a - tr(I_a)\tilde{\omega}_a\right]\right\} C_a \omega + M_t \ddot{\mathbf{q}}_o \\
&+ \sum_{a=1}^{N} \left\{2\Phi_{oa}^T C_a^T \left(\widetilde{\tilde{S}_a \omega_a}\right) + \Upsilon_{oa}^T C_a^T \left[I_a \tilde{\omega}_a + 3\tilde{\omega}_a I_a - 2tr(I_a)\tilde{\omega}_a\right]\right\} C_a \Upsilon_{oa} \dot{\mathbf{q}}_o \\
&+ \left(K_o + \dot{X}_{oa}\right) \mathbf{q}_o + \sum_{a=1}^{N} \Xi_a \ddot{\mathbf{q}}_a + 2\sum_{a=1}^{N} \left(\Phi_{oa}^T C_a^T \tilde{\omega} \overline{\Phi}_a + \Upsilon_{oa}^T C_a^T \Psi_a\right) \dot{\mathbf{q}}_a + \sum_{a=1}^{N} \dot{\hat{\Xi}}_a \mathbf{q}_a \\
&= \mathbf{Q}_o + \sum_{a=1}^{N} \left[\Phi_{oa}^T C_a^T \left(\tilde{\omega}_a \tilde{S}_a \omega_a + \tilde{S}_a \dot{\omega}_a\right) - \Upsilon_{oa}^T C_a^T \left(\tilde{\omega}_a I_a \omega_a + I_a \dot{\omega}_a\right)\right]
\end{aligned} \tag{4c}$$

$$\overline{\Phi}_a^T C_a \dot{\mathbf{V}}_O + \Pi_a^T \dot{\omega} - 2\left(\tilde{\Phi}_a^T \tilde{\omega}_a + \Psi_a^T\right) C_a \omega + \Xi_a^T \ddot{\mathbf{q}}_o - 2\left(\tilde{\Phi}_a^T \tilde{\omega}_a + \Psi_a^T\right) C_a \Upsilon_{oa} \dot{\mathbf{q}}_o + M_a \ddot{\mathbf{q}}_a$$
$$+ 2\tilde{H}_a \dot{\mathbf{q}}_a + \left(K_a + \dot{\tilde{H}}_a + \overline{H}_a\right) \mathbf{q}_a = \mathbf{Q}_a - \tilde{\Phi}_a^T \dot{\omega}_a + \Psi_a^T \omega_a, \ a = 1, 2, \ldots, N \tag{4d}$$

where K_o and K_a are stiffness matrices and

$$\mathbf{F} = \mathbf{F}_O + \sum_{i=1}^{n_o} \mathbf{F}_{oi} + \sum_{a=1}^{N} \sum_{i=1}^{n_a} C_a^T \mathbf{F}_{ai} \tag{5a}$$

$$\mathbf{M} = \mathbf{M}_O + \sum_{i=1}^{n_o} \tilde{r}_{oi} \mathbf{F}_{oi} + \sum_{a=1}^{N} \sum_{i=1}^{n_a} \left(\tilde{r}_{oa} C_a^T + C_a^T \tilde{r}_{ai}\right) \mathbf{F}_{ai} \tag{5b}$$

$$\mathbf{Q}_o = \sum_{i=1}^{n_o} \Phi_o^T (\mathbf{r}_{oi}) \mathbf{F}_{oi} + \sum_{a=1}^{N} \sum_{l=1}^{n_a} \left(\Phi_{oa}^T C_a^T + \Upsilon_{oa}^T C_a^T \tilde{r}_{ai}\right) \mathbf{F}_{ai} \tag{5c}$$

$$\mathbf{Q}_a = \sum_{i=1}^{n_a} \Phi_a^T (\mathbf{r}_{ai}) \mathbf{F}_{ai}, \quad a = 1, 2, \ldots, N \tag{5d}$$

are generalized force vectors, in which $\mathbf{F}_o$, $\mathbf{M}_o$ and $\mathbf{F}_{oi}$ are force and torque vectors acting on o and $\mathbf{F}_{ai}$ are force vectors acting on a, all in terms of components along local axes. The other various terms are as defined in [33].

Invoking the small-motions assumption, which implies that $\mathbf{V}_o \cong \dot{\mathbf{R}}_o$, $\omega \cong \dot{\theta}$, Eqs. (4) can be reduced to the form

$$[M_0 + M_1(t)]\ddot{\mathbf{x}} + G_1(t)\dot{\mathbf{x}} + [K_0 + K_1(t)]\mathbf{x} = [B_0^* + B_1^*(t)]\mathbf{f}(t) + \mathbf{d}(t) \tag{6}$$

where $\mathbf{x} = \left[\mathbf{R}_o^T \ \theta^T \ \mathbf{q}_o^T \ \mathbf{q}_1^T \ \mathbf{q}_2^T \ \ldots \ \mathbf{q}_N^T\right]^T$ is the configuration vector, M_0, K_0 and B_0^* are constant coefficient matrices, M_1, G_1, K_1 and B_1^* are time-varying coefficient matrices depending on the prescribed angular displacements, velocities and accelerations of bodies $a, \mathbf{f} = \left[\mathbf{F}_O^T \ \mathbf{M}_O^T \ \mathbf{F}_{o1}^T \ \ldots \ \mathbf{F}_{on_o}^T \ \mathbf{F}_{11}^T \ \ldots \ \mathbf{F}_{1n_1}^T \ \mathbf{F}_{21}^T \ \ldots \ \mathbf{F}_{Nn_N}^T\right]^T$ is recognized as the actual control vector and $\mathbf{d}$ is an inertial disturbance vector. The matrices M_0 and K_0 are symmetric. We note that quantities with the subscript 1 are one order of magnitude smaller than quantities with subscript 0. Explicit expressions for M_0, M_1, G_1, K_0, K_1, B_0^*, B_1^* and $\mathbf{d}$ can be obtained by comparing Eqs. (4) in conjunction with Eqs. (5) to Eq. (6).

SUBSTRUCTURE DECENTRALIZED CONTROL

The objective of the maneuver is to reorient the flexible substructures a relative to an inertial space. The strategy adopted here is to stabilize substructure o in an inertial space and to reorient the substructures a relative to substructure o, while any elastic vibration in all substructures is to be suppressed. The change in the orientation of substructures a relative to substructure o is to be carried out open loop. For a minimum-time maneuver, the control law is bang-bang, which implies constant angular acceleration during the first half of the maneuver time and constant deceleration during the second half, where the acceleration and deceleration are equal in magnitude.

As implied by Eqs. (3), every substructure is modeled independently. Consistent with this, we elect to control the elastic vibration of every substructure independently, which implies substructure decentralized control. This implies further that every substructure has its own sensors and actuators and the control gains relate the actuator forces on a given substructure to the signals from the sensors on the same substructure. The actuators on substructure o are charged with the control of the rigid-body perturbations and the elastic vibration of substructure o. The controls carrying out this task are $\mathbf{F}_o$, $\mathbf{M}_o$, $\mathbf{F}_{o1}$, $\mathbf{F}_{o2}$, $\dots$, $\mathbf{F}_{on_o}$. In addition, a different set of actuators change the orientation of substructures a relative to substructure o as if all substructures were rigid. This leaves the actuators on a typical substructure a in charge of the elastic vibration of that substructures alone. The controls assigned to this task are $\mathbf{F}_{a1}$, $\mathbf{F}_{a2}$, $\dots$, $\mathbf{F}_{an_a}$. Note that the sensors on substructure a sense only the elastic deformations. The idea of substructure decentralized control refers only to the design of control gains for each individual substructure separately. The control gains thus designed are used to generate actuator forces acting on the full structure.

To design the substructure decentralized controls described above, we treat the substructures as if they were independent. Hence, ignoring the coupling terms, Eqs. (1) can be written in the decentralized form

$$[M_{s0} + M_{s1}(t)]\ddot{\mathbf{x}}_s + G_{s1}(t)\dot{\mathbf{x}}_s + [K_{s0} + K_{s1}(t)]\mathbf{x}_s = [B^*_{s0} + B^*_{s1}]\mathbf{f}_s(t) + \mathbf{d}_s(t), \quad s = o, a; \quad a = 1, 2, \dots, N \tag{7}$$

in which $\mathbf{x}_o = \left[\mathbf{R}_O^T\ \boldsymbol{\theta}^T\ \mathbf{q}_o^T\right]^T$, $\mathbf{f}_o = \left[\mathbf{F}_O^T\ \mathbf{M}_O^T\ \mathbf{F}_{o1}^T\ \dots\ \mathbf{F}_{on_o}^T\right]^T$, $\mathbf{x}_a = \mathbf{q}_a^T$ and $\mathbf{f}_a = \left[\mathbf{F}_{a1}^T\ \mathbf{F}_{a2}^T\ \dots\ \mathbf{F}_{an_a}^T\right]^T$ $(a = 1, 2, \dots, N)$. The equations of motion can be simplified to some extent by expressing the configuration vector in a series of premaneuver eigenvectors corresponding to each substructure. To this end, we consider the eigenvalue problems $K_{s0}U_s = M_{s0}U_s\Lambda_s$ $(s = o, a;\ a = 1, 2, \dots, N)$, where $\Lambda_s = \text{diag}(\lambda_{si})$ $(s = o, a;\ a = 1, 2, \dots, N)$ is the matrix of eigenvalues and U_s is the matrix of eigenvectors. Because K_{s0} and M_{s0} are symmetric, the eigenvectors are orthogonal and can be normalized so as to satisfy the orthonormality relations $U_s^T M_{s0} U_s = I_s$, $U_s^T K_{s0} U_s = \Lambda_s$. Introducing the linear transformation $\mathbf{x}_s = U_s\boldsymbol{\eta}_s$ into Eq. (7), multiplying on the left by U_s^T and using the orthonormality relations, we obtain the pseudo-modal equations

$$\left(I + \overline{M}_{s1}\right)\ddot{\boldsymbol{\eta}}_s + \overline{G}_{s1}\dot{\boldsymbol{\eta}}_s + \left(\Lambda_s + \overline{K}_{s1}\right)\boldsymbol{\eta}_s = \mathbf{F}_s + \mathbf{D}_s, \quad s = o, a;\ a = 1, 2, \dots, N \tag{8}$$

where

$$\overline{M}_{s1} = U_s^T M_{s1} U_s,\ \overline{G}_{s1} = U_s^T G_{s1} U_s,\ \overline{K}_{s1} = U_s^T K_{s1} U_s, \tag{9a, b, c}$$

$$\mathbf{F}_s = U_s^T\left(B^*_{s0} + B^*_{s1}\right)\mathbf{f}_s, \quad \mathbf{D}_s = U_s^T\mathbf{d}_s \tag{9d, e}$$

CONTROL DESIGN

We are concerned here with the case in which the time-varying part of the coefficients in Eqs. (8) is of one order of magnitude smaller than the constant part. In this case, we can use a perturbation approach to compute the control gains [32]. To this end, we rewrite Eqs. (8) in the state form

$$\dot{\boldsymbol{\zeta}}_s(t) \cong [A_{s0} + A_{s1}(t)]\boldsymbol{\zeta}_s + [B_{s0} + B_{s1}(t)](\mathbf{F}_s + \mathbf{D}_s), \quad s = 0, a; \quad a = 1, 2, \dots, N \tag{10}$$

where $\boldsymbol{\zeta}_s = \left[\boldsymbol{\eta}_s^T\ \dot{\boldsymbol{\eta}}_s^T\right]^T$ $(s = o, a;\ a = 1, 2, \dots, N)$ is the pseudo-modal state vector; the coefficient matrices are obtained from Eqs. (7) by means of binomial expansions, recalling that quantities with the subscript 1 are of one order of magnitude smaller than quantities with the subscript 0.

Introducing the notation

$$\mathbf{g}_s = \mathbf{F}_s + \mathbf{D}_s, \quad s = o, a;\ a = 1, 2, \dots, N \tag{11}$$

and assuming that the disturbance vectors $\mathbf{D}_s$ are changing slowly during each half of the maneuver time interval, we can obtain the new state equations

$$\dot{\mathbf{w}}_s = \hat{A}_s\mathbf{w}_s + \hat{B}_s\dot{\mathbf{F}}_s, \quad s = o, a;\ a = 1, 2, \dots, N \tag{12}$$

in which $\mathbf{w}_s = \left[\boldsymbol{\zeta}_s^T \ \mathbf{g}_s^T\right]^T$ and

$$\hat{A}_s = \hat{A}_{s0} + \hat{A}_{s1}(t), \hat{B}_s = \begin{bmatrix} 0 \\ I \end{bmatrix} \tag{13a, b}$$

where in Eq. (13a)

$$\hat{A}_{s0} = \begin{bmatrix} A_{s0} + A_{s1}(T) & B_{s0} + B_{s1}(T) \\ 0 & 0 \end{bmatrix} \tag{14a}$$

$$\hat{A}_{s1}(t) = \begin{bmatrix} A_{s1}(t) - A_{s1}(T) & B_{s1}(t) - B_{s1}(T) \\ 0 & 0 \end{bmatrix} \tag{14b}$$

Note that the assumption that $\mathbf{D}_s$ change slowly is merely for the purpose of control design. This assumption is removed in the calculation of the response.

Next, we consider the performance indices

$$J_s = \frac{1}{2}\int_0^T \left(\mathbf{w}_s^T Q_s \mathbf{w}_s + \dot{\mathbf{F}}_s^T R_s \dot{\mathbf{F}}_s\right) dt, \quad s = o, a; \ a = 1, 2, \ldots, N \tag{15}$$

so that the optimal control law is given by

$$\dot{\mathbf{F}}_s = -R_s^{-1}\hat{B}_s^T P_s \mathbf{w}_s, \quad s = o, a; \ a = 1, 2, \ldots, N \tag{16}$$

where P_s is the solution of the matrix differential Riccati equation

$$-\dot{P}_s = P_s\hat{A}_s + \hat{A}_s^T P_s + Q_s - P_s\hat{B}_s R_s^{-1}\hat{B}_s^T P_s, \ P_s(T) = 0 \tag{17}$$

Consistent with the perturbation approach of [32], we divide P_s into a zero-order term and a small perturbing first-order term, or

$$P_s = P_{s0} + P_{s1} \tag{18}$$

Inserting Eq. (18) into the original matrix differential Riccati equation, Eq. (17), we obtain the zero-order matrix differential Riccati equation

$$-\dot{P}_{s0} = P_{s0}\hat{A}_{s0} + \hat{A}_{s0}^T P_{s0} + Q_s - P_{s0}\hat{B}_s R_s^{-1}\hat{B}_s^T P_{s0}, \ P_{s0}(T) = 0 \tag{19}$$

and the first-order matrix differential Lyapunov equation

$$-\dot{P}_{s1} = P_{s1}\hat{A}_{s0C} + \hat{A}_{s0C}^T P_{s1} + P_{S0}\hat{A}_{s1} + \hat{A}_{s1}^T P_{s0}, \ P_{s1}(T) = 0 \tag{20}$$

where

$$\hat{A}_{s0C} = A_{s0} - \hat{B}_s R_s^{-1}\hat{B}_s^T P_{s0} \tag{21}$$

denotes a closed-loop coefficient matrix.

If the final time T approaches infinity and the maneuver ends at time t_f, $t_f < T$, then during the postmaneuver period we can use the steady-state solution of the zero-order matrix differential Riccati equation, because the time-varying coefficients no longer exist after the termination of the maneuver. Then, the zero-order matrix differential Riccati equation reduces to a matrix algebraic Riccati equation. An exact solution of the first-order matrix differential Lyapunov equation can be obtained in integral form [32].

The control law given by Eq. (16) is in the form of a differential equation, so that it does not lend itself to ready implementation. Following the pattern established in

[32], Eq. (16) can be integrated to obtain the decentralized control law

$$
\begin{aligned}
\mathbf{F}_s(t) = & -\int_0^t \left[G_{si}^0 + G_{si}^1(\tau)\right] \boldsymbol{\eta}_s(\tau)\, d\tau - \int_0^t \left[G_{sp}^0 + G_{sp}^1(\tau)\right] \dot{\boldsymbol{\eta}}_s(\tau)\, d\tau \\
& - \int_0^t \left[G_{sd}^0 + G_{sd}^1(\tau)\right] \ddot{\boldsymbol{\eta}}_s(\tau)\, d\tau \\
= & - G_{si}^0 U_s^T M_{0s} \int_0^t \mathbf{x}_s(\tau)\, d\tau - G_{sp}^0 U_s^T M_{0s}\mathbf{x}_s - G_{sd}^0 U_s^T M_{0s}\dot{\mathbf{x}}_s \\
& - \int_0^t G_{si}^1(\tau) U_s^T M_{0s}\mathbf{x}_s(\tau)\, d\tau - \int_0^t G_{sp}^1(\tau) U_s^T M_{0s}\dot{\mathbf{x}}_s(\tau)\, d\tau \\
& - \int_0^t G_{sd}^1(\tau) U_s^T M_{0s}\ddot{\mathbf{x}}_s(\tau)\, d\tau, \quad s = 0, a;\ a = 1, 2, \ldots, n
\end{aligned} \tag{22}
$$

where G_{si}^j, G_{sp}^j and G_{sd}^j are obtained from the submatrices of P_s^j $(j = 0, 1)$. Equations (22) can be used to obtain the actual control vector

$$
\mathbf{f} = -\left[G_i^0 \ \ G_p^0 \ \ G_d^0\right] \begin{bmatrix} \int_0^t \mathbf{x}(\tau)\, d\tau \\ \mathbf{x}(t) \\ \dot{\mathbf{x}}(t) \end{bmatrix} - \int_0^t \left[G_i^1 \ \ G_p^1 \ \ G_d^1\right] \begin{bmatrix} \mathbf{x}(\tau) \\ \dot{\mathbf{x}}(\tau) \\ \ddot{\mathbf{x}}(\tau) \end{bmatrix} d\tau \tag{23}
$$

where $G_i^0 = \text{block} - \text{diag}\left\{\left[U_s^T (B_{s0}^* + B_{s1}^*)\right]^\dagger \ \ G_{si}^0 U_s^T M_{s0}\right\}$, etc, in which the dagger denotes the pseudo-inverse of the matrix.

Inserting Eq. (23) into Eq. (8), we obtain the closed-loop equations

$$
\begin{aligned}
\left[M_0 + M_1(t)\right]\ddot{\mathbf{x}} + \left\{G_1(t) + \left[B_0^* + B_1^*\right] G_d^0\right\}\dot{\mathbf{x}} + \left\{K_0 + K_1(t) + \left[B_0^* + B_1^*\right] G_p^0\right\}\mathbf{x} \\
+ G_i^0 \int_0^t \mathbf{x}(\tau)\, d\tau + B_0^* \int_0^t \left[G_i^1 \ \ G_p^1 \ \ G_d^1\right] \begin{bmatrix} \mathbf{x}(\tau) \\ \dot{\mathbf{x}}(\tau) \\ \ddot{\mathbf{x}}(\tau) \end{bmatrix} d\tau = \mathbf{d}(t)
\end{aligned} \tag{24}
$$

Equation (24) represents an integro-differential equation with time-varying coefficients and can be solved by a perturbation appraoch [32]. It should be pointed out that the decentralization scheme was merely for design of control gains. On the other hand, Eq. (24) is for the fully interacting structure.

NUMERICAL EXAMPLE

The substructure decentralized control theory developed above is applied to the structure shown in Fig. 2. The structure consists of a flexible central substructure and

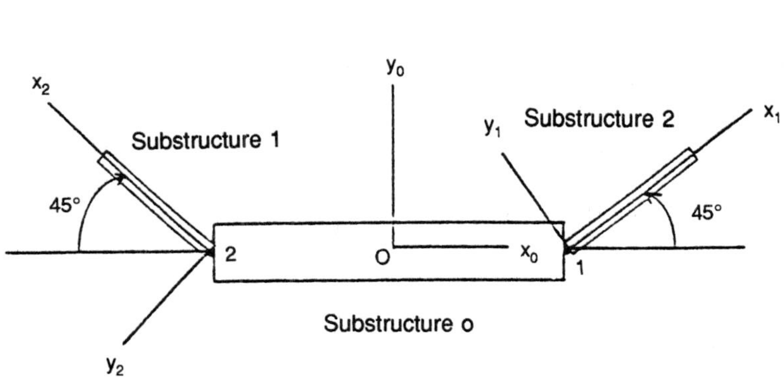

Figure 2. The Example Model

two flexible substructures, each one with one end hinged to the central substructure and the other end free. The central substructure and two maneuvering substructures of type a are uniform beams, as shown in Fig. 2, where the maneuvering beams are originally parallel to the x_o-axis of the platform. The maneuver consists of slewing each of the beams through a 45° angle in 3 s by using bang-bang control. Hence, we have

$$\ddot{\beta} = c,\ \dot{\beta} = ct,\ \beta = \frac{1}{2}ct^2 \text{ for } t \leq \frac{t_f}{2} \qquad (25a, b, c)$$

$$\ddot{\beta} = -c,\ \dot{\beta} = -c\left(t - t_f\right),\ \beta = -\frac{1}{2}c\left(t - t_f\right)^2 + \frac{1}{4}ct_f^2 \text{ for } \frac{t_f}{2} < t \leq t_f \qquad (25d, e, f)$$

where $c = 4\theta_f/t_f^2$, in which θ_f and t_f represent the final angle and time. Numerical data is as follows: $l_o = 10$ m, $l_1 = 5$ m, $l_1 = 6$ m, $m_o = 500$ kg, $m_1 = 25$ kg, $m_2 = 36$ kg, $EI_o = 10^5$ Nm^2, $EI_1 = 10^4$ Nm^2 and $EI_2 = 1.2\ 10^4$ Nm^2.

The elastic motion of the substructures of type a is represented by means of five cantilever modes. On the other hand, the elastic motion of the central substructure, is represented by quasi-comparison function [34] in the form of a combination of three free-free shape functions and two clamped-clamped shape functions.

The central structure is equipped with eight controllers to control the rigid-body translations and rotation and the elastic vibration. Each appendage has five actuators designed to suppress the elastic vibration along the beam. It should be noted that the actuators for a given substructure are activated by the state variables of the same substructure. The closed-loop equation, Eq. (24), is not feasible for computation in continuous time, so that the discrete-time approach of [32] is used for the calculation of the closed-loop response. Figures 3 through 5 show the uncontrolled and controlled rigid-body rotation, elastic displacement of the central substructure at joint 1 and the elastic displacement at the tip of beam 1. The conclusion is that the substructure decentralized control performs quite satisfactorily.

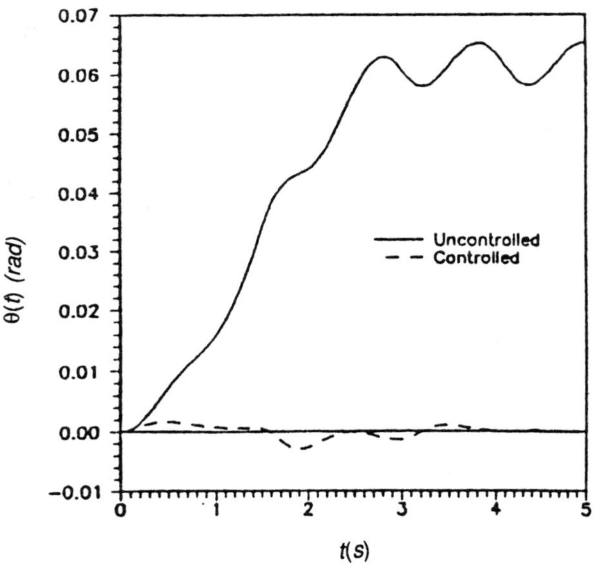

Figure 3. Time History of the Rigid-body Rotation

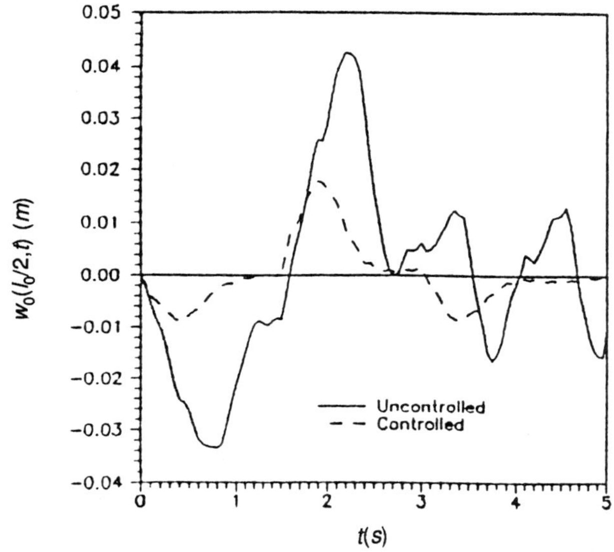

Figure 4. Time History of the Elastic Displacement of the Central Substructure at Joint 1

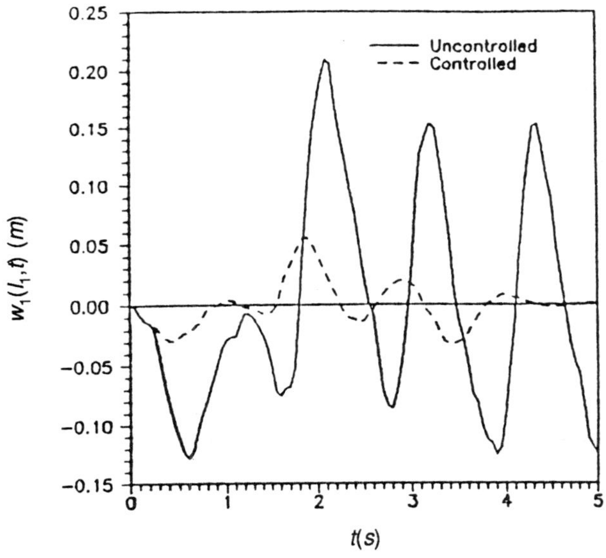

Figure 5. Time History of the Elastic Displacement at the Tip of Beam 1

SUMMARY

This paper is concerned with the problem of maneuvering and control of a space structure consisting of a central substructure and several articulated substructures. Using a perturbation approach, the problem is reduced to one characterized by time-varying coefficients and persistent disturbances. The modeling can be carried out conveniently by substructure synthesis. Consistent with this, the control design consists of substructure decentralized control, whereby the actuator forces acting on a given substructure depend only on the sensor output signals from the same substructure. An example involving a multibody system consisting of a central substructure and two maneuvering substructures illustrates the approach.

REFERENCES

1. W.J. Book and M. Majette, "Controller design for flexible, distributed parameter mechanical arms via combined state space and frequency domain techniques," *ASME Journal of Dynamic Systems, Measurement, and Control*, vol. 105, pp. 245-254, 1983.
2. S.B. Skaar, "Closed form optimal control solutions for continuous linear elastic systems," *Journal of Astronautical Sciences*, vol. 32, no. 4, pp. 447-461, 1984.
3. J.D. Turner and J.L. Junkins, "Optimal large-angle single axis rotational maneuvers of flexible spacecraft," *Journal of Guidance and Control*, vol. 3, no. 6, pp. 578-585, 1980.
4. J.A. Breakwell, "Optimal feedback slewing of flexible spacecraft," *Journal of Guidance, Control, and Dynamics*, vol. 4, no. 5, pp. 472-479, 1981.
5. J.D. Turner and H.M. Chun, "Optimal distributed control of a flexible spacecraft during a large-angle maneuver," *Journal of Guidance, Control, and Dynamics*, vol. 7, no. 3, pp. 257-264, 1984.
6. H.M. Chun, J.D. Turner and J.N. Juang, "Disturbance-accommodating tracking maneuvers of flexible spacecraft," *Journal of Astronautical Sciences*, vol. 33, no. 2, pp. 197-216, 1985.
7. J.N. Juang, L.G. Horta and H.N. Robertshaw, "A slewing control experiment for flexible structures," *Journal of Guidance, Control, and Dynamics*, vol. 9, no. 5, pp. 599-607, 1986.
8. A.L. Hale, R.J. Lisowski and W.E. Dahl, "Characteristic elastic systems of time-limited optimal maneuvers," *Journal of Guidance, Control, and Dynamics*, vol. 8, no. 1, pp. 86-93, 1985.
9. A.L. Hale and R.J. Lisowski, "Characteristic elastic systems of time-limited optimal maneuvers," *Journal of Guidance, Control, and Dynamics*, vol. 8, no. 5, pp. 628-636, 1985.
10. W.E. Vander Velde and J. He, "Design of space structure control systems using on-off thrusters," *Journal of Guidance, Control, and Dynamics*, vol. 6, no. 1, pp. 53-60, 1983.
11. H. Baruh and L.M. Silverberg, "Maneuver of distributed spacecraft," *Proceedings of the AIAA Guidance and Control Conference*, pp. 637-647, 1984.
12. L. Meirovitch, *Dynamics and Control of Structures*, Wiley-Interscience: New York, 1990.
13. S.B. Skaar, L. Tang and I. Yalda-Mooshabad, "On-off attitude control of flexible satellites," *Journal of Guidance, Control, and Dynamics*, vol. 9, no. 4, pp. 507-510, 1986.
14. R.C. Thompson, J.L. Junkins and S.R. Vadali, "Near-minimum time open-loop slewing of flexible vehicles," *Journal of Guidance, Control, and Dynamics*, vol. 12, no. 1, pp. 82-88, 1989.
15. G. Singh, P.T. Kabamba and N.H. McClamroch, "Planar, time-optimal rest-to-rest slewing maneuvers of flexible spacecraft," *Journal of Guidance, Control, and Dynamics*, vol. 12, no. 1, pp. 71-81, 1989.
16. J.Z. Ben-Asher, J.A. Burns and E.M. Cliffs, "Time optimal slewing of flexible spacecraft," *Journal of Guidance, Control, and Dynamics*, vol. 15, pp. 360-367, 1992.

17. L. Meirovitch and R.D. Quinn, "Equations of motion for maneuvering flexible spacecraft," *Journal of Guidance, Control and Dynamics*, vol. 10, no. 5, pp. 453-465, 1987.
18. L. Meirovitch and R.D. Quinn, "Maneuvering and vibration control of flexible spacecraft," *Journal of Astronautical Sciences*, vol. 35, no. 3, pp. 301-328, 1987.
19. R.D. Quinn and L. Meirovitch, "Maneuver and vibration control of SCOLE," *Journal of Guidance, Control, and Dynamics*, vol. 11, no. 6, pp. 542-553, 1988.
20. H. Soga, K. Hirako, Y. Ohkami, T. Kida and I. Yamaguchi, "Large-angle slewing of flexible spacecraft," *Joint AAS/JRS Symposium on Space Exploitation and Utilization*, Honolulu, HI, Paper AAS 85-671, December 1985.
21. L. Meirovitch and Y. Sharony, "Optimal vibration control of a flexible spacecraft during a minimum-time maneuver, *Journal of Optimization Theory and Applications*, vol. 69, pp. 31-54, 1991.
22. Y. Sharony and L. Meirovitch, "Accommodation of kinematic disturbances during a minimum-time maneuver of a flexible spacecraft," *Journal of Guidance, Control, and Dynamics*, vol. 14, pp. 268-277, 1991.
23. W.J. Book, O. Maizza-Neto and D.E. Whitney, "Feedback control of two beam, two joint systems with distributed flexibility," *ASME Journal of Dynamic Systems, Measurement and Control*, vol. 97, pp. 424-431, 1975.
24. H. Baruh and S.S.K. Tadikonda, "Issues in the dynamics and control of flexible robot manipulators," *Journal of Guidance, Control and Dynamics* vol. 14, no. 5, pp. 659-671, 1989.
25. L. Meirovitch, *Computational Methods in Structural Dynamics*, Sijthoff & Noordhoff: The Netherlands, 1980.
26. R.D. Quinn and R.Y.K. Chang, "Dynamic modeling of multi-jointed mechanical systems for space applications," *Proceedings of the Seventh VPI&SU Symposium on Dynamics and Control of Large Structures*, L. Meirovitch (Editor), pp. 507-521, 1989.
27. L. Meirovitch, "Hybrid state equations of motion for flexible bodies in terms of quasi-coordinates, *Journal of Guidance, Control, and Dynamics*, vol. 14, pp. 1008-1013, 1991.
28. L. Meirovitch and M.K. Kwak, "State equations for a spacecraft with flexible appendages in terms of quasi-coordinates, *Applied Mechanics Reviews*, vol. 42, no. 11, part 2, pp. S161-S170, 1989.
29. L. Meirovitch and M.K. Kwak, "Dynamics and control of a spacecraft with retargeting flexible antennas," *Journal of Guidance, Control, and Dynamics*, vol. 13, no. 2, pp. 241-248, 1990.
30. L. Meirovitch and M.K. Kwak, "Control of spacecraft with multi-targeted flexible antennas," *Journal of the Astronautical Sciences*, vol. 38, no. 2, pp. 187-199, 1990.
31. M.E.B. France and L. Meirovitch, "Discrete-time decentralized control of a spacecraft with retargetable flexible antennas, *Journal of Astronautical Sciences*, vol. 39, pp. 249-269, 1991.
32. L. Meirovitch and M.K. Kwak, "Control of flexible spacecraft with time-varying configuration," *Journal of Guidance, Control, and Dynamics*, vol. 15, pp. 314-324, 1992.
33. L. Meirovitch, M.K. Kwak and T. Stemple, "A substructure synthesis approach to the control of flexible multi-body systems," *ASME Winter Annual Meeting*, San Francisco, CA, Dec. 10-15, 1989.
34. L. Meirovitch and M.K. Kwak, "Rayleigh-Ritz based substructure synthesis for flexible multibody systems," *AIAA Journal*, vol. 29, pp. 1709-1719, 1991.

LOW SPEED MOTION CONTROL OF A HAMILTONIAN SYSTEM

Dan S. Necsulescu[1], Janislaw M. Skowronski[2], and Hamid Shaban-Zanjani[1]

[1]Department of Mechanical Engineering
University of Ottawa, Ottawa, Ont., K1N 6N5, Canada

[2]Department of Mechanical Engineering
University of Southern California, Los Angeles, CA 90089-1453, USA

ABSTRACT

This paper presents a solution to the low speed motion control problem of a mechanical system using self-adjusting sliding mode approach. First the Hamiltonian formulation is developed for obtaining power transfer equations and then power dissipation at low speed is analyzed. The example of a servomotor under proportional-derivative position controller is used for illustrating the accurate positioning problems in the presence of significant dry friction. Self-adjusting sliding mode control is proposed for improving positioning accuracy of this case and simulation results are presented for illustrating the efficiency of the proposed controller.

INTRODUCTION

The general control problem considered in this paper is the case of a system, described by Hamiltonian equations, having to position accurately in the presence of significant dry friction.

The development of the controller is based on modelling the power flow using the time derivative of the Hamiltonian subject to sufficient control conditions.

The paper focuses on the case of low Lagrangian velocities when the resulting sufficient conditions lead to very high magnitudes for the command signal, which often violate the saturation limits of the actuators. As Lagrangian velocities tend to zero, the controller cannot be defined any more. One solution proposed has been a variable structure (discontinuous) control law which, when the saturation limits are reached, switches to a relay controller dependent on Lagrangian speeds sign.

This paper presents new approaches of the low speed motion control problem of Hamiltonian systems. The case analyzed, the accurate positioning in a desired point when dry friction is present, has the particularity that when the speed changes sign a stick-slip process takes place. Ideally the system should be controlled in such a way as to behave as an efficiently damped virtual impedance defined in the operational space. This is achievable only if accurate positioning is obtained without any

Mechanics and Control, Edited by R.S. Guttalu
Plenum Press, New York, 1994

low speed sign change. Two solutions can be considered for this purpose: a first solution is based on computing a desired phase-plane path in reverse time starting at the desired point. A second solution is based on self-adjusting sliding mode approach. The switching surface is rotated in small increments toward higher speeds for given positions until sliding mode is first lost, at which time the switching surface rotation stops. For both solutions the desired point is reached without speed sign change.

The paper presents, in detail, the self-adjusting sliding mode approach to controlling precise motion of a Hamiltonian system at low Lagrangian speeds. Simulation results illustrate the advantages and the specificity of this approach.

HAMILTONIAN FORMULATION OF SYSTEM DYNAMICS

The standard format of Lagrange equations of motion for a mechanical nonlinear system is [9]

$$\frac{d}{dt}\left(\frac{\partial L}{\partial \dot{q}_i}\right) - \frac{\partial L}{\partial q_i} = -D_i(\mathbf{q}, \dot{\mathbf{q}}) - P_i(\mathbf{q}, \dot{\mathbf{q}}, \mathbf{U}) + F_i(\mathbf{q}, \dot{\mathbf{q}}, \mathbf{U}) \qquad i = 1, 2, \ldots, n \tag{1}$$

where the forces changing the total energy are included in the right hand side as generalized forces:

D_i - damping forces
F_i - applied forces as a result of the control, **U** , generated by the actuators,
P_i - perturbation forces (for example dry friction forces),
n - number of degree of freedom,
q, $\dot{\mathbf{q}}$, **U** - vectors of generalized positions, speeds and control.

The dynamically coupled Newtonian form is

$$\sum_{j=1}^{n} M_{ji}\ddot{q}_j + C_i(\mathbf{q}, \dot{\mathbf{q}}) + D_i(\mathbf{q}, \dot{\mathbf{q}}) + G_i(\mathbf{q}) + P_i(\mathbf{q}, \dot{\mathbf{q}}, t) = F_i(\mathbf{q}, \dot{\mathbf{q}}, \mathbf{U}) \tag{2}$$

where
C_i - the Coriolis and centrifugal forces,
G_i - the gravitational forces.

For the total energy of the system, T(**q**, $\dot{\mathbf{q}}$, t), the Legendre transformation introduces the generalized momentum

$$p_i(\mathbf{q}, \dot{\mathbf{q}}, t) = \frac{\partial T(\mathbf{q}, \dot{\mathbf{q}}, t)}{\partial \dot{q}_i} \tag{3}$$

The Hamiltonian function is defined as

$$H = \sum_{j=1}^{n} p_j \dot{q}_j - L \tag{4}$$

The Hamiltonian formulation of the system dynamic is given by

$$\dot{q}_i = \frac{\partial H(\mathbf{q},\mathbf{p},t)}{\partial p_i} \tag{5}$$

$$\dot{p}_i = -\frac{\partial H(\mathbf{q},\mathbf{p},t)}{\partial q_i} + Q_i \tag{6}$$

where

$$Q_i = -D_i - P_i + F_i \tag{7}$$

Assuming that the actuators are collocated with the degrees of freedom, we can write

$$F_i = B_i(\mathbf{q},\mathbf{p})\, U_i \tag{8}$$

where

$B_i(\mathbf{p},\mathbf{q})$ is the actuator transmission.

POWER TRANSFER EQUATIONS

The power transfer (inflow and outflow) in Hamiltonian formulation is given by:

$$\dot{H}(\mathbf{p},\mathbf{q}) = \sum_i \left[\frac{\partial H}{\partial q_i}\dot{q}_i + \frac{\partial H}{\partial p}\dot{p}_i\right] = \sum_i \left\{\frac{\partial H}{\partial q_i}\frac{\partial H}{\partial p_i} + \frac{\partial H}{\partial p_i}\left[-\frac{\partial H}{\partial q_i} - D_i(\mathbf{q},\mathbf{p})\right]\right\}$$
$$\{[-p_i(\mathbf{q},\dot{\mathbf{q}},t) + B_i(\mathbf{q},\mathbf{p})\,U_i]\} = \sum_i [-D_i(\mathbf{q},\mathbf{p}) - p_i(\mathbf{q},\dot{\mathbf{q}},t) + B_i(\mathbf{q},\mathbf{p})\,U_i]\dot{q}_i \tag{9}$$

Assuming Rayleigh dissipation functions we can write the dissipation forces as

$$D_i(\mathbf{q},\mathbf{p}) = d_i(q_i)\,\dot{q}_i \tag{10}$$

where the dissipation coefficients d_i are functions only of q_i . Also, we assume P_i resulting from dry friction only, $P_i(q_i)$. Equations (9) and (10) give:

$$\dot{H}(\boldsymbol{p},\boldsymbol{q}) = \sum_i \left[-P_i(\dot{q}_i)\,\dot{q}_i - d_i(q_i)\,\dot{q}_i^2 + B_i(\boldsymbol{q},\boldsymbol{p})\,U_i\dot{q}_i\right] \quad (11)$$

or in matrix form [9]

$$\dot{H}(\boldsymbol{p},\boldsymbol{q}) = \dot{\boldsymbol{q}}^T[-d(\boldsymbol{q})\,\dot{\boldsymbol{q}} - P(\boldsymbol{q}) + B(\boldsymbol{q},\boldsymbol{p})\,\boldsymbol{U}] \quad (12)$$

In case of Rayleigh dissipation functions assumption, the term $\dot{\mathbf{q}}^T\dot{\mathbf{d}}(\mathbf{q})\dot{\mathbf{q}}$ is positive and corresponds to a positive viscous type of damping. By a suitable choice of positive definite potential energy, the Hamiltonian can be used as Lyapanov function. The asymptotic stability condition for a passive system ($U_i=0$) is

$$\dot{H}(\boldsymbol{p},\boldsymbol{q}) < 0 \quad (13)$$

which is obviously verified by eq.(12) for $P(\mathbf{q})=0$. Active syst systems use actuator inputs $\dot{U}$ to produce the power transfer proportional to $\dot{\mathbf{q}}^T\mathbf{U}$. The inflow (outflow) of power increases (decreases) the total energy of the conservative part of the system.

POWER DISSIPATION AT LOW SPEED

In case that no (or reduced) passive damping exists in the system, $\mathbf{d}(\mathbf{q})$ is zero (small) and the power dissipation $\dot{\mathbf{q}}^T\mathbf{d}(\mathbf{q})\dot{\mathbf{q}}$ becomes impossible at low speeds. Active dissipation can be obtained using the last term in eq.(12), $\mathbf{B}\ \mathbf{U}$. In case that a speed error signal is used in creating active dissipation

$$\boldsymbol{U} = \boldsymbol{k}_p\,\dot{\boldsymbol{q}} \quad (14)$$

the power outflow from the system to the actuator is given by $\dot{\mathbf{q}}^T\mathrm{B}(\mathbf{q}\ \mathbf{p})\mathbf{k}_D\,\dot{\mathbf{q}}$ which increases the amount of power dissipation for positive definite $\mathrm{B}(\mathbf{q},\mathbf{p})\mathbf{k}_D$. However, the power dissipated $\dot{\mathbf{q}}^T\mathrm{B}(\mathbf{q},\mathbf{p})\mathbf{k}_D\,\dot{\mathbf{q}}$ decreases quadratically with decreasing $\dot{\mathbf{q}}$ and again, becomes insignificant at low speeds.

Low speed power dissipation problem has been often mentioned with regard to systems controlled for accurate positioning [4,1,6,11]. Classic control, using position-speed error feedback with constant gains, leads to a dissipation of virtual viscous friction type mentioned above. Dry friction and other discontinuous dynamics effects (for example of the type $P_i(\dot{q}_i) = p_i \operatorname{sgn} \dot{q}_i$ in eq. 11) problems, cannot be, in general, solved by this type of dissipation and limit cycles often limit the positioning accuracy [1,11]. Other solutions will be analyzed in this paper. Rewriting eq. (11) for the simplified case of $B_i=1$ and for Coulomb friction $P_i(\dot{q}_i) = p_i\mathrm{sgn}\ \dot{q}_i$, we obtain

$$\dot{H}(\boldsymbol{p},\boldsymbol{q}) = \sum_i \left[-d_i(q_i)\,\dot{q}_i^2 - p_i\,sgn(\dot{q}_i)\,\dot{q}_i + \dot{q}_i U_i\right] \quad (15)$$

Accurate positioning requires commands U_i which dissipate efficiently the energy at low speed U_i and lead to the equilibrium point (assumed here at $q_{di} = 0$).

POSITIONING UNDER PD CONTROL WHEN DRY FRICTION IS PRESENT

The problem of positioning under PD error control when dry friction is present has been extensively analyzed [3,6]. In order to compare new solutions with this classic approach, the main results will be reviewed here using a field controlled DC servomotor as an example [3].

The motor dynamics, when Coulomb friction is present, can be represented by an equation similar to eq. (2):

$$\dot{J}\theta + d\dot{\theta} + nT_c \, sgn(\dot{\theta}) = nT_M \tag{16}$$

where

$J = J_L + n^2 J_M$

$d = d_L + n^2 d_M$

$J_L(J_M)$ - moment of inertia of the load (motor),

$d_L(d_M)$ - viscous friction coefficient for the load (motor),

n - gear ratio

T_c - Coulomb friction torque,

T_M - motor torque

θ - angular position of the load shaft.

Assuming a PD position error controller, motor torque is given by

$$T_M = KK_{TG}\{(\theta_i - \theta) - nk\dot{\theta}/K\} \tag{17}$$

where

KK_TG - the gains of the position transducer, tachometer and power amplifier,

$k = K_T k_0$

k_0 - operational amplifier gain,

θ_i - input or desired position.

The canonical form of eq. (16), after replacing T_M by eq.(17), is given by

$$\ddot{\theta} + 2\xi\omega_0\dot{\theta} + \omega_0^2\theta + \omega_0^2 d\ sgn(\dot{\theta}) = \omega_0^2\theta_i \quad (18)$$

where

$\xi = \xi_v + \xi_T$
$\xi_v = F/2(nKK_TGJ)^{1/2}$
$\xi_{T2} = (n^3k^2K_TG/4KJ)^{1/2}$
$\omega_0^2 = nKK_TG/J$

The state-space representation of eq.(18) is given by

$$\begin{cases} \dot{x}_1 = x_2 \\ \dot{x}_2 = -\omega_0^2 x_1 - 2\xi\omega_0 x_2 - d\omega_0^2\ sgn(x_2) \end{cases} \quad (19)$$

where

$\theta_i = 0,\ x_1 = \theta$ and $x_2 = \dot{\theta}$

The state plane (x_1,x_2) representation of eq.(19) is given in Fig. 1, where $d=T_c/KK_TG$, corresponds to the static error [in radians] required for the motor to give a static torque equal to T_c[Nm]. Comparing Fig. 1 and 2 it can be seen that the effect of dry friction is a static error due to the torque $T_M < T_c$ in this region which results in "sticking" anywhere between -d and d. In the absence of Coulomb friction ($T_c=0$), accurate positioning is possible. Higher viscous damping coefficient will only increase the slope of the trajectory before reaching -d to d region, but will not solve the "sticking" problem. A type of solutions suggested by Fig. 1 is to achieve accurate positioning without any change of speed sign, so that Coulomb friction, even if present, will not affect the motion.

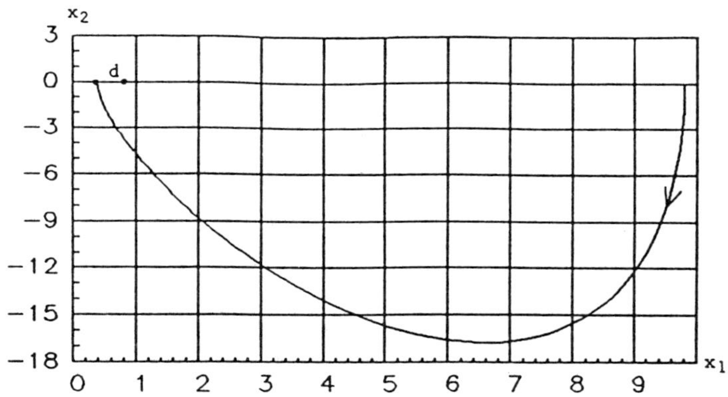

Fig. 1. State plane (x_1,x_2) representation of the servomotor dynamics under PD control in case that Coulomb friction is present ($Tc\neq0$, $d\neq0$) for $\omega_0=4$ rad./sec., $\xi=0.7$, $s=0.8$, $x_1(0)=9.8$ rad., sampling time $T=0.01$ sec.

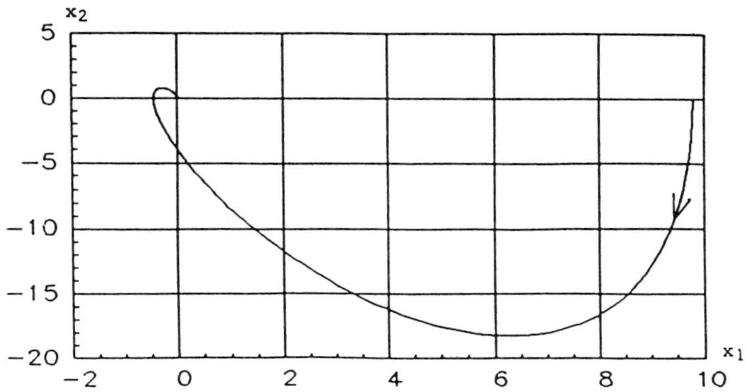

Fig. 2. Same as Fig. 1, except for $T_c=0$, i.e., no Coulomb friction ($d=0$).

ACCURATE POSITIONING WITHOUT CHANGING SPEED SIGN. SELF ADJUSTING SLIDING MODE CONTROL

A first solution for the dry friction problem can be the application of the switching curve strategy resulting from optimal (minimum time) control [2]. This switching curve strategy is based on precomputing in reverse time the state plane strategies which have a boundary at the initial position (i.e., final position in forward time) at the desired point. A minimum time trajectory that shows no speed change in the $\pm d$ range, will imply that the Coulomb friction does not affect final positioning. This approach is, consequently, an interesting solution to the dry friction problem, but requires exact knowledge of the parameters of the system, in particular of inertial parameters.

A more robust solution is proposed here, self-adjusting sliding mode approach, which requires longer time, but does not need exact knowledge of all parameters of the system. One parameter, for example inertial parameter, can actually be assumed unknown [5].

Let us assume a second order system

$$\dot{x}_i = x_2 \tag{20}$$

$$\dot{x}_2 = f(x_1, x_2) - bu \tag{21}$$

A time varying switching line is defined for the state plane:

$$\sigma(t) = c(t)\,x_1(t) + x_2(t) \tag{22}$$

This controller will increase continuously c(t) of the switching line as long as sliding mode is preserved and will stop the increase otherwise. Two sliding mode control laws will be analyzed:

$$u = (1/b)\,[f(x_1, x_2) + K\,sgn(\sigma)] \tag{23}$$

and

$$u = (1/b)\,[f(x_1, x_2) + cx_2 + k|x_1|\,sgn(\sigma)] \tag{24}$$

The nonlinear decoupling is achieved by the $(1/b)f(x_1,x_2)$ term, while the last term leads to a modified system dynamics given by

$$\dot{x}_2 = -K\,sgn(\sigma) \tag{25}$$

for the control law (23) and

$$\dot{x}_2 = -k|x_1|\,sgn(\sigma) - cx_2 \tag{26}$$

for the control law (24). The resulting trajectories show a behaviour which approximates an overdamped system in the last portion of the motion while the accuracy of positioning is achieved. Chattering is also present, due to the discontinuous control laws (21) or (22), and boundary layer solution can be applied for filtering the high frequency components of chattering [10].

Simulations were performed for the servomotor described by eq. (16) in order to illustrate the performance of these controllers in the presence of Coulomb friction. In this case, however, the compensation $(1/b)f(x_1,x_2)$ is normally limited the non-discontinuous terms of the servomotor dynamics (eq.(19)) i.e.

$$f(x_1, x_2) = -\omega_0^2 x_1 - 2\xi\omega_0 x_2 \tag{27}$$

and consequently, the modified dynamics for controller (23) is given by

$$\dot{x}_2 = -K\,sgn(\sigma) - d\omega_0^2\,sgn(x_2) \tag{28}$$

Simulation results in Fig. 3 show the trajectory which reaches the desired point (origin) without changing the speed x_2 sign, i.e., without being influenced by the presence of Coulomb friction or the "sticking region" $\pm$d. The dotted line represents the motion of the self-adjusting switching line, eq.(22). It can be seen that the dynamics is similar to the case of a nonlinear time constant switching line, but the determination of such a nonlinear switching line requires the exact knowledge of the system parameters, (as for the case of switching curve strategy) while self-adjusting sliding mode can be applied even in the case of an unknown inertial parameter [5] and, thus, is more robust.

The position, speed and acceleration errors versus time, $x_1(t)$, $x_2(t)$ and $\dot{x}_2(t)$, respectively, are shown in Fig. 4. The smooth approaching stage during the first apr. 0.7 sec is followed by the chattering during the sliding mode stage, when rapid acceleration error,[$\dot{x}_2(t)$] changes occur, as predictable from eq. (28) as a result of the K sgn(σ) term. Fig. 5 shows the details A-A and B-B of Fig.4 for the final stage of the motion.

Similar good performance is shown in Fig. 6 for the controller given by eq. (24).

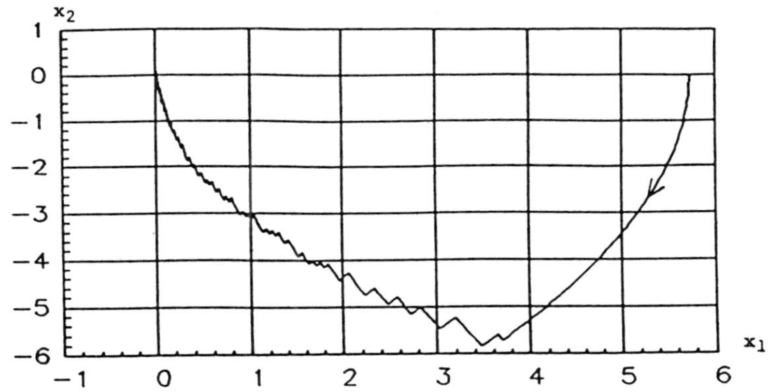

Fig.3. State plane representation of the servomotor dynamics under sliding mode controller eq. (23), in case that Coulomb friction is present for K = 10, $x_1(0) = 5.74$ rad, d = 2 rad.

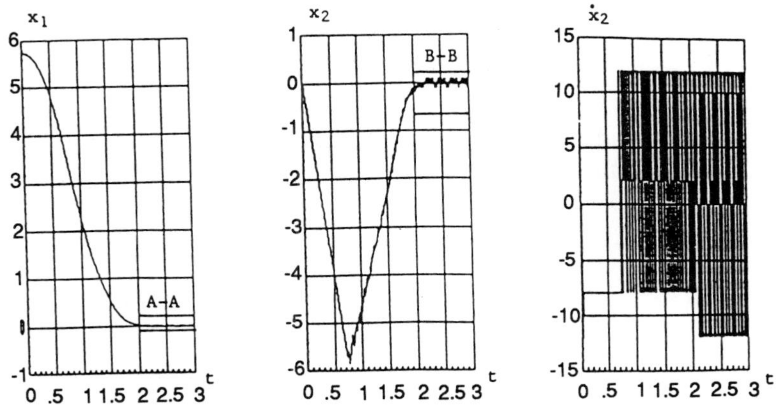

Fig. 4. $x_1(t)$, $x_2(t)$ and $\dot{x}_2(t)$ for the simulation shown in Fig. 3.

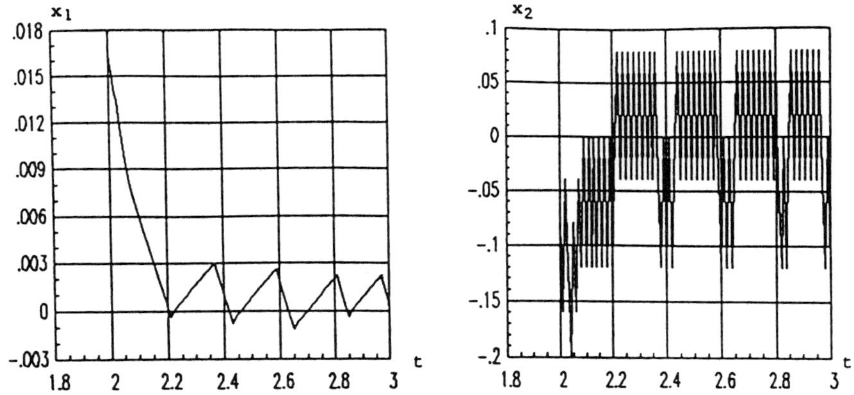

Fig. 5. The details A-A and B-B for $x_1(t)$ and $x_2(t)$ in Fig. 4.

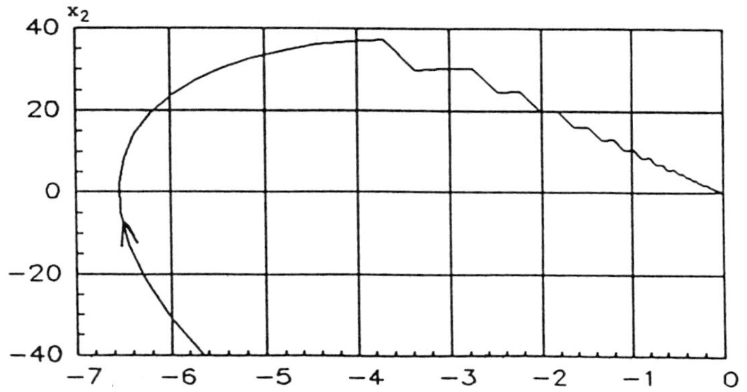

Fig. 6. State plane representation of the servomotor dynamics under sliding mode controller eq. (24), in case that Coulomb friction is present, for $k=1$, $x_1(0)=$ -5.74 rad., $x_2(0)=$ -40 rad/sec.

FINAL REMARKS

The controllers given by eq. (23) and (24) can be analyzed from the point of view of their power dissipation performance using eq.(11) for $B_i=1$. The power dissipated by the discontinuous part of the controllers is $K\,\mathrm{sgn}(\sigma)\dot{q}$ and $K|q|\,\mathrm{sgn}(\sigma)\dot{q}$ respectively for $q = \theta_i-\theta$. It can be seen that the power dissipated decreases linearly with the decrease of Lagrangian speed $\dot{q}$, while in the PD error controller case, the dissipative part, given by eq.(14), decreases quadratically with decreasing speed $\dot{q}$. We can conclude that self-adjusting sliding mode controllers can be a very interesting candidate for solving the problem of accurate positioning in the presence of dry friction at low speed.

REFERENCES

1. B. Armstrong-Hélouvry, "Control of Machines with Friction", Kluver Academic Publishers, (1991).
2. M. Athans, P.L. Falb, "Optimal Control", McGraw Hill, (1966).
3. D.P. Atherton, "Nonlinear Control Engineering", Van Nostrand,(1975).
4. H. Flashner, J.M. Skowronski, Model Tracking Control of Hamiltonian Systems, in: "J. Dyn. Syst., Measurement and Control", Trans. ASME, pp. 656-660, Dec. 1989.
5. U. Itkis, "Control Systems of Variable Structures", J. Wiley, (1976).
6. N.E. Leonard, P.S. Krishnaprasad, Comparative Study of Compensating Control Strategies for Servomechanisms, Technical Research Report, Systems, Research Center, University of Maryland, TR 91-88.
7. L. Meirovitch, "Methods of Analytical Dynamics", McGraw Hill,(1970).
8. H. Nijmeijer, A. Van der Schaft,"Nonlinear Dynamical Control Systems", Springer Verlag,(1990).
9. J.M. Skowronski,"Control of Nonlinear Mechanical Systems", Plenum Press,(1991).
10. J.J. Slotine, W. Li, "Applied Nonlinear Control", Prentice Hall,(1991).
11. G.D. Warshaw, A. Jnifene, D. Necsulescu, Nonlinear Friction Effects on Precise Motion Control of a Manipulator, Proc. 13th Can. Congress of Applied Mechanics, Winnipeg, pp. 660-661, June 2-6, (1991).

BOUNDARY STABILIZATION OF NONLINEAR FLEXIBLE SYSTEMS

Nasir U. Ahmed[1] and Janislaw M. Skowronski[2]

[1]Department of Electrical Engineering, and Mathematics
University of Ottawa
Ottawa ON K1N 6N5, Canada
[2]Department of Mechanical Engineering
University of Southern California
Los Angeles, CA 90089-1453

ABSTRACT

We present some results on feed back stabilization of nonlinear beams and space craft with flexible appendages by use of controls applied to the boundary of flexible members. Semigroup theory and basic concepts of Lyapunov functions are used to prove stability, asymptotic stability, and exponential stability.

INTRODUCTION

In recent years, deployment of large mechanical structures in space vehicles, such as, space shuttle, satellites and space stations has generated great interest in the study of their stability and control [1-3,5,6,9]. These systems are governed by partial differential equations or a combination of partial and ordinary differential equations. In many of these papers distributed controls have been used [1,2,5,8]. In [6] the system considered is linear with boundary feedback controls proving exponential stability. In [5] optimal control of space structures has been considered. In [1,2], modeling and control of large space structures such as the proposed U.S space station has been considered where distributed controls have been used. In [9], stabilization by use of boundary control of a flexible structure attached to space shuttle has been studied. Our interest here is to study the stability of nonlinear flexible structures by use of boundary feedback controls. We present some recent results on feedback stabilization of nonlinear beams, and space craft with flexible appendages. Semigroup theory along with the basic concepts of Lyapunov functions are used to prove asymptotic stability of nonlinear beams with boundary feedback controls. We prove a similar result for a satellite with appendages, in particular, flexible masts carrying antennas. We also show some intimate relation between stability and observability of cantilever beams. We have presented results on stability, asymptotic stability and in some cases exponential stability.

Mechanics and Control, Edited by R.S. Guttalu
Plenum Press, New York, 1994

FEEDBACK STABILIZATION OF A NONLINEAR BEAM

We consider the question of stabilizability of certain nonlinear beams with feedback controls applied to the boundary. Particularly, we consider control of the bending moment and shear forces of the free end of the beam by use of linear feedback of the displacement and bending rates of the free end (tip) of the beam.

The complete dynamics of the system is given by the following set of equations and boundary conditions:

$$\begin{aligned}
&\rho\left(\frac{\partial^2 y}{\partial t^2}\right) + \frac{\partial^2}{\partial x^2}\left(EI\frac{\partial^2 y}{\partial x^2}\right) + f(y_t) + g(y) = 0, t > 0, x \in (0, \ell) \\
&y(t,0) = 0, \qquad y_x(t,0) = 0 \\
&EID^2 y(t,\ell) = u_1, \quad D\big(EI(D^2 y(t,\ell))\big) = u_2,
\end{aligned} \tag{S1}$$

where $D\phi \equiv \frac{\partial\phi}{\partial x}; D^2\phi \equiv \frac{\partial^2\phi}{\partial x^2}$. The functions u_1 and u_2 are the boundary controls to be provided as feedback controls for controlling the bending moments and the shear forces at the tip of the beam. Of course this equation must be solved with respect to certain initial data given by,

$$y(0,x) = y_0, y_t(0,x) = y_1.$$

Note that both the mass density ρ and the flexural rigidity EI may be functions of x but not t.

In the absence of the damping term f and with controls identically zero the system is conservative. We will show that for appropriate choice of controls, the system is stable in the Lyapunov sense. More precisely we have the following result.

Theorem 2.1. Consider the system $(S1)$ with the feedback controls as given by

$$\begin{aligned}
u_1(t) &= -\alpha(Dy)_t(t,\ell) \\
u_2(t) &= \beta y_t(t,\ell)
\end{aligned} \tag{1}$$

for suitable constants $\alpha, \beta \in R$. Suppose $Inf\{\rho(x), EI(x), x \in (0,\ell)\} > \sigma > 0$.

Then

(1) the system $(S1)$ is stable in the Lyapunov sense if the following conditions hold:

(i) $G(\xi) \equiv \int_0^\xi g(\theta)\, d\theta \geq 0$, $f(\xi)\,\xi \geq 0$ for $\xi \in R$, and (ii) $\alpha, \beta \geq 0$.

(2) if $f(\xi)\xi > 0$ for $\xi \neq 0$, and $g(\xi)\xi \geq 0$ for $\xi \in R$, then the system is asymptotically stable with respect to the zero state for any α, $\beta \geq 0$

(3) In case f and g merely satisfy the condition (i) of part (1), the system is still asymptotically stable with respect to the origin of the energy space provided $\alpha + \beta > 0$, and it is exponentially stable in every neighborhood of the origin of energy space contained in the region where viscous damping dominates over stiffness.

Proof. We present here a brief outline of the proof.

(1) We introduce the energy function:

$$V(t) \equiv V(y, y_t) \equiv (1/2)\int_0^\ell \{\rho(y_t)^2 + EI\left(\frac{\partial^2 y}{\partial x^2}\right)^2 + G(y)\}\, dx. \tag{2}$$

and show that it is a Lyapunov function. Differentiating V with respect to time and using the differential equation (S1), and the boundary controls as given by (1), it is easy to verify that

$$(d/dt)V = \int_0^\ell -f(y_t)y_t \; dx - \alpha \; (Dy_t(t,\ell))^2 - \beta \; (y_t(t,\ell))^2. \tag{3}$$

This shows that $(d/dt)V \;\le 0$, and consequently $V(t) \;\le V(0)$. Thus the trajectories remain bounded for bounded initial disturbances. Hence the system $(S1)$ is stable in the Lyapunov sense proving (1).

(2) For (2), we prove that $(d/dt)V \;< 0$ by showing that if $(d/dt)V \equiv 0$ on any interval of time $(s, \tau]$ of positive length then V also vanishes on the same interval of time. For details see Ahmed [3].

(3) Case (3) is more subtle. In this case apparently there may be at times no damping and no restraining force. The proof of this is based on semigroup (the solution operator) theory.

For this it is convenient to write the beam equation as an ordinary differential equation on the energy space,

$$X \equiv H_0^2 \times L_2(0,\ell),$$

where

$$H_0^2 \equiv \{\varphi \in L_2 : \frac{\partial \varphi}{\partial x} \equiv D\varphi \in L_2, \; D^2\varphi \in L_2 \text{ and } \varphi(0) = 0, D\varphi(0) = 0\}.$$

Let A denote the formal beam operator given by

$$A\psi \equiv (1/\rho)\frac{\partial^2}{\partial x^2}\left(EI\frac{\partial^2}{\partial x^2}\psi\right) \equiv (1/\rho)D^2\left(EID^2\psi\right).$$

Define the state as

$$w \equiv \begin{pmatrix} w_1 \\ w_2 \end{pmatrix} \equiv \begin{pmatrix} y \\ y_t \end{pmatrix}.$$

Then the system $(S1)$ can be written as

$$(d/dt)w = \mathcal{A}w + \mathcal{F}(w) \tag{4}$$

where the operator $\mathcal{A}$ is given by the restriction of the formal differential operator

$$\mathcal{L} \equiv \begin{pmatrix} 0 & 1 \\ -A & 0 \end{pmatrix},$$

to the domain $D(\mathcal{A})$ given by

$$D(\mathcal{A}) \equiv \{w \in X : \mathcal{L}w \in X \text{and } EID^2w_1(\ell) + \alpha Dw_2(\ell) = 0, \\ D\left(EID^2w_1(\ell)\right) - \beta w_2(\ell) = 0\}.$$

The nonlinear operator $\mathcal{F}$ is given by

$$\mathcal{F}(w) \equiv -(1/\rho)\begin{pmatrix} 0 \\ g(w_1) + f(w_2) \end{pmatrix}.$$

Since the modulus of rigidity EI and the mass density ρ are strictly positive, the space X is a Hilbert space with respect to the scalar product,

$$\langle \varphi, \psi \rangle \equiv (EID^2\varphi_1, D^2\psi_1) + (\rho\varphi_2, \psi_2),$$

where the first term represents elastic potential energy and the second term is a measure of kinetic energy. For the proof of exponential stability, essentially one shows that the semigroup is uniformly bounded and that the corresponding frequency domain transfer function $H(\omega) \equiv (i\omega - \mathcal{A})^{-1}$ is also uniformly bounded for all $\omega \in R$. Hence it follow from a result of Chen [6,Theorem 4, pp. 82] that there exists a $\mu > 0$ such that the semigroup corresponding to $\mathcal{A}$ is a contraction and exponentially stable, that is, $\|T_{\mathcal{A}}(t)\| \leq e^{-\mu t}$ for all $t \geq 0$. Thus the linear part is exponentially stable. Since the nonlinear terms are dissipative, this implies that the system $(S1)$ is asymptotically stable. For $\gamma > 0$ define the set

$$\Omega_\gamma \equiv \{w \in X : (f(w_2), w_2) - \gamma(G(w_1), 1) \geq 0\}.$$

This is the region where relatively viscous damping dominates stiffness. Clearly this contains the origin. Suppose for some $\gamma > 0$ the set is nontrivial. Let $\Sigma \subset \Omega_\gamma$ be a neighborhood of the origin and let $w^o \in \Sigma$ and $w(t) = w(t, w^o)$ be the solution of (4) starting from w^o. Then taking the scalar product (in X) of (4) with w and defining

$$E(t) \equiv \|w(t)\|^2 + (G(w_1(t)), 1)$$

one finds that

$$(d/dt)E(t) \leq -min\{\mu, \gamma\}E(t) \quad \text{for } t \geq 0.$$

Hence the system is exponentially stable in Σ. This completes the proof. ■

Stability vs Observability. It is interesting to note the connection between stability and observability. According to the preceding result the operator $(\mu I + \mathcal{A})$ is maximal dissipative in the Hilbert space X. This implies that the linear system

$$(d/dt)w = \mathcal{A}w, \quad \text{with the output } z \equiv Cw \equiv \begin{pmatrix} \sqrt{\beta}w_2(t,\ell) \\ \sqrt{\alpha}Dw_2(t,\ell) \end{pmatrix}$$

is completely observable in the sense that the output $z \equiv 0$ if and only if $w \equiv 0$. More specifically one can verify that

$$\mu\|w(t)\|_X^2 \leq \|z(t)\|_{R^2}^2 \quad \text{for } t \geq 0.$$

This means that the operator C is injective and C^*C is coercive. The physical implication of this is that the tip (free end) of the beam can not stay in the equilibrium state for any positive length of time without having the entire body of the beam in the rest state for the same period of time. In other words the free end of any cantilever beam is the signature of the state of the entire beam.

An important question is, is it possible to determine the state of the beam from the observation of z. If the range of the operator C^*C is all of X, it follows from it's strict positivity that C^*C is invertible and hence one can determine $w(t) = (C^*C)^{-1}C^*z(t)$. In any case the operator C^*C is certainly invertible on it's range $R(C^*C)$. One can also introduce an observability operator

$$\Gamma_\tau y \equiv \int_0^\tau T^*(t)C^*CT(t)\, y\, dt, \quad y \in X,$$

where $\{T(t), t \geq 0\}$ is the semigroup corresponding to the generator $\mathcal{A}$. Thus, if for some $\tau < \infty$, the observability operator is invertible, then one can, in principle, determine the state of the beam, say at time $t = 0$, from the record of $z(t), t \in [0, \tau]$ from the expression

$$w_0 = \Gamma_\tau^{-1} \int_0^\tau T^*(t)C^*z(t)\, dt.$$

As in the finite dimensional case Ahmed [4, pp.229], one can determine the operator Γ_τ by solving the following operator equation

$$\begin{aligned} &d/dt(M(t), \xi, \eta) + (M(t)A\xi, \eta) + (M(t)\xi, A\eta) + (C^*C\xi, \eta) = 0, \\ &(M(\tau)\xi, \eta) = 0, \quad \text{for all } \xi, \eta \in D(\mathcal{A}); \end{aligned}$$

giving $\Gamma_\tau = M(0)$.

If the nonlinear terms f and g in the system $(S1)$ are replaced by a more general expression say $h(y, y_t)$ and the operator $\mathcal{F}$ is accordingly modified we obtain a more general nonlinear beam. In this case we have the following result.

Theorem 2.2. Consider the system (4) with the nonlinear operator $\mathcal{F}$ as described above. Suppose the function $h \in C^1$ vanishing at zero with it's derivative being uniformly bounded on R^2 and also vanishing at zero. Then for any $0 < \nu < \mu$ the system is exponentially stable with exponent $-\nu$ in the neighborhood of the origin of X.

Proof. For detailed proof see Ahmed [3]. ■

Remark. If the beam is clamped at both ends and a control $u(t) = -\delta y_t(t, \ell/2)$ is applied at the mid point of the span, the function V given by (2) is also a Lyapunov function for this system. In this case

$$(d/dt)V(t) \leq -\delta\, (y_t(t, \ell/2))^2 - \int_0^\ell f(y_t)y_t\, dx,$$

and the system is stable.

STABILIZATION OF BEAMS SUBJECT TO LARGE VIBRATIONS

To allow for moderately large vibrations, the following model has been considered to be more appropriate [11,pp.99].

$$\begin{aligned} &\rho\left(\frac{\partial^2 y}{\partial t^2}\right) + \frac{\partial^2}{\partial x^2}\left(EI\frac{\partial^2 y}{\partial x^2}\right) - N(y)\left(\frac{\partial^2 y}{\partial x^2}\right) + f(y_t) = 0, t > 0, x \in (0, \ell) \\ &y(t, 0) = 0, \qquad y_x(t, 0) = 0 \\ &EID^2y(t, \ell) = u_1, \quad D\big(EI(D^2y(t, \ell))\big) - Ny_x(t, \ell) = u_2 \end{aligned} \qquad (S2)$$

where, in general, the nonlinear operator N is given by

$$N \equiv a + b\int_0^\ell \left(\frac{\partial y}{\partial x}\right)^2 dx.$$

The nonlinear term represents membrane force. Here a and b are constants. Again we wish to find a feedback control law that stabilizes the system. The following result is similar to that of Theorem 1.

Theorem 3.1. Consider the system ($S2$) and suppose the constants $a, b \geq 0$ and

$$f(\xi)\xi \geq 0, \text{ for } \xi \in R,$$

and let the feedback controls be given by

$$u_1(t) = -\delta D y_t(t, \ell)$$
$$u_2(t) = \gamma y_t(t, \ell)$$

where $\gamma, \delta \geq 0$.

Then

(1) the system is stable in the Lyapunov sense.

(2) if $f(\xi)\xi > 0$ for $\xi \neq 0$, then the system is asymptotically stable.

Proof. The proof follows from similar arguments as in Theorem 1, with

$$V \equiv (1/2) \int_0^\ell \left\{ \rho(\frac{\partial y}{\partial t})^2 + EI(\frac{\partial^2 y}{\partial x^2})^2 + (1/2)N(\frac{\partial y}{\partial x})^2 \right\} dx$$

as the Lyapunov function. For detailed proof see Ahmed [10] ■

Remark. Note that the constants appearing in the nonlinear term N are not required to be positive for small perturbations. We prove this in the next theorem.

In the absence of damping ($f \equiv 0$) it is difficult to prove asymptotic stability of the nonlinear beam equation ($S2$) by simple Lyapunov argument. Below , we give such a result based on semigroup arguments.

Theorem 3.2. Consider the system ($S2$) with no external damping ($f \equiv 0$), and the constants $a = 0$, and $b \in R$ not necessarily positive. Suppose the feedback controls are as given in theorem 3 with $\delta,\ \gamma \geq 0,\ \delta + \gamma \neq\ 0$.

Then the system is exponentially stable in the neighborhood of the origin of the energy space $X \equiv H_0^2 \times L_2$. That is, the system is locally exponentially stable.

Proof. Recall the operator $\mathcal{A}$ as given in equation (4). By virtue of Chen's result, we already know that $\mathcal{A}$ is strongly dissipative and the corresponding semigroup $T_{\mathcal{A}}(t), t \geq 0$, is exponentially stable, that is, there exist $1 \geq M > 0$ and $\mu > 0$ such that

$$\|T_{\mathcal{A}}(t)\|_{\mathcal{L}(X)} \leq Me^{-\mu t} \text{ for all } t \geq 0.$$

Define

$$B(w) \equiv \begin{pmatrix} 0 \\ N(w_1)D^2 w_1 \end{pmatrix}.$$

It is easy to verify that B is a locally bounded Lipschitz continuous (nonlinear) map from X to X and

$$Lim_{\|z\|_X \to 0} \|B'(z)\|_{\mathcal{L}(X)} = 0.$$

Using the variation of constants formula, the solution of equation ($S2$) is then given by the solution of the integral equation

$$w(t) = T_{\mathcal{A}}(t)w^0 + \int_0^t T_{\mathcal{A}}(t-\theta)\ B(w(\theta))\ d\theta.$$

Following Lasiecka's procedure [7], choosing $\nu \in (0, \mu)$ and defining $v = we^{\nu t}$ with $w^0 \equiv w(0) = v(0)$, one obtains an integral equation for v,

$$v(t) = e^{\nu t}\; T_A(t)w^0 + \int_0^t e^{\nu t}\; T_A(t-\theta)\; B(e^{-\nu\theta}\; v(\theta))\; d\theta. \tag{5}$$

Define the operator G by

$$G(v)(t) \equiv e^{\nu t}\; T_A(t)w^0 + \int_0^t e^{\nu t}\; T_A(t-\theta)\; B(e^{-\nu\theta}\; v(\theta))\; d\theta, t \geq 0.$$

Then the question of existence of solution of the integral equation is equivalent to the question of existence of a fixed point for the operator G, that is an element $v \in C([0, \infty), X)$ such that $v = Gv$. For $r > 0$, let $D_r \equiv \{z \in X : \|z\|_X \leq r\}$ denote the ball of radius r centered at the origin in X. Let

$$C_r \equiv \{v \in C(([0, \infty), X) : v(t) \in D_r \;\text{ for all } t \geq 0\}.$$

Since $Lim_{\|z\|\to 0}\|B'(z)\|_{\mathcal{L}(X)} = 0$, it is easy to verify that , for sufficiently small $r > 0$, the operator G is a contraction in C_r. Hence by Banach fixed point theorem, there exists a $v^* \in C_r$ such that $v^* = Gv^*$. In other words, for each initial perturbation $w^0 \in D_r$, the equation (9) has a solution w such that $\|w(t, w^0)\|_X \leq re^{-\nu t}$ for all $t \geq 0$, proving exponential stability. This completes the proof. ■

Remark. If the parameter a appearing in the nonlinear expression for the membrane force N is sufficiently small so that $|a|M < (\mu - \nu)$, then there exists an $r > 0$ such that the system is exponentially stable in the ball $D_r \equiv \{w \in X : \|w\| < r\}$.

BOUNDARY STABILIZATION OF A FLEXIBLE SPACE CRAFT

In this section we consider the problem of boundary stabilization of large flexible structures in space. In particular, we consider a satellite with a flexible mast whose far end is free though it may be attached to a pay load such as an antenna. It is this free end where we wish to apply controls, possibly feedback controls, in addition to a control torque applied to the body of the space craft. We shall use the model proposed in Biswas and Ahmed [5]:

$$\begin{aligned} &I\dot{\omega} + \int_0^\ell (x+r)\rho\; y_{tt}\; dx = u_1(t) \\ &\rho\; y_{tt} + \frac{\partial^2}{\partial x^2}\left(EI\frac{\partial^2 y}{\partial x^2}\right) + (x+r)\;\rho\;\dot{\omega} + f(y_t) = 0, \;\text{ for } x\; \in (0, \ell). \end{aligned} \tag{S3}$$

Here ω denotes the attitude rate of the rigid body and y denotes the deflection of the mast from its rest position, r is the radius of the satellite and ℓ is the length of the mast. These equations are subject to the boundary conditions as given below:

$$\begin{aligned} &y(t, 0) = 0,\; y_x(t, 0) \equiv Dy(t, 0) = 0 \\ &EI(D^2y)(t, \ell) = u_2,\; D(EID^2y)(t, \ell) = u_3, \end{aligned} \tag{6}$$

where u_2 and u_3 are the boundary controls controlling the bending moment and the shear forces respectively. The problem is to find a feedback control law $\{u_1, u_2, u_3\}$ such that the system is stable with respect to the rest state $\{\omega, y, y_t\} = \{0, 0, 0\}$.

The system $(S3)$ is a simplified model of those given in [1,2]. However, it contains the essentially important couplings between the rigid body and the flexible boom. In this problem we may take $Y \equiv R \times X$, X as defined in the previous problems, as the state space. Here we are interested in the stabilization problem as stated above.

Theorem 4.1. Consider the system $(S3)$ and suppose that $I \geq I_0$ where

$$I_0 \equiv \int_0^\ell \rho\,(x+r)^2\,dx$$

is the mass moment of inertia of the beam. That is, the rigid body is at least as massive as the mast. Suppose that $f(\xi)\;\xi \geq 0$ for $\xi \in R$ and let the controls be given by

$$u_1(t) = -d\,\omega,\quad u_2(t) = -\alpha\left((Dy)_t(t,\ell)\right),\quad u_3 = \beta(y_t(t,\ell)), \tag{7}$$

for $d \geq 0$, $\alpha, \beta \geq 0$. Then

(1) the system $(S3)$ is stable in the Lyapunov sense.

(2) if $f(\xi)\;\xi > 0$ for $\xi \neq 0$, and $d > 0$, then the system is asymptotically stable with respect to the zero state for any $\alpha, \beta \geq 0$.

Proof. For this system we can prove that the appropriate Lyapunov function is given by

$$V \equiv (1/2)(I - I_0)\,\omega^2 + (1/2)\int_0^\ell \left\{\rho\,(y_t + (x+r)\,\omega)^2 + EI(D^2y)^2\right\}\,dx. \tag{8}$$

For details see Ahmed [3]. ■

The most important question now is; what if $f(\xi)\;\xi \geq 0$, or more importantly, what if there is no viscous damping at all as in the case of geosynchronous satellite. Are the boundary controls good enough for exponential stability or at least asymptotic stability. We present below a result that shows that the system without viscous damping is exponentially stable if certain conditions are satisfied.

Theorem 4.2. Consider the system $(S3)$ with the feedback control laws as given in equation (7). Suppose d, α, $\beta > 0$ and $I > I_0$. Then the system $(S3)$ is exponentially stable with respect to zero state even in the absence of viscous damping, that is, $f \equiv 0$.

Proof. For detailed proof see Ahmed [3]. Here we give only an outline. The proof is based on semigroup theory after the system has been written as an abstract differential equation in the Hilbert space Y:

$$\mathcal{E}(d/dt)z = \mathcal{A}z + \mathcal{F}(z). \tag{9}$$

Here we have used the fact that, under the given assumptions, the operator $\mathcal{A}$ is strictly m-dissipative, and $\mathcal{E}$ is a strictly positive self adjoint operator. This later fact allows us to define a norm equivalent to the energy norm and prove that in this new norm the system is exponentially stable and, therefore, by equivalence, it is exponentially stable in the energy norm. ■

The interesting difference between the above results is that, according to Theorem 4.2, for exponential stability, one needs $d > 0$ and $I > I_0$; where as, for mere asymptotic stability, according to Theorem 4.1, this is not essential. The question that remains open is whether or not it is possible to achieve asymptotic stability with only boundary feed back control with $f \equiv 0$ (for example geosynchronous orbit) and $d = 0$.

REFERENCES

[1] N. U. Ahmed and S. S. Lim, "Modeling and control of flexible space stations (slew maneuvers)", *Proceedings of the 3rd Annual Conference on Aerospace Computational Control* (D.E.Bernard and G.K.Mann, Eds.), JPL, NASA Pub. 89-45, Vol. 2, pp. 900-914, 1989.

[2] N. U. Ahmed and S. K. Biswas, "Mathematical modeling and control of large space structures with multiple appendages", *J.Math. Comput. Modeling*, Vol. 10, pp. 891-900, 1988.

[3] N. U. Ahmed, "Stability and control of nonlinear flexible systems", *Dynamic Systems and Applications* (submitted).

[4] N. U. Ahmed, *Elements of Finite Dimensional Systems and Control Theory*, Longman Scientific and Technical, UK, 1988.

[5] S. K. Biswas and N.U. Ahmed, "Optimal control of large space structures governed by a coupled system of ordinary and partial differential equations", *J. Math. Control, Signals, and Systems*, Vol. 2, pp. 1-18, 1989.

[6] G. Chen, S. G. Krantz, D. W. Ma and C. E. Wayne, "The Euler-Bernoulli beam equation with boundary energy dissipation", in *Operator Methods in Optimal Control Problems* (Sung J.Lee, Ed.), Marcel Dekker lecture notes, Vol. 108, pp. 67-96, 1987.

[7] I. Lasiecka, "Stabilization of hyperbolic and parabolic systems with nonlinearly perturbed boundary conditions", *J. Diff. Equations*, Vol. 75, pp. 53-87, 1988.

[8] P. Li and N. U. Ahmed, "On exponential stability of infinite dimensional systems with bounded or unbounded perturbations", *Applicable Analysis*, Vol. 30, pp. 175-187, 1988.

[9] P. Li, N. U. Ahmed and S. S. Lim, "Robust boundary control of flexible structures governed by hybrid dynamics", *Canadian Conference on Electrical and Computer Engineering*, pp. 18.3.1-18.3.4., 1990.

[10] J. M. Skowronski, H. Flashner and R. S. Gutalu (Eds.), *Mechanics and Control*, Proceedings of the 3rd Workshop on Control Mechanics, Springer-Verlag, Berlin, Vol. 151, 1991.

[11] S. K. Srinivasan and R. Vasudevan, *Introduction to Random Differential Equations and Their Applications*, Elsevier, New York, 1971.

DIFFERENTIAL DYNAMIC PROGRAMMING AND STATE VARIABLE INEQUALITY CONSTRAINED PROBLEMS

David J.W. Ruxton

Department of Mathematics and Computing
University of Central Queensland
Rockhampton, Queensland 4702
Australia

ABSTRACT

The differential dynamic programming (DDP) algorithm is shown to be readily adapted to handle state variable inequality constrained continuous optimal control problems. A new approach, using multiplier penalty functions implemented in conjunction with the DDP algorithm, is introduced and shown to be effective. The new DDP and multiplier penalty function algorithm is compared with the gradient-restoration method before being applied to solve a problem involving control of a constrained robot arm in the plane.

INTRODUCTION

Research by a number of workers including [1,2,3,4,5], has resulted in the development of a variety of numerical methods for solving the optimal control problem subject to state variable inequality constraints (SVIC).

In particular, and of interest here, the differential dynamic programming (DDP) algorithm developed by Jacobson and Mayne [6], has been modified to cater for the SVIC problem by Martensson [7], Jarmark [8] and Tun and Dillon [9]. Whilst these DDP algorithm variants can give encouraging results, they all involve considerable implementation and computational effort, particularly for highly non-linear problems.

Although penalty function schemes for handling constraints of SVIC form are recognised as having stability and accuracy problems they are nevertheless considered useful by a number of workers [5,8,9,10,11,12,13]. This popularity of penalty function methods may be due to their flexibility and ease of implementation when used in conjunction with established numerical methods for the unconstrained optimal control problem.

Apart from work by Jarmark [8], there appears to have been scant interest, or at least little reported, on the use of penalty function schemes with DDP applied to the SVIC problem. Chen and Chang et al [14,15], have recently reported on the effectiveness of incorporating a multiplier penalty function scheme, as first proposed by Bertsekas [16], with the DDP algorithm applied to *discrete* SVIC optimal control problems.

The intent here is to illustrate the ease of implementation and effectiveness of penalty function schemes when combined with the DDP algorithm and applied to *continuous* SVIC optimal control problems. In particular, following the lead of Chen and Chang et al, a new DDP algorithm variant using a multiplier penalty function scheme is shown to compare favourably with the gradient-restoration algorithm before being applied to solve a non-linear problem involving the control of a constrained planar robot arm.

PROBLEM STATEMENT

Solution of the continuous optimal control problem involves finding a control function $u(t)$ which minimises a cost functional

$$V(x_0, t_0) = \int_{t_0}^{t_f} L(x, u; t)\, dt + F(x(t_f); t_f) \tag{1}$$

subject to dynamic system constraints described by a set of ordinary differential equations

$$\dot{x} = f(x, u; t); \quad x(t_0) = x_0, \quad x(t_f) = x_{t_f} \tag{2}$$

where

- $x(t)$ is an n-dimensional vector function of time which describes the state of the dynamic system at any time $t \in [t_0, t_f]$;

- $u(t)$ is an m-dimensional vector function of time which describes the control available for adjustment at any time $t \in [t_0, t_f]$;

- $f(x, u; t)$ is an n-dimensional vector function which describes the dynamical structure of the system. The notation adopted means that f is a function of $x(t)$ and $u(t)$ explicitly and possibly also of time explicitly.

- L and F are scalar functions of their arguments

- V is a scalar which describes the value of some performance index or cost functional associated with the dynamical system.

The object of the SVIC optimal control problem is to solve equation 2 for a control $u(t)$; $t \in [t_0, t_f]$, such that the cost functional given by equation 1 is minimised and the following scalar SVIC is satisfied

$$g(x; t) \leq 0 \quad \forall \quad t \in [t_0, t_f]. \tag{3}$$

DDP AND PENALTY FUNCTIONS

A full description of the DDP algorithm can be found in Jacobson and Mayne [6]. Here only those sections of the algorithm that need significant alteration in order to accommodate penalty function schemes are discussed.

The essence of the iterative DDP method involves the repeated backward integration of the following matrix differential equations:

$$\begin{aligned} -\dot{V_x} &= H_x + V_{xx}(f - f(\bar{x}, \bar{u}\,; t)) \\ -\dot{V_{xx}} &= H_{xx} + {f_x}^T V_{xx} + V_{xx} f_x - (H_{ux} + {f_u}^T V_{xx}) {H_{uu}}^{-1} (H_{ux} + {f_u}^T V_{xx}) \end{aligned}$$

where the dot notation means differentiation with respect to time and where the Hamiltonian H is defined in the usual manner as

$$H = H(x, u, V_x; t) = L(x, u; t) + < V_x, f(x, u; t) >$$

In addition to requiring partial derivatives of H with respect to state and control variables the DDP algorithm requires a complete minimisation of H with respect to current control values $\bar{u}$ at each time step in a current iteration to calculate improved controls u^*. All terms in the above matrix differential equations are then evaluated at $\bar{x}$ and u^* unless otherwise stated.

It will be shown that exterior quadratic penalty function (EQPF) and multiplier penalty function (MPF) schemes require simple modifications to the form of the Hamiltonian and some minor modifications to the control of flow in the DDP algorithm. This is in marked contrast when compared with other DDP algorithm variants previously developed for the SVIC problem.

A DDP/EQPF SCHEME

The augmented integrand of the cost functional in equation (1) becomes

$$L^*(x,u,w;t) = L(x,u;t) + w\sum_{i=1}^{n} A_i g_i{}^2(x;t)$$

where $A_i = 1$ if the $g_i > 0$, $A_i = 0$ otherwise and n is the number of constraints g.

This results in an augmented Hamiltonian

$$H^*(x,u,V_x,w\,;t) = H(x,u,V_x\,;t) + w\sum_{i=1}^{n} A_i g_i{}^2(x;t).$$

Repeated iteration of the DDP algorithm, using H^* in place of H and using a high fixed value for, or successively increasing values of, the penalty weighting parameter w between iterations, usually results in satisfaction of the constraints of SVIC form to within some specified tolerance.

The only required modifications to the DDP algorithm are in the backward integration phase and involve simple changes to the partial derivatives of the augmented Hamiltonian with respect to the state variables. The incorporation of a weighting parameter updating scheme is a minor addition to the main flow of the algorithm. That is,

$$H^* = H + w\sum_{i=1}^{n} A_i g^2, \quad H_x^* = H_x + 2w\sum_{i=1}^{n} A_i g g_x$$

and

$$H_{xx}^* = H_{xx} + 2w\sum_{i=1}^{n} A_i (g g_{xx} + g_x^2)$$

where $g = g_i(x;t)$ is used for notational convenience.

A NEW DDP/MPF APPROACH

Essentially, the MPF method merges the penalty function idea with a primal-dual Lagrange multiplier scheme, producing an algorithm which largely overcomes the ill conditioning and rate of convergence difficulties associated with EQPF schemes.

Bertsekas [16,17], has analysed the use of multiplier methods for general constrained minimisation problems and concluded that they have significant advantages over traditional penalty function methods in terms of reliability and speed of convergence. In particular, he points out that convergence for the method of multipliers can usually be obtained without the need to increase the penalty weighting parameter to the point where ill conditioning becomes a problem.

Chen and Chang et al [14,15], recently used ideas suggested by Bertsekas [17], to incorporate the MPF method with the DDP algorithm, giving encouraging results for discrete optimal control problems subject to constraints of SVIC form. Proceeding along similar lines, it is now shown that the DDP algorithm is readily modified to incorporate the MPF method in order to effectively deal with continuous optimal control problems subject to constraints of SVIC form.

Dual Problem Formulation

The constrained optimal control problem specified by equations 1, 2, and 3 involves two sets of constraints: the system dynamics as specified by equation 2 and the explicit constraint on state given by equation 3. A dual problem may be constructed, relaxing all but the constraints on the system dynamics, by applying a MPF scheme along the lines proposed by Bertsekas [17], as follows

Let

$$\alpha_j(x, u; t) = w g_j(x, u; t)$$

where w is a weighting parameter associated with a series of constraints $\{g_j\}$.

Note that x and u are functions of time with $g_j(x, u; t) = g_j(x)$ if the constraint does not involve u or t explicitly.

$P(\alpha, \mu; t)$ is introduced as a MPF associated with the constraints $g_j(x, u; t)$ as follows

$$P(\alpha, \mu; t) = \sum_{j=1}^{n} P_j(\alpha_j, \mu_j\,; t)$$

where $\mu_j(t)$ is a multiplier and n is the total number of constraints.

The integrand of the cost functional given by equation 1 is then modified to accommodate this MPF, thus forming an augmented cost integrand

$$L^*(x, u, \alpha, \mu; t) = L(x, u; t) + 1/wP(\alpha, \mu; t).$$

The dual problem corresponding to equations (1), (2) and (3) is then

$$\max_{\mu} D(\alpha, \mu; t) \tag{4}$$

with

$$D(\alpha, \mu; t) = \min_{u} \int_{t_0}^{t_f} L^*(x, u, \alpha, \mu; t)\, dt + F(x(t_f), t_f) \tag{5}$$

subject to equation 2.

Solution of this dual problem, and thus also of the constrained primal problem given by equations (1), (2) and (3) is simply a matter of using a numerical method, suitable for unconstrained optimal control problems, to minimise equation 5. This algorithm would start with nominal values for $\mu(t)$ and w and then iteratively generate improved values until both equations 4 and 5 are satisfied to the required accuracy. In other words 4 and 5 are alternated in an iteration sequence until convergence.

The updated multipliers at each unconstrained problem iteration sequence are generated for each of the j constraints from

$$\begin{aligned}\mu^{i+1}(t) &= \nabla_\alpha P(\alpha, \mu^i; t)\\ \alpha_j &= w^i g_j[x(\mu^i, w^i; t)\;,\; u(\mu^i, w^i; t)\;; t]\end{aligned}$$

where $\nabla_\alpha P = \partial P/\partial \alpha_j$, i is an iteration index and $\{w^i\}$ is a positive non-decreasing sequence chosen such that successive solutions of (5) converge whilst constraint (3) is satisfied.

Selection of A Multiplier Penalty Function

The following MPF form that is both smooth and numerically well conditioned was suggested by Bertsekas [17], then subsequently modified and used by Chen and Chang et al [14,15], in solving discrete SVIC problems

$$\begin{aligned}P(\alpha, \mu; t) &= \mu\alpha + \mu\alpha^2\,, \quad for\ \alpha\ \geq\ ;0\\ &= \frac{\mu\alpha}{1-\alpha}\,, \quad for\ \alpha\ <\ 0.\end{aligned} \tag{6}$$

The MPF Scheme and DDP

In what follows it is assumed, for notational simplicity and clarity, that only one SVIC is active, however the rationale extends to the case of multiple constraints.

The DDP algorithm is readily modified to accommodate the SVIC problem by incorporating the MPF given by equation 6. The augmented cost integrand becomes either:

(a) Constraint active, that is $g(x) \geq 0$

$$\begin{aligned} P(\alpha, \mu; t) &= \mu\alpha + \mu\alpha^2 \\ L^* &= L + \mu g + \mu w g^2 \end{aligned}$$

note the similarity of the last term to the exterior quadratic penalty form;

(b) Constraint inactive, that is $g(x) < 0$

$$\begin{aligned} P(\alpha, \mu; t) &= \mu\alpha(1-\alpha)^{-1} \\ L^* &= L + \mu g(1 - wg)^{-1}. \end{aligned}$$

These modifications result in an augmented Hamiltonian

$$\begin{aligned} H^* &= H + 1/wP(\alpha, \mu; t) \\ &= H + A\mu g(1 + wg) + B\mu g(1 - wg)^{-1} \end{aligned}$$

where $g = g(x)$ and $A = 1, B = 0$ for $g \geq 0$ or $A = 0, B = 1$ for $g < 0$.

Successive iterations of the DDP algorithm using H^* in place of H will, under convexity assumptions already noted, converge to a solution of the relaxed problem given by equation 5 for current values of $\mu(t)$ and w.

As is the case for the EQPF method, the only modifications required in the backward integration phase of the DDP algorithm are changes to the partial derivatives of the Hamiltonian. These partial derivatives become

$$\begin{aligned} H_x^* &= H_x + A\mu(g_x + 2wgg_x) + B\mu g_x(1 - wg)^{-2} \\ H_{xx}^* &= H_{xx} + A\mu(g_{xx} + 2w(gg_{xx} + {g_x}^2)) + B\mu(g_{xx} - wgg_{xx} + 2w{g_x}^2))(1 - wg)^{-3}. \end{aligned}$$

The DDP algorithm is readily modified, along the lines suggested by Bertsekas [17], updating multipliers and penalty function weights after each minimisation of (5) with respect to $u(t)$ as follows

$$\begin{aligned} \mu^{i+1}(t) = \frac{\partial P}{\partial \alpha}(\alpha, \mu^i; t) &= \mu^i(t)(1 + 2w^i g)\ ,\ g \geq 0 \\ &= \mu^i(t)(1 - w^i g)^{-2}\ ,\ g < 0 \end{aligned}$$

and $w^{i+1} = a.w^i$, $a > 0$ and $a = 1$ for a fixed weighting scheme.

COMPARISONS

The following "evergreen" problem, considered by [1,5,7,18,19,20], is chosen as a convenient basis for comparing select algorithms. The problem is to minimise

$$V(x_0, t_0) = \int_0^1 u^2 dt$$

subject to

$$\begin{aligned} \dot{x_1} = x_2 \ &;\ x_1(0) = 0 \ ;\ x_1(1) = 0 \\ \dot{x_2} = u \ &;\ x_2(0) = 1 \ ;\ x_2(1) = -1 \end{aligned}$$

$$g(x_1) = x_1 - 0.15 \leq 0$$

Results

Results of applying the DDP/EQPF, DDP/MPF algorithm variants and the gradient-restoration algorithm to the problem are presented in Tables 1 and 2.

When interpreting these results the following points should be noted:

- The mnemonics pf, gr and m refer to the DDP/EQPF, the gradient -restoration and the DDP/MPF algorithms respectively;
- Gradient-restoration results are as published by Miele [18];
- The optimal cost is known to be 5.926 [19];
- Results for the MPF scheme are given for both 16 and 18 iterations;
- Note especially that $u^{gr} > 0$ at $t = 0.5$ whereas u^m and $u^{opt} < 0 \; \forall \; t \in [0, 1]$;
- Constraint satisfaction by either state or control trajectories may be measured by

$$E(p) = \frac{\sum_{i=1}^{n} | (p_a - p_c) |}{n}$$

where p_a is the actual optimal value of the trajectory and p_c is the computed value of the trajectory, measured for the n discrete time steps used over the entire interval $[0, 1]$. Here $E(u)$ is considered, rather than the more natural $E(x_1)$, for convenience in matching and presenting results from the various methods.

- The gradient-restoration algorithm involves transformation of the original problem into one of higher dimension. Although the first-order gradient-restoration algorithm is computationally more efficient, being of $\mathcal{O}(n^2)$, than the second-order DDP algorithm, at $\mathcal{O}(n^3)$, it has more work to do for a given problem since the state space dimension n increases according to the number of inequality constraints present. Thus, at least for problems of low dimension, the two algorithms have similar work to do per iteration.

Table 1. State and control trajectories

t	x_1^{gr}	$x_1^{pf\&m}$	u^{gr}	u^{pf}	u_{16}^m	u_{18}^m	x_1^{opt}	u^{opt}
0.4	0.1496	0.1498	-0.4730	-0.4954	-0.4982	-0.4961	0.1498	-0.4938
0.5	0.1499	0.1500	+0.0375	-0.0514	-0.0416	-0.0206	0.1500	0.0000
0.6	0.1497	0.1498	-0.4405	-0.4938	-0.4923	-0.4940	0.1498	-0.4938

Table 2. Summary of results

Method	Cost	Its.	$E(u)$	w	a	μ
gr	5.927	16	0.019	-	-	-
pf	5.925	13	0.007	10^4 *to* 10^6	-	-
m_{16}	5.926	16	0.006	100	5	50
m_{18}	5.926	18	0.004	100	5	50

Comment on The Comparisons

The results for the single problem considered here are indicative of those obtained by Ruxton [20], where the DDP/MPF algorithm variant is tested on a number of linear and non-linear problems including a problem involving a SVIC which is non-linear and time dependent. It would seem that both the DDP/MPF and DDP/EQPF algorithm variants give more accurate results, and especially better constraint boundary adhesion, than the gradient-restoration algorithm albeit for increased computational effort. Furthermore the DDP/MPF algorithm appears to produce significantly better constraint boundary adhesion than the DDP/EQPF algorithm. This is to be expected given the modus operandi of the two schemes. For example, it is a simple matter to show that as the trajectory sequence evolves, with $g(x_1)$ tending ever closer to zero, H_x is continuous across the constraint boundary $g(x_1) = 0$ for the DDP/MPF formulation. This is not the case for the DDP/EQPF formulation.

A PROBLEM FROM ROBOTICS

An extendable robot arm rotating in the plane may be represented as shown in Figure 1.

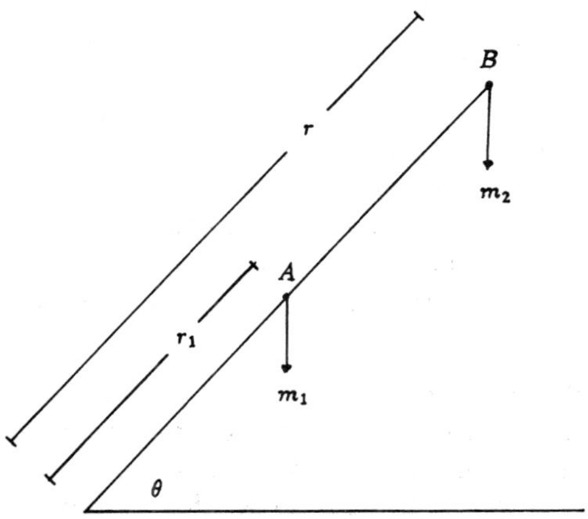

Figure 1. The robot arm in the plane

where

$$
\begin{aligned}
r_1 &= \text{fully retracted arm length} \\
r &= \text{variable arm extension} \\
m_1, m_2 &= \text{mass of manipulator arm with payloads at A and B}
\end{aligned}
$$

Lagrangian analysis yields:

$$m_1 r_1^2 \ddot{\theta} + m_2 r^2 \ddot{\theta} + 2 m_2 r \dot{r} \dot{\theta} + g \cos\theta (m_1 r_1 + m_2 r) = T_\theta$$

$$m_2 \ddot{r} - m_2 r \dot{\theta}^2 + m_2 g \sin\theta = F_r$$

where T_θ is the torque about the rotational actuator and F_r is the force applied by the linear actuator.

In the horizontal plane, $g = 0$, and the state and control variables are defined as

$$x_1(t) = \theta \ , \ x_2(t) = \dot{\theta} \ , \ x_3(t) = r \ , \ x_4(t) = \dot{r} \ , \ u_1(t) = T_\theta \ , \ u_2(t) = F_r,$$

yielding the state equations

$$\begin{aligned} \dot{x}_1 &= x_2 \\ \dot{x}_2 &= (u_1 - 2m_2 x_3 x_4 x_2)/(m_1 r_1^2 + m_2 x_3^2) \\ \dot{x}_3 &= x_4 \\ \dot{x}_4 &= (u_2 + m_2 x_3 x_2^2)/m_2 \end{aligned}$$

subject to boundary conditions $x(t_0) = x_{t_0}$, $x(t_f) = x_{t_f}$.

Limits may be placed on the arm extension, velocity and acceleration as follows:

$a < x_3 < b$	(m) ,	extension limits
$\mid \dot{x}_2 \mid < m$	(rad/sec) ,	radial velocity limits
$\mid \dot{x}_4 \mid < n$	(m/sec) ,	extension velocity limits
$\mid \ddot{x}_2 \mid < o$	(rad/sec^2) ,	radial acceleration limits
$\mid \ddot{x}_4 \mid < p$	(m/sec^2) ,	extension acceleration limits

This general control problem can be converted to an optimal control problem by the addition of a performance criterion. The performance criterion chosen here involves the minimisation of control effort. Other criteria may be considered but may not produce a form amenable to application of the standard DDP algorithm. For example, the minimisation of arm travel time results in a singular optimal control problem.

A choice of parameters $m_1 = 7kg$, $m_2 = 1kg$, $r_1 = 1m$, $a = 1$, $b = m = n = o = p =$ *unbounded* [1], yields the optimal control problem

$$\begin{aligned} \dot{x}_1 &= x_2 \\ \dot{x}_2 &= (u_1 - 2x_3 x_4 x_2)/(7 + x_3^2) \\ \dot{x}_3 &= x_4 \\ \dot{x}_4 &= (u_2 + x_3 x_2^2) \end{aligned}$$

for some $x(t_0) = x_{t_0}$ and $x(t_f) = x_{t_f}$, with the performance index

[1] It can be shown [20], conveniently, that the only constraint that affected the optimal result for a reasonable range of acceleration and velocity boundary conditions is that involving the arm retraction, x_3.

$$J = \int_{t_0}^{t_f} (u_1^2 + u_2^2)\, dt$$

subject to the single constraint of SVIC form:

$$g_2(x_3) \quad = \quad -x_3 + 1 \leq 0 \tag{7}$$

The problem may now be considered for a range of boundary conditions. Here the boundary conditions are chosen as:

$$x(0) = [3.14 \ , \ 0.1 \ , \ 1.5 \ , \ 0]^T$$

$$x(10) = [6.28 \ , \ 0 \ , \ 1.5 \ , \ 0]^T.$$

The costs, number of iterations and maximum arm retraction resulting from use of the DDP/EQPF and DDP/MPF schemes are:

scheme	cost	total iterations	x_3 max. retraction
EQPF	5.694	66	1.0020
MPF	5.686	50	1.0001

Figures 2 to 5 show select control and state trajectory evolutions and the optimal trajectory in the r, θ plane.

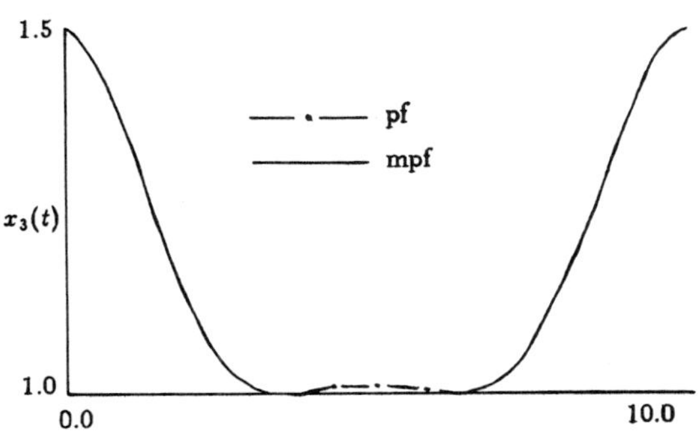

Figure 2. State trajectory $x_3(t)$

Comments on The Results

A marked softness in constraint boundary adhesion when using the EQPF scheme is evident on examination of the results. The optimal costs for both schemes differ with the MPF scheme clearly requiring fewer iterations to achieve the optimal solution with better constraint boundary adhesion. The softness in constraint boundary adhesion using the EQPF scheme is best illustrated by Figures 2 and 3, which show that the state trajectory $x_4(t)$ is not always zero, as it should be, whilst trajectory $x_3(t)$ lies on the constraint boundary $x_3(t) = 1$. Figure 4 shows a slight variation in control $u(t)$ between the two schemes over the constraining interval as would be expected.

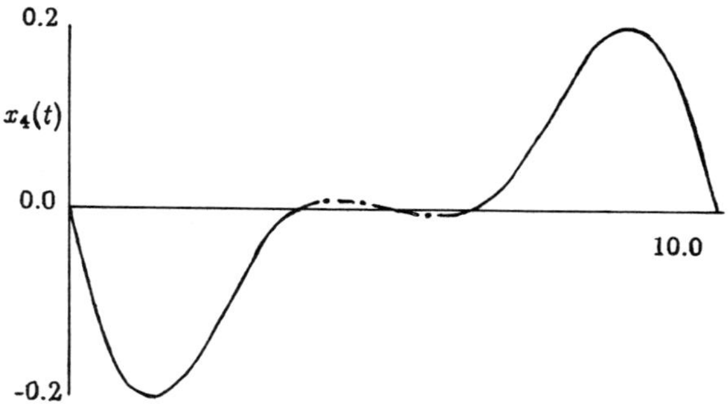

Figure 3. State trajectory $x_4(t)$

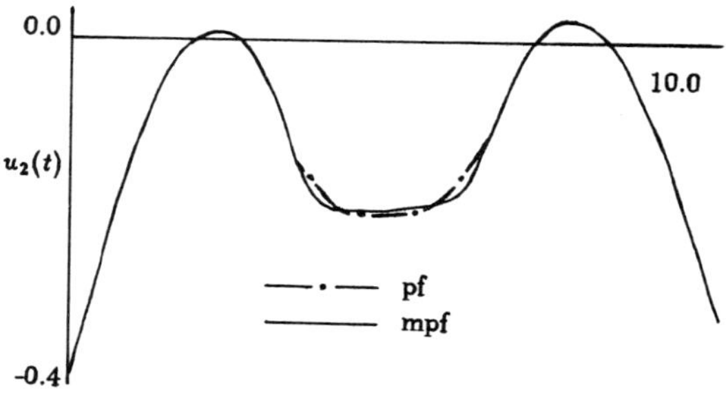

Figure 4. Control trajectory $u_2(t)$

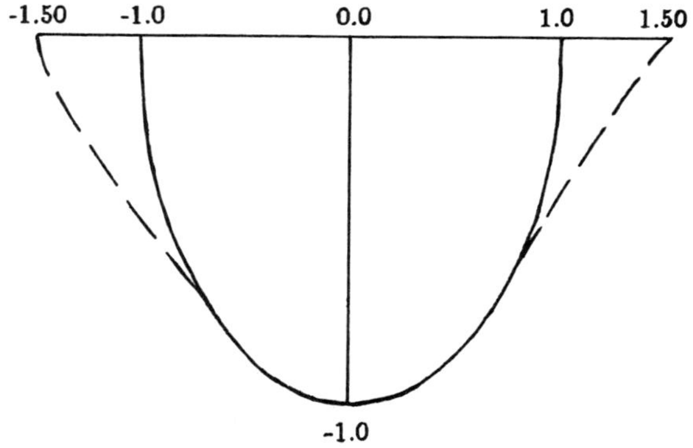

Figure 5. Robot arm trajectory in the r, θ plane

The results show that the EQPF scheme can produce optimal controls which look plausible and which produce a cost close to the optimal cost, but which may be inaccurate in satisfying constraints. In examples considered it is clear that the MPF scheme yields results at least as good as the EQPF scheme and gives more reliable constraint satisfaction.

CONCLUSION

The results presented here indicate that the DDP/MPF algorithm produces more accurate results than the DDP/EQPF algorithm and compares favourably with the well known gradient-restoration algorithm.

The DDP/MPF algorithm would seem to be a relatively easily implemented and effective alternative numerical method suitable for application to non-trivial optimal control problems subject to SVIC constraints.

REFERENCES

1. Bryson, A.E. Jr., Denham, W.F. and Dreyfus, S.E., "Optimal programming problems with inequality constraints I: necessary conditions for extremal solutions," *AIAA J.*, Vol. 1, pp. 2544-2550, 1963.

2. Jacobson, D.H., Lele, M.M. and Speyer, J.L., "New necessary conditions of optimality for control problems with state-variable inequality constraints," *J. Math. Anal. Appl.*, Vol. 35, pp. 255-284, 1971.

3. McIntyre, J. and Paiewonsky, B., "On optimal control with bounded state variables," in *Adv. Cont. Sys.*, Vol. 5 (C. T. Leondes, ed.), Academic Press, New York, 1967.

4. Speyer, J.L. and Bryson, A.E. Jr., "Optimal programming problems with a bounded state space", *AIAA J.*, Vol. 6, pp. 1488-1491, 1968.

5. Lasdon, L.S., Warren, A.D. and Rice, R.K., "An interior penalty method for inequality constrained optimal control problems," *IEEE Trans. Auto. Cont.*, Vol AC-12, pp. 388-395, 1967.

6. Jacobson, D.H. and Mayne, D.Q., *Differential Dynamic Programming*, American Elsevier, New York, 1970.

7. Martensson, K., *New approaches to the numerical solution of optimal control problems*, Studentlitteratur (Ph.D. thesis), Lund, Sweden, 1972.

8. Jarmark, B., "Calculation aspects on an optimisation program," Report R82-02, School of Electrical Engineering, Chalmers University of Technology, Goteborg, Sweden, 1982.

9. Tun, T. and Dillon, T.S., "Extensions of the differential dynamic programming method to include systems with state dependent control constraints and state variable inequality constraints," *J. Appl. Sci. Engng. A*, Vol. 3, pp. 171-192, 1978.

10. Mayorga, R.V., Quintana V.H. and Gerez, V., "A numerical solution for state constrained continuous optimal control problems using improved penalty functions," in *Proc. 22nd IEEE Conf. on Decision and Control*, USA, Vol. 1, pp. 432-434, December 1983.

11. Lele, M.M. and Jacobson, D.H., "A proof of the convergence of the Kelley-Bryson penalty function technique for state-constrained control problems," *J. Math. Anal. Appl.*, Vol. 26, pp. 163-169, 1969.

12. Xing, A.Q. and Wang, C.L., "Applications of the exterior penalty method in constrained optimal control problems," *Opt. Cont. Appl. Methods*, Vol. 10, pp. 333-345, 1989.

13. Polak, E., "An historical survey of computational methods in optimal control," *SIAM Rev.*, Vol. 15, pp. 553-584, 1973.

14. Chen, C.H., Chang S.C. and Fong, I.K., "An effective differential dynamic programming algorithm for constrained optimal control problems," in *Proc. American Control Conference*, Pittsburgh, Vol. 2, pp. 1763-1764, June 1989.

15. Chang, S.C., Chen, C.H., Fong, I.K. and Luh, P.B., "Hydroelectric generation scheduling with an effective differential dynamic programming algorithm," *IEEE Trans. on Power Systems*, Vol. 5, pp. 737-743, 1990.

16. Bertsekas, D.P., *Multiplier methods: a survey*, *Automatica*, Vol. 12, pp. 133-145, 1976.

17. Bertsekas, D.P., *Constrained Optimization and Lagrange Multiplier Methods*, Academic Press, New York, 1982.

18. Miele, A., "Gradient algorithms for the optimisation of dynamic systems," *Control and Dynamic Systems*, Vol. 16, pp. 3-52, 1980.

19. Bryson, A.E. Jr. and Ho, Y.C., *Applied Optimal Control— optimisation, estimation and control*, Halsted Press, New York, 1975.

20. Ruxton, D.J.W., *Differential dynamic programming and optimal control of inequality constrained continuous dynamic systems*, M.Sc. thesis, University of Central Queensland, Australia, 1991.

DYNAMICAL PULSE-WIDTH-MODULATION CONTROL OF RIGID AND FLEXIBLE MANIPULATORS

Hebertt Sira-Ramírez and Orestes Llanes-Santiago

Departamento Sistemas de Control
Universidad de Los Andes
Mérida
Venezuela

ABSTRACT

Due to the associated chattering effects, Pulse-Width-Modulation (PWM) control strategies have been of limited theoretical interest in the investigations about regulation of mechanical systems, such as robotic manipulators. This article proposes dynamical PWM control as a means of circumventing the inconveniences caused by the discontinuities associated to the PWM control strategies and still retain the classical robustness features associated to this practical discontinuous control technique. As examples, the results are applied to, both, rigid and flexible joint single link robotic manipulators.

INTRODUCTION

The design of PWM control policies for the regulation of dynamical systems is generally based on approximating an average (i.e., smooth) desired behavior by means of high-frequency discontinuous control signals which, somehow, emulate the effects of the designed average feedback control inputs. In spite of its well known robustness features, the PWM control approach induces undesirable bang-bang control inputs causing noticeable high frequency discontinuities, or "chattering", in the controlled responses. This kind of discontinuous behavior severely limits, for a large class of mechanical and electro-mechanical systems, the applicability of PWM stabilizing feedback control policies, as well as that of some other closely related discontinuous feedback control techniques (such as Sliding Mode control and Pulse-Frequency-Modulation control). For mechanical systems, in general, and robotic manipulators, in particular, low limits to possible wear and tear effects, and high accuracy of the regulated variables, are important design concerns when choosing the regulating feedback control scheme.

Mechanics and Control, Edited by R.S. Guttalu
Plenum Press, New York, 1994

PWM control constitutes a robust feedback control policy due to its insensitivity to external disturbance inputs, certain certain immunity to model parameter variations, within known bounds, and enhanced performance in spite of the effects of ever present modeling errors. For basic work related to PWM control schemes, in linear and nonlinear dynamical systems, the reader is referred to Polak [1], Skoog [2], Skoog and Blankenship [3], Tsypkin [4], La Cava et al [5], Csaki [6] and also Sira-Ramirez [7]-[12]. A large collection of mechanical nonlinear systems belong to the class of *systems exactly linearizable by state coordinate transformations and static state feedback.* The *generalized state models* (see Fliess [13], [14]) of such systems do not include input signal derivatives. Thus, static, instead of dynamical, controllers naturally arise from the exact linearization procedure introduced in [14]. In this article, wer present a general design method for synthesizing *static*, as well as *dynamical*, PWM feedback control laws stabilizing to a constant equilibrium point the state of any nonlinear dynamical system belonging to the above class. PWM controllers are proposed which effectively stabilize to zero a suitably designed auxiliary scalar output function of the system. The obtained restricted dynamics results, in turn, in a linearization and local asymptotic stabilization of the original system state to its equilibrium value. However, the obtained static PWM controller is shown to include undesirable chattering effects on the generated input signals. A dynamical PWM controller alternative is then proposed, and synthesized, on the basis of *Fliess's generalized controller canonical form* (FGCCF) (see [14]) of the associated *extended system* (see Nijmeijer and Van der Schaft [15]). Continuous, instead of bang-bang, feedback control input signals are, thus, obtained which robustly stabilize to a constant operating point the closed loop system, while effectively smoothing out the chattering effects on the control signals and regulated state variables.

Next section presents a fundamental result on the PWM control of an elementary scalar dynamical system. It is shown that, based on this scalar result, a general PWM controller design procedure can be proposed for higher order nonlinear controlled plants. This section also reviews FGCCF on which, both, the static, and dynamical, PWM controller design procedures are based. The dynamical PWM control design scheme for linearizable systems, proposed in this section, utilizes the concept of the *extended system.* Two illustrative examples are presented later. The first one is constituted by a purely mechanical system, representing a single link rigid robotic manipulator, on which the performances of a static and a dynamical PWM feedback controller are tested by means of computer simulations. The second example deals with the more delicate issue of using static and dynamical PWM feedback control policies for a single link flexible robotic manipulator.

PWM CONTROL OF NONLINEAR SYSTEMS

PWM Control of a Scalar System

Consider the scalar PWM controlled dynamical system with state variable s, in which the constants a, W and r are all strictly positive quantities.

$$\begin{aligned}\dot{s} &= -as - Wv \\ v = PWM_t(s) &= \begin{cases} sgn[s(t_k)] & \text{for } \; t_k \le t < t_k + \tau[s(t_k)]T \\ 0 & \text{for } \; t_k + \tau[s(t_k)]T \le t < t_k + T \end{cases}\end{aligned} \tag{1}$$

$$\tau[s(t)] = \begin{cases} 1 & \text{for } |s(t)| > \frac{1}{r} \\ r|s(t)| & \text{for } |s(t)| \le \frac{1}{r} \end{cases}$$

$$k = 0, 1, 2, \cdots; \quad t_{k+1} = t_k + T.$$

The t_k's represent *regularly* spaced sampling instants. The sampling interval $[t_k, t_{k+1}]$ is known as the *duty cycle* and its width is here represented by the constant T, i.e., $T = t_{k+1} - t_k$. At each sampling instant, t_k, the value of the width of the next sign-modulated, fixed amplitude, control pulse v is determined by the sampled value of the *duty ratio function*, represented by $\tau[s(t_k)]$. The function "sgn" stands for the *signum* function:

$$sgn(s) = \begin{cases} +1 & \text{if } s > 0 \\ 0 & \text{if } s = 0 \\ -1 & \text{if } s < 0 \end{cases}$$

The pulse width, $\tau[s(t_k)]T$, saturates to the value of the duty cycle, T, i.e., $\tau[s(t_k)]$ saturates to 1, as long as the value of the controlled state s is larger, in absolute value, than a given prespecified threshold $1/r$ (see Figure 1). When the absolute value of the state, s, of the scalar system (1) is driven below the value $1/r$, the duty ratio, $\tau[s(t_k)]$, also starts decreasing, in a linear fashion, with respect to $|s|$. When the scalar system

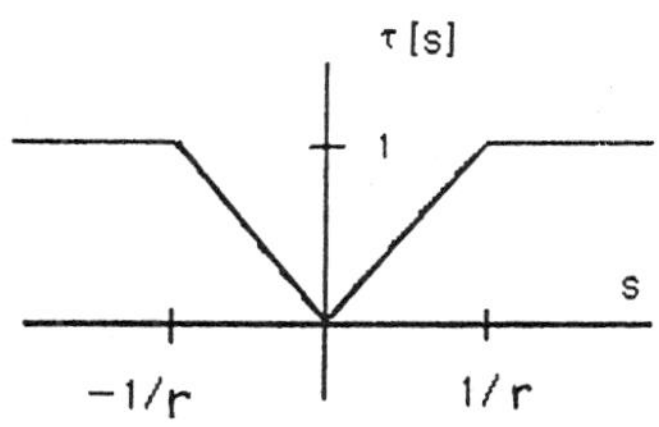

Figure 1. Duty Ratio function for PWM controlled Scalar System.

is, eventually, at rest in $s = 0$, no control pulses are then applied to the system. The basic idea behind the above discontinuous PWM control law (1) is that large errors in the scalar state s should require larger corrective pulse widths during the fixed inter-sampling periods. Small errors, on the other hand, should be driven to zero, on the basis of the adopted sampling frequency, with corrective pulse widths decreasing to zero until stabilization. In a sense, PWM policies are "proportional" feedback policies subject to saturation. The following theorem establishes a sufficient condition for the asymptotic stability to zero of the PWM controlled system (1). The same theorem is also found in [1] and also in [2]. In reference [2] it is further established that the proposed convergence condition is also necessary. We furnish, however, a different proof of the sufficiency result.

Theorem 1: The PWM controlled system (1) is asymptotically stable to $s = 0$, if

$$rW < a \tanh\left(\frac{aT}{2}\right) \tag{2}$$

Proof: Due to the piecewise constant nature of the control inputs and the linearity of the underlying continuous system (1), it suffices to study the stability of the discretized version of (1) at the sampling instants. An exact discretization of the PWM controlled system (1) thus yields

$$s(t_k + T) = e^{-aT} s(t_k) - \frac{We^{-aT}}{a} \left(e^{a\tau[s(t_k)]T} - 1 \right) sgn[s(t_k)] \tag{3}$$

Suppose the initial condition $s(t_0)$ is chosen deep into the region $|s| > 1/r$. The evolution of the sampled values of $s(t)$ obey, according to (1):

$$s(t_{k+1}) = e^{-aT} \left\{ s(t_k) - \frac{W}{a}(e^{aT} - 1)\, sgn[s(t_k)] \right\} \tag{4}$$

The absolute value of the incremental step, $\Delta s(t_k) := s(t_{k+1}) - s(t_k)$, is readily obtained from (4), from where it is easily found that

$$\begin{aligned} |\Delta s(t_k)| &= \left(1 - e^{-aT}\right) \left| \left(s(t_k) + \frac{W}{a} \right) \right| \\ &< \left(1 - e^{-aT}\right) \left[|s(t_k)| + \frac{W}{a} \right] \end{aligned} \tag{5}$$

The condition $|\Delta s(t_k)| < 2/r$ is sufficient to ensure that the value of $s(t_k)$ will be eventually found within the bounded region $|s| < 1/r$, irrespectively of the initial condition value $s(t_0)$ given in $|s| > 1/r$. Sufficiency is clear from the facts that, in the region $|s| > 1/r$, the $|\Delta s(t_k)|$ decrease at each step and that one must guarantee that $s(t_k)$ does not persistently "jump" over the band $|s| < 1/r$, thus falling into a possible limit cycle behavior. ¿From (5) and the fact that $|(t_k)| > 1/r$, the condition $|\Delta s(t_k)| > 1/r$, the condition $|\Delta s(t_k)| < 2/r$ is guaranteed if we let

$$\left(1 - e^{-aT}\right) \left[1 + \frac{rW}{a} \right] < 2 \tag{6}$$

which is just the expression (2), after some straightforward manipulations.

Suppose now that the initial state, $s(t_0)$, of the scalar system (1), is found in the region $|s| < 1/r$. The exact discretization of the PWM controlled system is now given by

$$s(t_{k+1}) = e^{-aT} s(t_k) - \frac{We^{-aT}}{a} \left(e^{ar|s(t_k)|T} - 1 \right) sgn[s(t_k)] \tag{7}$$

The absolute values of the incremental steps $\Delta s(t_k) = s(t_k + T) - s(t_k)$ are thus given by

$$\begin{aligned} |\Delta s(t_k)| &= \left(1 - e^{-aT}\right) \left| s(t_k) + \frac{W}{a} \frac{\left(e^{a(r|s(t_k)|-1)T} - e^{-aT} \right)}{\left(1 - e^{-aT}\right)} \right| \\ &< \left(1 - e^{-aT}\right) \left[|s(t_k)| + \frac{W}{a} \frac{\left(e^{a(r|s(t_k)|-1)T} - e^{-aT} \right)}{\left(1 - e^{-aT}\right)} \right] \end{aligned} \tag{8}$$

A sufficient condition for asymptotic stability of (7) to zero is given by $|\Delta s(t_k)| < 2|s(t_k)|$. Notice however that from the fact that $r|s(t_k)| < 1$, (8) implies that

$$|\Delta s(t_k)| < \left(1 - e^{-aT}\right) \left[|s(t_k)| + \frac{W}{a} \right]$$

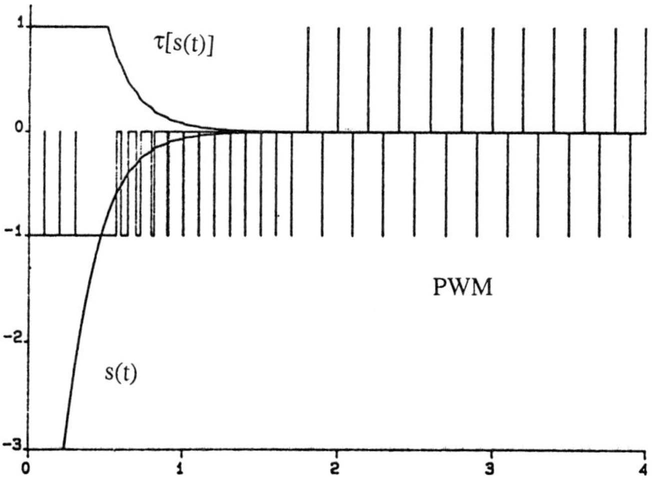

Figure 2. State Response, duty ratio function and PWM control signal for first order PWM controlled scalar system.

Hence, the above convergence condition is fully guaranteed if we let

$$|\Delta s(t_k)| < \left(1 - e^{-aT}\right)\left[|s(t_k)| + \frac{W}{a}\right] < 2|s(t_k)| \tag{9}$$

i.e., from the second inequality one has

$$\frac{W}{a} < \frac{2}{(1 - e^{-aT})}|s(t_k)| - |s(t_k)| = |s(t_k)| \tanh\left(\frac{aT}{2}\right)$$

which after multiplication of by r, and the fact that $r|s(t_k)| < 1$, results in:

$$\frac{rW}{a} < r|s(t_k)| \tanh\left(\frac{aT}{2}\right) < \tanh\left(\frac{aT}{2}\right)$$

The result follows.

A computer simulation of system (1) is shown below in Figure 2. This figure depicts the state s(t), the PWM control input signal v(t) and the duty ratio function $\tau(s(t))$. The values of the involved constants were chosen as: $a = 5.4, r = 0.01, T = 0.1, W = 10$. Since $rW/a = 0.0185 < \tanh(0.27) = 0.2636$, asymptotic stability of s(t) to zero is guaranteed by Theorem 1.

Fliess' Generalized Controller Canonical Form

Consider the analytical n-dimensional state variable representation of a nonlinear system:

$$\dot{x} = F(x, u) \tag{10}$$

It is assumed that the nonlinear system (8) exhibits a constant equilibrium point of interest characterized by $F(X(U), U) = 0$. We refer to this equilibrium point as

$(X(U), U)$. Associated to the system (10) is FGCCF given by (see [14] for further details)

$$\begin{aligned} \dot{z}_1 &= z_2 \\ \dot{z}_2 &= z_3 \\ &\cdots \\ \dot{z}_{n-1} &= z_n \\ \dot{z}_n &= c(z, u, \dot{u}, \cdots, u^{(\alpha)}) \end{aligned} \tag{11}$$

Implicit in this representation is the assumption of the existence of an element, $z_1 = h(x)$, called the *differential primitive element,* which generates the *generalized state representation* (11) of system (10). Under these circumstances, system (10) is transformed into system (11) by means of an invertible input-dependent state coordinate transformation of the form

$$z = \Phi(x, u, \dot{u}, \cdots, u^{(\alpha-1)}) \tag{12}$$

given by

$$\Phi(x, u, \dot{u}, \cdots, u^{(\alpha-1)}) = [h(x), \dot{h}(x), \cdots, h^{(n-1)}(x, u, \dot{u}, \cdots, u^{(\alpha-1)})]' \tag{13}$$

We assume that the equilibrium point $(X(U), U)$ of (8) is transformed, by (10), into the vector $z = 0$, i.e., $\Phi(X(U), U, 0, \cdots, 0) = 0$. ¿From this assumption, it readily follows that $c(0, U, \cdots, 0) = 0$. Suppose that for a suitably designed feedback control policy the state of (9) is asymptotically driven to zero. The autonomous dynamics described by

$$c(0, u, \dot{u}, \cdots, u^{(\alpha)}) = 0 \tag{14}$$

constitutes the *zero dynamics* (see Fliess [16]). In order to guarantee an overall stable performance of the controlled system, it is of crucial importance to assess the stability of such a dynamics around its possible equilibrium points. Around the equilibrium point of interest $u = U$ we assume that the dynamical system (14) is asymptotically stable, i.e., the system $dx/dt = F(x, u)$ with output $y = h(x)$ is *minimum phase. Non-minimum phase* systems can also be controlled by PWM control policies by means of suitable change of the differential primitive element. This topic, however, will be explored elsewhere (see also Fliess and Messager [17] for related aspects in connection with sliding mode control of linear time invariant systems). Motivated by the fact that for a large class of mechanical and electromechanical systems the integer α, in the corresponding GCCF (11), is equal to zero, we shall concentrate our developments, from now on, on such a particular class of systems. This class corresponds to *systems exactly linearizable by means of state coordinate transformations and static state feedback* i.e., those systems in which the function c, in (11), is of the form $c(z, u) = c(\Phi(x), u)$, and for which the transformation (12) only involves state variables. We further assume that $\partial c/\partial u$ is not identically zero, at least, locally around the equilibrium point.

Static PWM Control for Exactly Linearizable Systems

Let $P_{n-1}(\lambda)$ be an $(n-1)$th order Hurwitz polynomial with constant coefficients:

$$P_{n-1}(\lambda) = \lambda^{n-1} + a_{n-1}\lambda^{n-2} + \cdots + a_2\lambda + a_1 \tag{15}$$

Consider now the following auxiliary output function of the system (11):

$$s(z) = z_n + a_{n-1}z_{n-1} + \cdots + a_2 z_2 + a_1 z_1 \tag{16}$$

If the condition $s = 0$ is achieved by means of suitable controls, the restricted motions of the system (11) satisfy the following asymptotically stable linear time-invariant dynamics:

$$\begin{aligned} \dot{z}_1 &= z_2 \\ \dot{z}_2 &= z_3 \\ &\cdots \\ \dot{z}_{n-1} &= -a_{n-1}z_{n-1} - \cdots - a_2 z_2 - a_1 z_1 \end{aligned} \tag{17}$$

The following proposition is a direct consequence of the preceding considerations and of Theorem 1.

Proposition 1: A system of the form (10) is locally asymptotically stabilizable to the equilibrium point $(U, X(U))$ if the control action u is specified as a PWM control policy given by the solution of the following implicit (algebraic) equation

$$c(\Phi(x), u) = -\sum_{i=1}^{n}(a_{i-1} + aa_i)h^{(i-1)}(x) - W\ PWM_\tau\left[\sum_{i=1}^{n} a_i h^{i-1}(x)\right] \tag{18}$$

where $a_0 = 0$, and $a_n = 1$.

Proof: Imposing on the auxiliary output function $s(z)$, given in (16), the asymptotically stable discontinuous PWM controlled dynamics defined by (1), one immediately obtains, in terms of the transformed coordinates z, an *implicit* nonlinear equation for the required control input u. Rewriting the obtained expression in original state and input coordinates the static controller adopts the form (18). ¿From the assumption that $\partial c/\partial u$ is locally non zero, it follows that (18) can be explicitly solved for u. We denote, in general, the solution for u in an equation of the form: $c(\Phi(x), u) = z_{n+1}$, as $u = g(x, z_{n+1})$, i.e., $c(\Phi(x), g(x, z_{n+1})) \equiv z_{n+1}$ for some given indeterminate z_{n+1}.

As it can be easily seen from (18), the case of exactly linearizable systems results in a *static* PWM controller and, hence, the proposed scheme yields discontinuous control actions. Hence, "bang-bang" feedback control signals are generated in the closed loop system.

Dynamical PWM Control of Nonlinear Systems

Consider now the *extended system*, associated to system (8) (see [15]):

$$\begin{aligned} \dot{x} &= F(x, u) \\ \dot{u} &= v \end{aligned} \tag{19}$$

It is easy to see that if $z_1 = h(x)$ is a differential primitive element for (10), z_1 also qualifies as a differential primitive element for (19). Letting $c(z, u)$ become a new state variable z_{n+1}, it is follows readily that the GCCF of (19) is written as

$$\begin{aligned} \dot{z}_1 &= z_2 \\ \dot{z}_2 &= z_3 \end{aligned} \tag{20}$$

$$\cdots$$

$$\dot{z}_n = z_{n+1}$$

$$\dot{z}_{n+1} = \frac{\partial c(\Phi(x),u)}{\partial x} F(x,u)\Bigg|_{\substack{x=\Phi^{-1}(z)\\ u=g(z,z_{n+1})}} + \frac{\partial c(\Phi(x),u)}{\partial u}\Bigg|_{\substack{x=\Phi^{-1}(z)\\ u=g(z,z_{n+1})}} v$$

$$:= f(z,z_{n+1}) + g(z,z_{n+1})v$$

Hence, if the original system is exactly linearizable by means of state coordinates transformations, and static state feedback, so is the extended system with respect to the new auxiliary input v. An equilibrium point of (20) is evidently given by $v=0, u=U, x=X(U)$. We denote this equilibrium point by $((X(U),U),0)$.

Notice that the state coordinate transformation taking (19) into the linearizable form (20) is given by

$$\hat{z} := \begin{bmatrix} z \\ z_{n+1} \end{bmatrix} = \begin{bmatrix} \Phi(x) \\ c(\Phi(x),u) \end{bmatrix} := \hat{\Phi}(x,u) \tag{21}$$

which is evidently invertible whenever $\Phi(x)$ is invertible and $\partial c/\partial u$ is non zero. It is easy to see that $z=0$ is an equilibrium point of (20). Let $p_n(\lambda)$ be an $n-th$ order Hurwitz polynomial with constant coefficients:

$$p_n(\lambda) = \lambda^n + a_n\lambda^{n-1} + \cdots + a_2\lambda + a_1 \tag{22}$$

If one considers now the following auxiliary output function of the system (20)

$$\sigma(z,z_{n+1}) = z_{n+1} + a_n z_n + \cdots + a_2 z_2 + a_1 z_1 \tag{23}$$

then the condition $\sigma=0$ implies that the restricted motions of the system (20) satisfy the following asymptotically stable linear time-invariant dynamics

$$\begin{aligned} \dot{z}_1 &= z_2 \\ \dot{z}_2 &= z_3 \\ &\cdots \\ \dot{z}_n &= -a_n z_n - \cdots - a_2 z_2 - a_1 z_1 \end{aligned} \tag{24}$$

The following proposition is a direct consequence of the preceding considerations and of Theorem 1.

Proposition 2: A system of the form (10) is locally asymptotically stabilizable to the equilibrium point $((X(U),U),0)$ if the control action u is specified as a dynamical PWM control policy given by the solution of the following explicit first order nonlinear differential equation with discontinuous right hand side:

$$\begin{aligned} f(\hat{\Phi}(x,u)) + g(\hat{\Phi}(x,u))\dot{u} &= -\sum_{i=1}^{n}(a_{i-1}+aa_i)h^{(i-1)}(x) - (a_n+a)h^{(n)}(x,u) \\ &\quad -W\ PWM_\tau\left[\sum_{i=1}^{n} a_i h^{(i-1)}(x) + h^{(n)}(x,u)\right] \end{aligned} \tag{25}$$

where $a_0 = 0$.

Proof: As in Proposition 1, above, imposing on the auxiliary output function $\sigma(z, z_{n+1})$, given in (23), the asymptotically stable discontinuous PWM controlled PWM controlled dynamics, defined by (1), one immediately obtains, in terms of the transformed coordinates (z, z_{n+1}), a nonlinear algebraic equation for the required control input v. Rewriting the obtained static controller expression in terms of the original state and input coordinates (x, u), the controller adopts the dynamical form of equation (25).

Notice that since $g = \partial c/\partial u$ is assumed to be nonzero, the controller (2.25) is locally well defined and no *impasse points* needs to be considered (see Fliess and Hassler [18]).

APPLICATIONS TO SINGLE-LINK RIGID AND FLEXIBLE JOINT MANIPULATORS

PWM Control of a Single Link Rigid Robotic Manipulator

Consider the following nonlinear dynamical model of a single link robotic manipulator (Khalil [19]):

$$\begin{aligned} \dot{x}_1 &= x_2 \\ \dot{x}_2 &= \frac{g}{L}\sin x_1 - \frac{k}{M}x_2 + \frac{1}{ML^2}u \end{aligned} \tag{26}$$

where x_1 is the link angular position, x_2 is the angular velocity and u represents the applied torque. The mass M is assumed to be concentrated at the tip of the manipulator. The constant k is the viscous damping coefficient.

It is desired to synthesize, both, a static and a dynamical PWM feedback control policy which drives the angular position of the system to a constant desired angular position x_{1d}.

Static PWM Controller Design. Let $z_1 = x_1 - x_{1d}$. It is easy to see that z_1 qualifies as a differential primitive element for (26) and that to obtain the FGCCF of (26) it simply requires the use of the following (trivial) state coordinate transformation $z_1 = x_1 - x_{1d}, z_2 = x_2$:

$$\begin{aligned} \dot{z}_1 &= z_2 \\ \dot{z}_2 &= -\frac{g}{L}\sin(z_1 + x_{1d}) - \frac{k}{M}z_2 + \frac{1}{ML^2}u \end{aligned} \tag{27}$$

Let $s(z) = z_2 + a_1 z_1$ with $a_1 > 0$, be an auxiliary output function for system (27). Notice that if $s(x)$ is stabilized to zero, the constrained dynamics evolves according to $dz_1/dt = -a_1 z_1$ (or $dx_1/dt = -a_1(x_1 - x_{1d})$) thus achieving the desired regulation task. Imposing on the auxiliary output function $s(z)$ the asymptotically stable dynamics of the PWM controlled system (1) one obtained the following static PWM controller

$$\begin{aligned} u &= ML^2[-aa_1 z_1 + \left(\frac{k}{M} - a - a_1\right) z_2 \\ &\quad + \frac{g}{L}\sin(z_1 + x_{1d}) - W\ PWM_\tau(z_2 + a_1 z_1)] \end{aligned} \tag{28}$$

which, in original coordinates is rewritten as

$$u = ML^2 \left\{ -aa_1(x_1 - x_{1d}) + \left(\frac{k}{M} - a - a_1\right) x_2 \right. \tag{29}$$

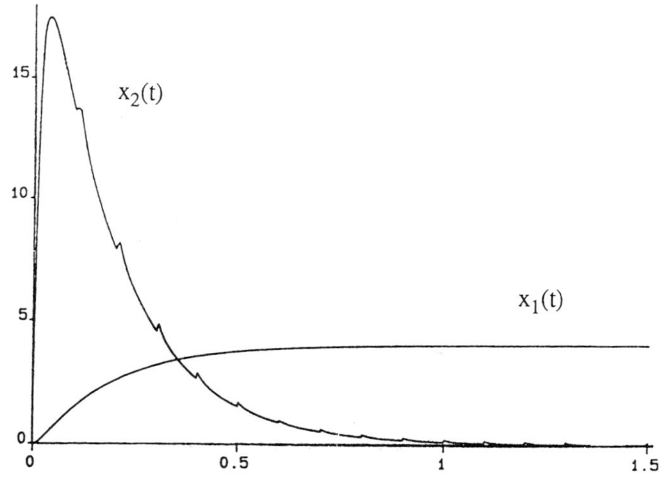

Figure 3. Angular position and angular velocity of static PWM controlled rigid robotic manipulator.

$$+\frac{g}{L}\sin(x_1) - WPWM_\tau[x_2 + a_1(x_1 - x_{1d})]\Big\}$$

Simulations were run for the above PWM controlled manipulator (26), (29), with the following parameters: $M = 0.01068\ [Kg], L = 0.5\ [m], k = 0, x_{1d} = 4\ [rad], g = 9.8\ [m/s^2]$. The auxiliary output function was synthesized with $a_1 = 60\ [s^{-1}]$. The static PWM controller parameters were chosen as: $a = 5.4\ [s^{-1}], W = 100\ [rad/s], T = 0.1\ [s], r = 0.01\ [s/rad]$. In this case the sufficient condition of Theorem 1 is verified as $rW/a = 0.185 < \tanh(0.27) = 0.2636$. Figure 3 depicts the state trajectories of the controlled system clearly showing convergence to the desired angular position.

The angular velocity is also shown to converge to zero with significant chattering. Figure 4 shows a magnified view of a portion of the *discontinuous* applied torque input signal u, as generated by (29), and of the PWM signal.

In spite of the fact that the discontinuities associated to the input torque variable u do not noticeably propagate towards the angular position variable (due to the two natural integration steps existing between the angular acceleration and the angular position) the bang-bang input behavior is deemed as highly undesirable. These discontinuities not only cause wear and tear, but, also, they represent an opportunity for unnecessary excitation of high-frequency unmodelled dynamics of the mechanical system.

Dynamical PWM Controller Design. Consider now the extended system of (26):

$$\begin{aligned} \dot{x}_1 &= x_2 \\ \dot{x}_2 &= \frac{g}{L}\sin x_1 - \frac{k}{M}x_2 + \frac{1}{ML^2}u \\ \dot{u} &= v \end{aligned} \tag{30}$$

Taking again the angular position error $x_1 = x_{1d}$ as a differential primitive element

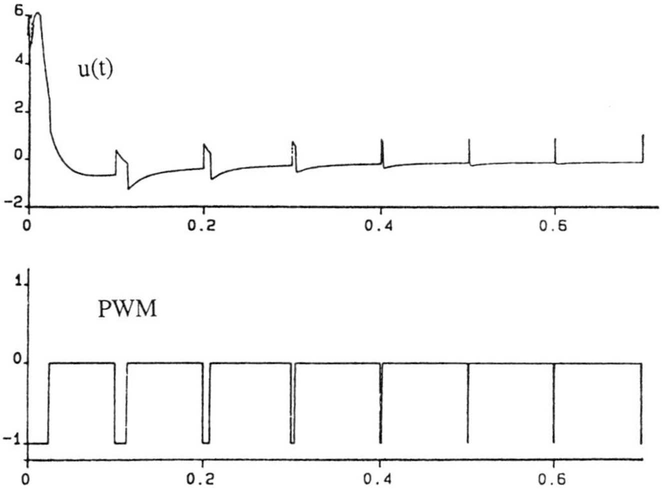

Figure 4. Discontinuous control input torque signal (magnified view of only a portion) and PWM control signal for static PWM controlled robotic manipulator.

x_1. The resulting FGCCF of the extended system is now obtained as

$$\begin{aligned} \dot{z}_1 &= z_2 \\ \dot{z}_2 &= z_3 \\ \dot{z}_3 &= -\frac{g}{L} z_2 \cos(z_1 + x_{1d}) - \frac{k}{M} z_3 + \frac{1}{ML^2} v \end{aligned} \tag{31}$$

with $z_1 = x_1 - x_{1d}, z_2 = x_2, z_3 = -(g/L)\sin x_1 - (k/M)x_2 + [(1/ML^2)]u$. Let the auxiliary output function $\sigma(z)$ be defined as: $z_3 + a_2 z_2 + a_1 z_1$, with a_2 and a_1 positive constants, chosen in the standard second order system form, with damping factor ζ and natural frequency ω_n : $a_2 = 2\zeta\omega_n$ and $a_1 = \omega_n^2$. Notice that if $\sigma(z)$ is stabilized to zero, the constrained dynamics evolves in accordance to the asymptotically stable second order dynamics: $dz_1/dt = z_2$; $dz_2/dt = -2\zeta\omega_n z_2 - \omega_n^2 z_1$ thus achieving the desired control task: $z_2 = x_2 \to 0$ and $z_1 = x_1 - x_{1d} \to 0$.

Imposing on $\sigma(z)$ the same asymptotically stable dynamics of the PWM controlled system (1) one obtains the following static PWM controller for the extended system

$$\begin{aligned} v &= ML^2 \left[-\left(a + \frac{k}{M} + 2\zeta\omega_n \right) z_3 - (2\zeta\omega_n a + \omega_n^2) z_2 - a\omega_n^2 z_1 \right. \\ & \left. + \frac{g}{L} z_2 \cos(z_1 + x_{1d}) - W\ PWM_\tau (z_3 + 2\zeta\omega_n z_2 + \omega_n^2 z_1) \right] \end{aligned} \tag{32}$$

which, in original coordinates, is rewritten as a dynamical PWM controller given by the solution of the following time-varying ordinary differential equation for the control input u with discontinuous (PWM) right hand side

$$\begin{aligned} \dot{u} &= ML^2 \left\{ -\left(a + \frac{k}{M} + 2\zeta\omega_n \right) \left(-\frac{g}{L} \sin x_1 - \frac{k}{M} x_2 + \frac{1}{ML^2} u \right) \right. \\ & - \left(2\zeta\omega_n a + \omega_n^2 \right) x_2 - a\omega_n^2 (x_1 - x_{1d}) + \frac{g}{L} x_2 \cos(x_1) \\ & \left. - W\ PWM_\tau \left[-\frac{g}{L} \sin x_1 - \frac{k}{M} x_2 + \frac{1}{ML^2} u + 2\zeta\omega_n x_2 + \omega_n^2 (x_1 - x_{1d}) \right] \right\} \end{aligned} \tag{33}$$

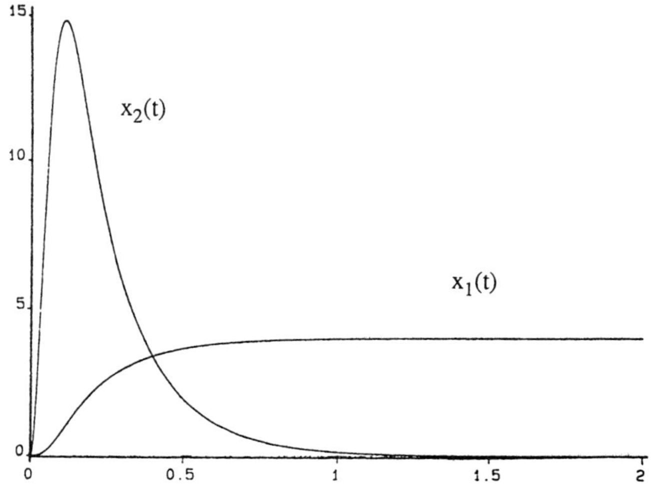

Figure 5. Angular Position and angular velocity of dynamical PWM controlled rigid robotic manipulator.

Simulations were run for the dynamically PWM controlled manipulator (26), (33) with the same physical parameter values as before. The auxiliary output function σ was synthesized such that the corresponding characteristic polynomial of the linearized system is $p_2(\lambda) = \lambda^2 + 2\zeta\omega_n\lambda + \omega_n^2$, with $\zeta = 0.8$ and $\omega_n = 28.0$ $[rad/s]$. The static PWM controller parameters were chosen in the same manner as in the previous example. Figure 5 depicts the state trajectories of the controlled system clearly showing convergence to the desired angular position and to zero angular velocity with no chattering being exhibited now.

Figure 6 shows the PWM signal and the substantially smoothed out (chattering-free) applied torque input signal u, as generated by the dynamical PWM controller (33). The effect of adding an integrator to the original input u of the system results in a low pass filtering effect on the generated auxiliary input v of the extended system.

PWM Control of a Single Link Flexible Joint Robotic Manipulator

Consider the following damping-free nonlinear dynamical model of a single link robotic manipulator with a flexible joint (Spong and Vidyasagar [20], Khalil [19]):

$$\begin{aligned} I\ddot{q}_1 + MgL\sin q_1 + \kappa(q_1 - q_2) = 0 \\ J\ddot{q}_2 - \kappa(q_1 - q_2) = u \end{aligned} \tag{34}$$

where q_1 is the angular position of the link, q_2 is the motor shaft's angular position and u represents the generated torque applied to the shaft (see [20] for details). The constants I and M represent, respectively, the moment of inertia and the mass of the link. J is the moment of inertia of the link and the motor shaft. The constant κ is the elasticity coefficient associated to the flexible joint.

A state space model of the above system is readily obtained by defining $x_1 = q_1, x_2 = dx_1/dt, x_3 = q_2, x_4 = dx_3/dt$ as shown below. (See also Korasani [21], for a different state space model).

$$\dot{x}_1 = x_2 \tag{35}$$

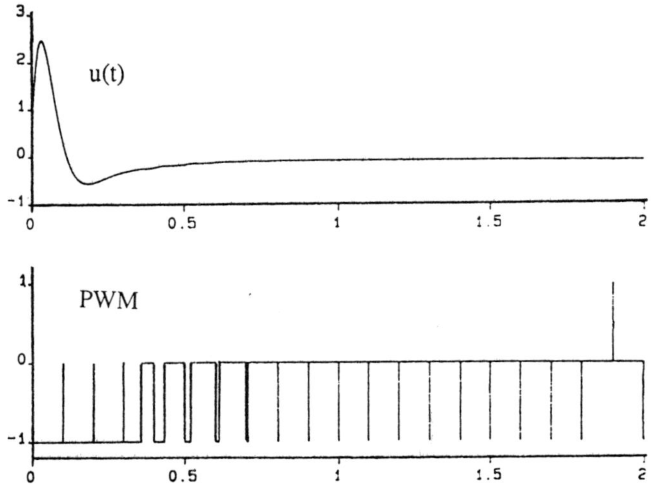

Figure 6. Chattering free control input torque signal and PWM signal for dynamical PWM controlled rigid robotic manipulator.

$$\begin{aligned}\dot{x}_2 &= -\frac{MgL}{I}\sin(x_1) - \frac{\kappa}{L}(x_1 - x_3) \\ \dot{x}_3 &= x_4 \\ \dot{x}_4 &= \frac{\kappa}{J}(x_1 - x_3) + \frac{1}{J}u\end{aligned}$$

It is desired to synthesize, both, a static and a dynamical PWM feedback control policy which drives the angular position of the system to a desired constant angular position x_{1d}. We summarize below the steps leading to both controllers designs.

Static PWM Controller Design

State coordinate transformation and its inverse to obtain FGCCF of the flexible link manipulator model:

$$\begin{aligned}z_1 &= x_1 - x_{1d} \\ z_2 &= x_2 \\ z_3 &= -\frac{MgL}{I}\sin(x_1) - \frac{\kappa}{I}(x_1 - x_3) \\ z_4 &= -\frac{MgL}{I}x_2\cos(x_1) - \frac{\kappa}{I}(x_2 - x_4)\end{aligned} \tag{36}$$

$$\begin{aligned}x_1 &= z_1 + x_{1d} \\ x_2 &= z_2 \\ x_3 &= \frac{1}{\kappa}[Iz_3 + MgL\sin(z_1 + x_{1d})] + z_1 + z_{1d} \\ x_4 &= \frac{1}{\kappa}[Iz_4 + MgLz_2\cos(z_1 + x_{1d})] + z_2\end{aligned} \tag{37}$$

Fliess generalized controller canonical form:

$$\begin{aligned} \dot{z}_1 &= z_2 \\ \dot{z}_2 &= z_3 \\ \dot{z}_3 &= z_4 \\ \dot{z}_4 &= \frac{MgL}{I}\left[z_2^2 + \frac{MgL}{I}\cos(z_1 + z_{1d}) + \frac{\kappa}{I}\right]\sin(z_1 + x_{1d}) \\ &\quad - \left[\frac{\kappa}{I} + \frac{\kappa}{J} + \frac{MgL}{I}\cos(z_1 + x_{1d})\right]\left[z_3 + \frac{MgL}{I}\sin(z_1 + x_{1d})\right] + \frac{\kappa}{IJ}u \end{aligned} \tag{38}$$

Auxiliary output function:

$$s(z) = z_4 + a_3 z_3 + a_2 z_2 + a_1 z_1 \tag{39}$$

with the a's chosen so that $p_3(\lambda) = \lambda^3 + a_3\lambda^2 + a_2\lambda + a_1$ is a Hurwitz polynomial (for instance $p(\lambda) = (\lambda + a)(s^2 + 2\zeta\omega_n\lambda + \omega_n^2)$, with $a > 0$ and $1 > \zeta > 0, \omega_n > 0$). In original coordinates:

$$\begin{aligned} s(x) &= -\frac{MgL}{I}x_2\cos(x_1) - \frac{\kappa}{I}(x_2 - x_4) - \frac{MgLa_3}{I}\sin(x_1) \\ &\quad - \frac{\kappa a_3}{I}(x_1 - x_3) + a_2 x_2 + a_1(x_1 - x_{1d}) \end{aligned}$$

Static PWM controller in transformed coordinates:

$$\begin{aligned} u &= -\frac{JMgL}{\kappa}\left[z_2^2 + \frac{MgL}{I}\cos(z_1 + x_{1d}) + \frac{\kappa}{I}\right]\sin(z_1 + x_{1d}) \\ &\quad + \left[1 + \frac{I}{J} + \frac{MgL}{\kappa}\cos(z_1 + x_{1d})\right]\left[z_3 + \frac{MgL}{I}\sin(z_1 + x_{1d})\right] \\ &\quad - \frac{IJ}{\kappa}\left[(a + a_3)z_4 + (a_2 + aa_3)z_3 + (a_1 + aa_2)z_2 + aa_1 z_1\right] \\ &\quad - W\ PWM_\tau[z_4 + a_3 z_3 + a_2 z_2 + a_1 z_1] \end{aligned} \tag{40}$$

Static PWM controller in original coordinates:

$$\begin{aligned} u &= -\frac{JMgL}{\kappa}\left[x_2^2 + \frac{MgL}{I}\cos(x_1) + \frac{\kappa}{I}\right]\sin(x_1) + \left[1 + \frac{I}{J} + \frac{MgL}{\kappa}\cos(x_1)\right] \times \\ &\quad \left[-\frac{MgL}{I}\sin(x_1) - \frac{\kappa}{I}(x_1 - x_3) + \frac{MgL}{I}\sin(x_1)\right] \\ &\quad - \frac{IJ}{\kappa}\left\{(a + a_3)\left[\frac{MgL}{I}x_2\cos(x_1) + \frac{\kappa}{I}(x_2 - x_4)\right]\right. \\ &\quad \left. + (a_2 + aa_3)\left[\frac{MgL}{I}\sin(x_1) + \frac{\kappa}{I}(x_1 - x_3)\right] + (a_1 + aa_2)x_2 + aa_1[x_1 - x_{1d}]\right\} \\ &\quad - W\ PWM_\tau\left[-\frac{MgL}{I}x_2\cos(x_1) - \frac{\kappa}{I}(x_2 - x_4)\right. \end{aligned} \tag{41}$$

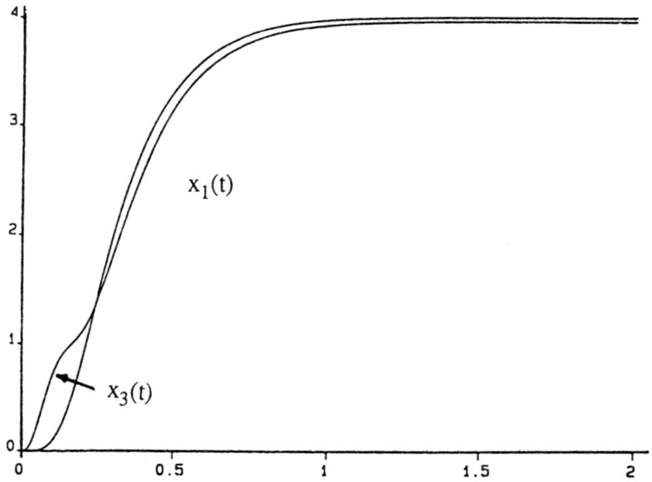

Figure 7. Angular position of link and motor shaft for static PWM controlled flexible robotic manipulator.

$$-a_3\frac{MgL}{I}\sin(x_1) - a_3\frac{\kappa}{I}(x_1 - x_3) + a_2x_2 + a_1(x_1 - x_{1d})\Big]$$

Simulations were run for the static PWM controlled manipulator (35)-(40) with the following parameters: $M = 0.01068\ [Kg], L = 0.5\ [m], \kappa = 1.0\ Nm/rad, x_{1d} = 4\ [rad], g = 9.8\ [m/s^2], J = 0.07\ [Nms^2/rad]$. The auxiliary output function was synthesized such that the corresponding characteristic polynomial is given by $p_2(\lambda) = (\lambda+b)(\lambda^2+2\zeta\omega_n\lambda+\omega_n^2)$ with $b = 30, \zeta = 0.8, \omega_n = 28.0$. The static PWM controller parameters chosen again as: $a = 5.4\ [s^{-1}], W = 10^4\ [rad/s], T = 0.1\ [s], r = 10^{-4}\ [s/rad]$.

Figure 7 depicts the controlled trajectories of the angular positions x_1 and x_3 clearly showing convergence of x_1 to the desired angular position. Due to the assumed flexibility, the shaft's angular position x_3 exhibits a small steady state error of 0.052 $[rad]$ with respect to x_1. Figure 8 shows the link and shaft's angular velocities x_2 and x_4. The effects of the PWM discontinuities are clearly portrayed in the shaft's angular velocity response.

Dynamical PWM Controller Design

The extended system:

$$\begin{aligned}
\dot{x}_1 &= x_2 \\
\dot{x}_2 &= -\frac{MgL}{I}\sin(x_1) - \frac{\kappa}{L}(x_1 - x_3) \\
\dot{x}_3 &= x_4 \\
\dot{x}_4 &= \frac{\kappa}{J}(x_1 - x_3) + \frac{1}{J}u \\
\dot{u} &= v
\end{aligned} \tag{42}$$

The (PWM) signal and the discontinuous applied torque input u, as generated by (41), are shown in Figure 9.

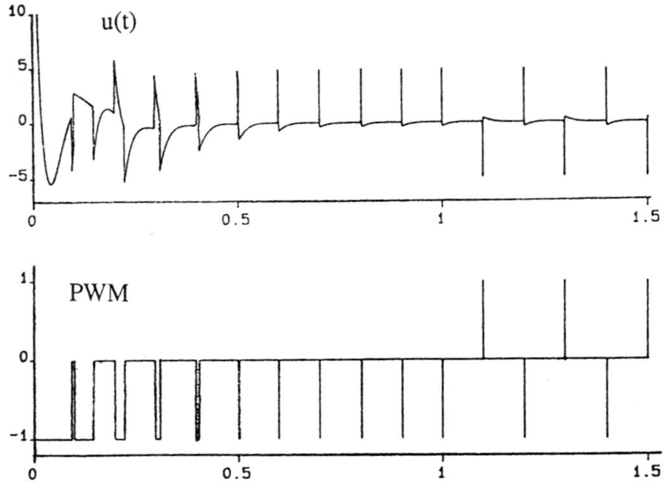

Figure 8. Angular velocities of link and motor shaft for static PWM controlled flexible robotic manipulator.

State coordinate transformation and its inverse to obtain FGCCF of the extended manipulator system:

$$
\begin{aligned}
z_1 &= x_1 - x_{1d} \\
z_2 &= x_2 \\
z_3 &= -\frac{MgL}{I}\sin(x_1) - \frac{\kappa}{I}(x_1 - x_3) \\
z_4 &= -\frac{MgL}{I}x_2\cos(x_1) - \frac{\kappa}{I}(x_2 - x_4) \\
z_5 &= \left(\frac{MgL}{I}\right)^2 \sin(x_1)\cos(x_1) + \frac{MgL\kappa}{I^2}(x_1 - x_3)\cos(x_1) \\
&\quad + \frac{MgL}{I}x_2^2\sin(x_1) + \frac{MgL\kappa}{I^2}\sin(x_1) + \left(\frac{\kappa}{I}\right)^2(x_1 - x_3) \\
&\quad + \frac{\kappa^2}{IJ}(x_1 - x_3) + \frac{\kappa}{IJ}u
\end{aligned}
\tag{43}
$$

$$
\begin{aligned}
x_1 &= z_1 + x_{1d} \\
x_2 &= z_2 \\
x_3 &= \frac{1}{\kappa}[Iz_3 + MgL\sin(z_1 + x_{1d})] + z_1 + x_{1d}
\end{aligned}
\tag{44}
$$

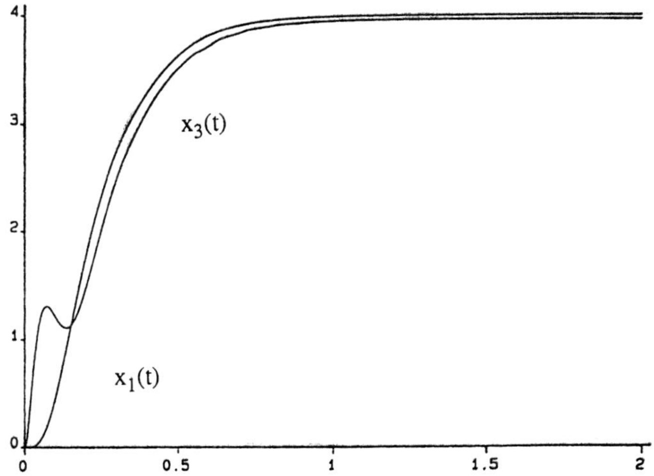

Figure 9. Discontinuous control input torque signal (magnified view of only a portion) and PWM control signal for static PWM controlled flexible robotic manipulator.

$$x_4 = \frac{1}{\kappa}[Iz_4 + MgLz_2 \cos(z_1 + x_{1d})] + z_2$$

$$\begin{aligned} u &= z_5 - \left(\frac{MgL}{I}\right)^2 \sin(z_1 + x_{1d}) \cos(z_1 + x_{1d}) \\ &+\frac{1}{\kappa}[Iz_3 + MgL \sin(z_1 + z_{1d})] \left[\frac{MgL\kappa}{I^2} + \left(\frac{k}{I}\right)^2 + \frac{k^2}{IJ}\right] \cos(z_1 + x_{1d}) \\ &-\frac{MgL}{I} z_2^2 \sin(z_1 + x_{1d}) - \frac{MgL\kappa}{I^2} \sin(z_1 + x_{1d}) \end{aligned}$$

Fliess generalized controller canonical form of the extended system:

$$\begin{aligned} \dot{z}_1 &= z_2 \\ \dot{z}_2 &= z_3 \\ \dot{z}_3 &= z_4 \\ \dot{z}_4 &= z_5 \\ \dot{z}_5 &= c(z, v) \end{aligned} \tag{45}$$

The function $c(z, v)$ is easily obtained from (42), (43) and it only requires long but straightforward manipulations. It is not presented here in the interest of brevity.

Auxiliary output function for the extended system:

$$\sigma(z) = z_5 + a_4 z_4 + a_3 z_3 + a_2 z_2 + a_1 z_1 \tag{46}$$

with the a's chosen so that $p_4(\lambda) = \lambda^4 + a_4\lambda^3 + a_3\lambda^2 + a_2\lambda + a_1$ is a Hurwitz polynomial (for instance $p(\lambda) = (\lambda^2 + 2\zeta_1\omega_{1n}\lambda + \omega_{1n}^2)(\lambda^2 + 2\zeta_2\omega_{2n}\lambda + \omega_{2n}^2)$, with $1 > \zeta_1, \zeta_2 > 0, \omega_{1n}, \omega_{2n} > 0$). In original coordinates:

$$\begin{aligned}\sigma(x) &= \left(\frac{MgL^2}{I}\right)\sin(x_1)\cos(x_1) + \frac{MgL\kappa}{I^2}(x_1 - x_3)\cos(x_1) \\ &+\frac{MgL}{I}x_2^2\sin(x_1) + \frac{MgL}{I}\left(\frac{\kappa}{I} - a_3\right)\sin(x_1) + \left[\left(\frac{\kappa}{I}\right)^2 - \frac{\kappa}{I}a_3\right](x_1 - x_3) \\ &+\frac{\kappa^2}{IJ}(x_1 - x_3) + \frac{\kappa}{IJ}u - \frac{MgLa_4}{I}x_2\cos(x_1)\frac{\kappa a_4}{I}(x_2 - x_4) \\ &+a_2x_2 + a_1(x_1 - x_{1d})\end{aligned} \tag{47}$$

Dynamical PWM controller in original coordinates:

$$\begin{aligned}\dot{u} &= \frac{IJ}{\kappa}\Big\{-W\,PWM[\sigma(x)] - a_4\frac{MgL}{I}\left[\frac{MgL}{I}\sin(x_1) + \frac{\kappa}{I}(x_1 - x_3)\cos(x_1)\right] \\ &-a_4x_2^2\sin(x_1) + \frac{\kappa}{I}\left[\frac{MgL}{I}\sin(x_1) + (x_1 - x_3)\left(\frac{\kappa}{I} + \frac{\kappa}{J}\right) + \frac{u}{J}\right] \\ &+a_3\left[\frac{MgL}{I}x_2\cos(x_1)\frac{\kappa}{I}(x_2 - x_4)\right] + a_2\left[\frac{MgL}{I}\sin(x_1) + \frac{\kappa}{I}(x_1 - x_3)\right] - a_1x_2 \\ &-\frac{MgL}{I}\Big\{x_2^3\cos(x_1) - 2x_2\left[\frac{MgL}{I}\sin(x_1) + \frac{\kappa}{I}(x_1 - x_3)\right]\sin(x_1) \\ &-\left[\frac{MgL}{I}x_2\cos(x_1) + \frac{\kappa}{I}(x_2 - x_4)\right]\cos(x_1) \\ &\qquad\qquad +\left[\frac{MgL}{I}\sin(x_1) + \frac{\kappa}{I}(x_1 - x_3)\right]x_2\sin(x_1)\Big\} \\ &-\frac{\kappa}{I}\left[\frac{MgL}{I}x_2\cos(x_1) - \left(\frac{\kappa}{J} + \frac{\kappa}{I}\right)(x_2 - x_4)\right] + \frac{MgL}{I}x_2^2\sin(x_1) \\ &+\frac{MgL\kappa}{I^2}\sin(x_1) + \frac{MgL\kappa}{I^2}\sin(x_1) + \left(\frac{\kappa}{I}\right)^2(x_1 - x_3) + \frac{\kappa^2}{IJ}(x_1 - x_3) + \frac{\kappa}{IJ}u\Big\}\end{aligned} \tag{48}$$

Simulations were run for the dynamically PWM controlled manipulator (42), (48) with the same parameter values as before. The auxiliary output function was synthesized such that the corresponding characteristic polynomial is given by $p_4(\lambda) = (\lambda^2 + 2\zeta_1\omega_{1n}\lambda + \omega_{1n}^2)(\lambda^2 + 2\zeta_2\omega_{2n}\lambda + \omega_{2n}^2)$ with $\zeta_1 = 0.8, \zeta_2 = 0.9, \omega_{1n} = 28.0\ [rad/s], \omega_{2n} = 15.25\ [rad/s]$. The PWM controller parameters were chosen as before. Figure 10 depicts the controlled trajectories of the angular positions x_1 and x_3 clearly showing convergence of x_1 to the desired reference value.

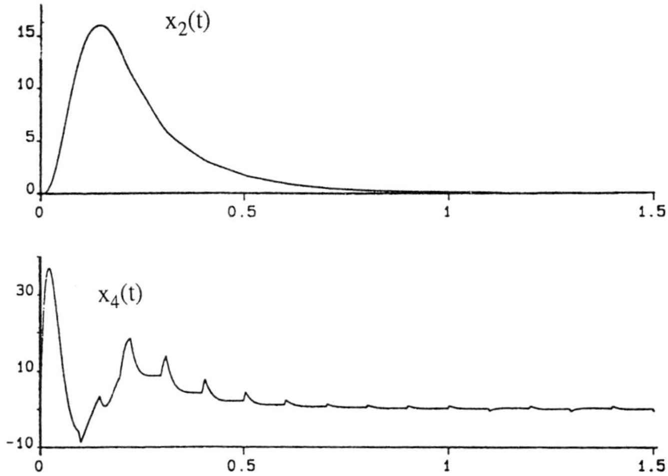

Figure 10. Angular position and angular velocity of dynamical PWM controlled flexible robotic manipulator.

Due to the assumed flexibility, the shaft's angular position x_3 exhibits a small steady state error of 0.052 $[rad]$ with respect to x_1. Figure 11 shows the link and shaft's angular velocities x_2 and x_4.

The (chattering-free) dynamically generated torque input signal u, as generated by (48), is shown in Figure 12.

CONCLUSIONS

A general stabilizing design procedure, based on dynamical PWM feedback control policies, has been presented for nonlinear systems describing some linearizable mechanical systems. A stabilizing discontinuous static controller of the PWM type is proposed for an elementary scalar system. Based on this simple result, a static PWM controller design can be obtained, for general high order nonlinear systems, by the zeroing of a suitably chosen auxiliary scalar output function. Zeroing of such an auxiliary output function induces an asymptotically stable motion for the constrained dynamics characterized by a linear time-invariant system with eigenvalues placeable at will. The results are easily implemented on a dynamical extension of the original nonlinear system which now results in a dynamical PWM feedback controller. In the dynamical controller case, the discontinuities, associated to the PWM regulator, take place on the one-dimensional state space of the dynamical controller, and not in the state space of the system. The resulting integrated control actions are, thus, *continuous* with substantially reduced (smoothed out) chattering.

In order to establish the salient stability features of the actual closed loop PWM controlled system, no need arises to resort to *average* controlled system considerations, nor high sampling frequency assumptions. As a matter of fact, the sampling frequency plays no crucial role in the stabilization features of the system, aside from the verification of a simple algebraic condition. New applications areas, such as nonlinear chemical process control, nonlinear electromechanical systems control etc., in which PWM con-

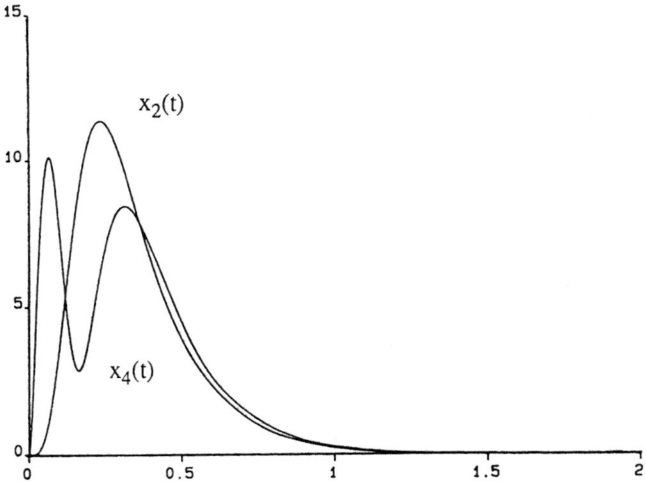

Figure 11. Angular velocities of link and motor shaft for dynamical PWM controlled flexible robotic manipulator.

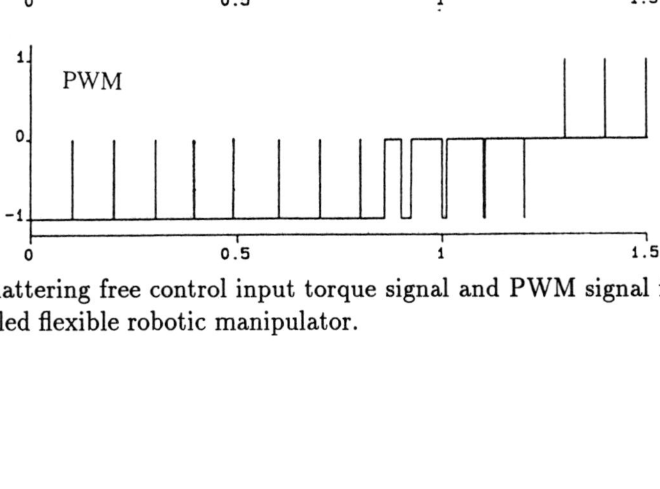

Figure 12. Chattering free control input torque signal and PWM signal for dynamical PWM controlled flexible robotic manipulator.

trol was not traditionally feasible, can now benefit from the inherent robustness and high performance characteristics of this class of discontinuous control strategies.

REFERENCES

1. F. Czaki, *Modern Control Theories*, Akademia Kiadó, Budapest, 1972.

2. M. Fliess, "Nonlinear control theory and differential algebra," in *Modeling and Adaptive Control,* Byrnes and A. Kurzhanski (Eds.), Lecture Notes in Control and Information Sciences, Vol. 105, pp. 134-145, Springer-Verlag, New York, 1988.

3. M. Fliess, "Generalized controller canonical form for linear and nonlinear dynamics," *IEEE Trans. Auto. Cont.,* Vol. AC-35. pp. 994-1001, 1990.

4. M.Fliess, "What the Kalman state representation is good for," *Proc. 29th IEEE Conference on Decision and Control*, Vol. 3, pp. 1282-1287, Honolulu, Hawaii, December 1990.

5. M. Fliess and M. Hasler, "Questioning the classic state-space description via circuit examples," in *Realization and Modelling in System Theory*, M. A. Kashoek, J. H. Van Schuppen and A. C. M. Ran (eds.), Birkhäuser, Boston, 1990.

6. M. Fliess and F. Messager, "Sur la commande en régime Glissant," *C. R. Academie Sciences Paris*, Series I. Vol. 313, pp. 951-956, 1991.

7. H. Khalil, *Nonlinear Systems*, MacMillan, New York, 1992.

8. K. Khorasani, "Nonlinear feedback control of flexible joint manipulators: A single link case study," *IEEE Tran. Auto. Cont.*, Vol. 35, No. 10, pp. 1145-1149, 1990.

9. M. LaCava, G. Paletta and C. Piccardi, "Stability analysis of PWM control systems with PID regulators," *Int. J. Cont.*, Vol. 39, No. 5, pp. 987-1005, 1984.

10. H. Nijmeijer and A. Van der Schaft, *Nonlinear Dynamical Control Systems,* Springer-Verlag, New York, 1990.

11. E. Polak, "Stability and graphical analysis of first order pulse-width-modulated sampled-data regulator systems," *IRE Tran. Auto. Cont.,* Vol. AC-6, pp. 276-282, 1961.

12. H. Sira-Ramírez, "Nonlinear pulse-width-modulation controller design," in *Variable Structure Control for Robotics and Aerospace Systems,* K. K. D. Young (ed.), Elsevier Science, The Netherlands, 1993.

13. H. Sira-Ramírez, "A Geometric approach to pulse-width-modulated control in nonlinear dynamical systems," *IEEE Trans. Auto. Cont.,* Vol. AC-34. pp. 184-187, 1989.

14. H. Sira-Ramírez, "Invariance conditions in nonlinear PWM controlled systems," *Int. J. Systems Sci.*, Vol. 20, pp. 1679-1690, 1989.

15. H. Sira-Ramírez and P. Lischinsky-Arenas, "Dynamical discontinuous feedback control of nonlinear systems," *IEEE Trans. Auto. Cont.*, Vol. AC-35, pp. 1373-1378, 1990.

16. H. Sira-Ramírez, "Nonlinear dynamical discontinuous feedback controlled descent on a non atmosphere-free planet: A differential algebraic Approach," *Control-Theory and Advanced Technology*, Vol 7, No. 2, pp. 301-320, 1991.

17. H. Sira-Ramírez, "Dynamical pulse-width-modulation control of nonlinear systems," *Systems and Control Letters*, Vol. 18, No. 2, pp. 223-231, 1992.

18. R. A. Skoog, "On the stability of pulse-width-modulated feedback systems," *IEEE Tran. Auto. Cont.,* Vol. AC-13, No. 5, pp. 532-538, 1968.

19. R. A. Skoog and G. Blankenship, "Generalized pulse-modulated feedback systems: Norms, gains, Lipschitz constants and stability," *IEEE Tran. Auto. Cont.,* Vol. AC-15. pp. 300-315, 1970.

20. M.Spong and M. Vidyasagar, *Robot Dynamics and Control,* John Wiley and Sons, New York, 1988.

21. Y. Z. Tsypkin, *Relay Control Systems,* Cambridge University Press, Cambridge, 1984.

A CHAOTIC CONTROLLER FOR THE DOUBLE PENDULUM

Thomas L. Vincent[1], Thomas J. Schmitt[2], and Tyrone L. Vincent[3]

[1]Aerospace and Mechanical Engineering
University of Arizona, Tucson, AZ 85745
[2]Optical Storage Development
IBM AdStar, Tucson, AZ 85744
[3]Electrical and Computer Engineering
University of Arizona, Tucson, AZ 85745

ABSTRACT

Here we consider a two link pendulum attached at the top link to a fixed DC motor. A second DC motor provides a bearing between the two links. The torque provided by both motors is limited so that each motor operating under maximum current at steady state will lift each link to an angle of less than 20°. Chaotic motion may be easily produced in this system by driving the system with the first motor under a sine input with no input to the second motor. This paper examines the potential usefulness of using such chaotic motion as a part of an overall control system design strategy in order to achieve target states which would otherwise require more complicated control algorithms (e.g. optimal control theory). In particular we examine the problem of obtaining an inverted position for the second link starting the system with the two links at rest hanging straight down. We demonstrate that an open loop chaotic controller in conjunction with a closed loop controller can indeed result in a useful control design. The key to using such a design depends on having available a simple and accurate method for determining the domain of attraction for the closed loop controller. Results are presented both in terms of a computer simulation of this system and data from an actual two link pendulum.

INTRODUCTION

There are many studies that deal with the control of pendulum systems. In their introduction, Meier Zu Farwig and Unbegauen [6] provide a short list of papers dealing with the control of single, double, and triple pendulums. These studies are generally concerned with stabilizing the pendulums in an inverted position. Other studies, such as Richter and Scholz [8] which describe the possible chaotic nature of a conservative double pendulum having no control inputs reveals a deeper complexity to these problems than are traditionally dealt with by assuming motion only in the vicinity of a desired equilibrium point. Our objective is to use the chaotic motion in such a way that it can be used to achieve a desired design objective.

Ott, Grebogi, and Yorke [7] and Vincent and Yu [11] have already shown that chaos can be suppressed by stabilizing one of the unstable equilibrium points embedded in the chaotic attractor. One can start with any given initial condition and wait until the trajectory approaches a sufficiently small region about one of the equilibrium points and then apply the feedback control which will capture the system at that equilibrium point. This idea as extended to transient chaos is also discussed by Tél [10]. To use these ideas in a practical way for nonlinear control design, three essential components are required: (1) a control law which will produce chaotic motion in state variables of interest (2) a control law for stabilizing the system about the desired operating point and (3) knowledge of the domain of attraction for the stabilizing controller. The double pendulum provides an ideal test bed for examining the practicality of such an approach. Chaos can be easily produced, it can be easily suppressed and methods exist for determining domains of attraction for a given control law.

We will begin with a differential equation model that can accurately simulate the behavior of a double pendulum subject to motor inputs at its pivot points. The motors will be used to produce stability in the vicinity of an unstable equilibrium point and to produce chaotic motion elsewhere. Ideally the control algorithm will produce chaotic motion via open loop control, to move the system into a local domain of attraction for a particular closed loop controller. The closed loop controller is then switched on to "trap" or stabilize the system.

Our model will be based on a physical system which exists in the University of Arizona Controls Laboratory. We use the equations of motion for a *general* double pendulum applied to this system. Integration of the equations reveal the chaotic nature of the system. We use the model to develop our control algorithm and then apply the results to the physical laboratory system.

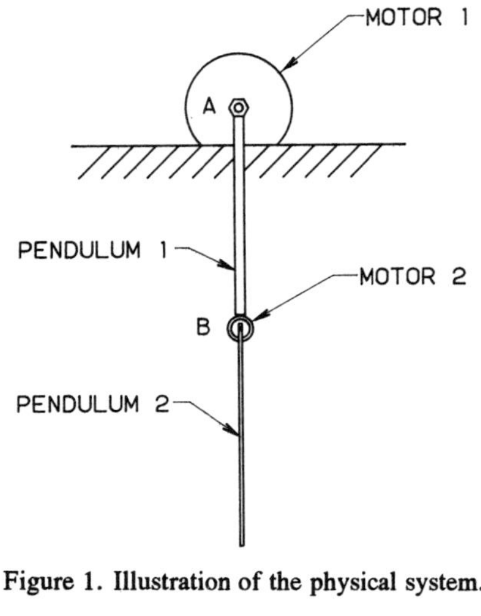

Figure 1. Illustration of the physical system.

THE DOUBLE PENDULUM SYSTEM

Figure 1 shows an illustration of the physical system. DC motors are attached to each pendulum. The larger motor is attached to a table top which represents an inertial frame of reference. The smaller motor is rigidly attached to the end of the large pendulum. The bearings of the motors become the pivots A and B, and motion is confined to a vertical plane. Neither motor can produce enough torque (within rated current) to rotate their respective pendulums much beyond fifteen or twenty degrees from the stable vertical positions. A digital, rotary encoder is attached to the shaft of each motor to measure angular position. Angular velocities are determined by a digital filter. A 16 MH AT personal computer with a Data Translation A/D, D/A board is used to monitor and control the system. Measured system constants for the laboratory system as needed for the simulation are given in Table 1.

Table 1. Double pendulum system constants

Symbol	Units	Description
$g = 32.17$	ft/sec^2	Gravitational constant
Pendulum #1		
$\omega_n = 5.026$	rad/sec	Undamped natural frequency
$r_1 = .95833$	feet	Distance to center of mass
$r_3 = 1.3542$	feet	Distance to pivot B
$I_A = .05$	$slug\text{-}ft^2$	Moment about center of rotation
$m_1 = 1.327 / g$	slugs	Mass of pendulum #1
Pendulum #2		
$\omega_n = 6.28$	rad/sec	Undamped natural frequency
$r_2 = .65104$	feet	Distance to center of mass
$I_B = .0035$	$slug\text{-}ft^2$	Moment about center of rotation
$m_2 = .1684 / g$	slugs	Mass of Pendulum #2
Motor #1		
$K_{\gamma 1} = .194$	ft-lb/amp	Motor torque constant
$K_{\beta 1} = .263$	Volt-sec/rad	Back EMF constant ($K_{\gamma 1} = .7375 K_{\beta 1}$)
$R_1 = 7.7$	Ohms	Resistance of armature windings
$\mu_1 = .017$	ft-lb	Friction damping
$\eta_1 = .000455$	ft-lb/rad/sec	Velocity damping
$U1_{max} = 19.9$	Volts	Maximum supply voltage
Motor #2		
$K_{\gamma 2} = .0218$	ft-lb/amp	Motor torque constant
$R_2 = 2.8$	Ohms	Resistance of armature windings
$\mu_2 = .0043$	ft-lb	Friction damping
$\eta_2 = .000455$	ft-lb/rad/sec	Velocity damping
$U2_{max} = 1.7$	Amps	Maximum supply current

EQUATIONS OF MOTION

Equations of motion for a double pendulum system confined to planar motion are easy to find in literature for pendulums with simple shapes. The laboratory double pendulum does not have lumped masses or slender, uniform rod pendulums so a more general form for the equations of motion are needed. These may be derived using the fundamental form of Lagrange's equations.

$$\frac{d}{dt}\left(\frac{\partial T}{\partial \dot{\boldsymbol{q}}}\right) - \frac{\partial T}{\partial \boldsymbol{q}} = \boldsymbol{Q}^{T}$$

where T is the systems total kinetic energy, $\boldsymbol{q}$ is a vector of the generalized coordinates, $\boldsymbol{Q}^{T}$ is a vector of the generalized forces, and the dot over $\boldsymbol{q}$ represents the time derivative. Details of the derivation for the general case depicted in Figure 2 are given by Schmitt [9]. The two motors provide torque inputs to the system. These are lumped together with any friction or damping torques applied at the pivots into the total equivalent torques U_1 and U_2.

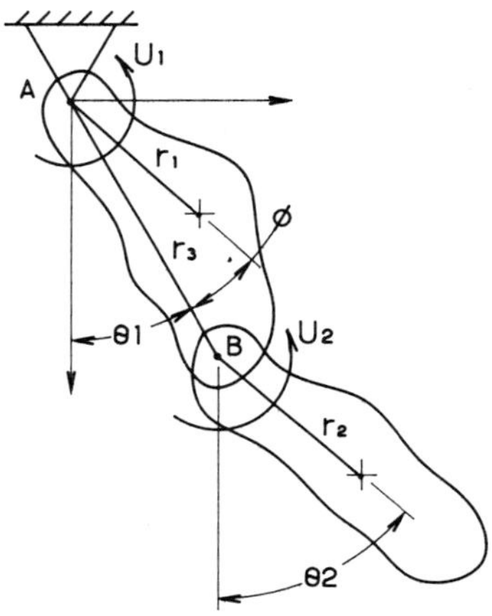

Figure 2. A general double pendulum system.

In terms of the angles defined in Figure 2 and the system constants defined in Table 1, the equations of motion can be written as:

$$(I_A + m_2 r_3^2)\ddot{\theta}_1 + m_2 r_2 r_3 \cos(\theta_2-\theta_1)\ddot{\theta}_2 - m_2 r_2 r_3 \sin(\theta_2-\theta_1)\dot{\theta}_2^2 + m_1 r_1 g \sin(\theta_1+\phi) + m_2 r_3 g \sin(\theta_1) = U_1 - U_2$$

$$I_b\ddot{\theta}_2 + m_2 r_2 r_3 \cos(\theta_2-\theta_1)\ddot{\theta}_1 + m_2 r_2 r_3 \sin(\theta_2-\theta_1)\dot{\theta}_1^2 + m_2 r_2 g \sin(\theta_2) = U_2$$

For the laboratory system, the center of mass of the first pendulum is approximately in line between A and B. We will assume that $\phi = 0$.

The first motor was driven by supplying a known voltage to the motor leads. Making the assumption that the winding inductance is small enough to be negligible, the equation for the motor torque M_1 as a function of the control voltage input u_1 can be written as

$$M_1 = \frac{K_{\gamma 1} u_1}{R_1} - \frac{K_{\gamma 1} K_{\beta 1}}{R_1} \dot{\theta}_1$$

where the second term is the back EMF produced by rotation. The second motor was driven by controlling the current through the motor windings. The motor torque equation is simply

$$M_2 = K_{\gamma 2} u_2$$

The Coulomb damping was modeled as kinetic friction. The coefficient was multiplied by the sign of the relative angular velocity at the pivots.

$$f_i = -\mu_i \, \mathrm{SGN}(\omega_i)$$

Where SGN(x) is the BASIC language sign function. From experiment it was determined that Coulomb friction was the largest contributor to the damping moment at the pivot points. Viscous damping was not included in the simulation.

For each pendulum, the Coulomb friction can be written:

$$f_1 = -\mu_1 \, \mathrm{SGN}(\dot{\theta}_1)$$
$$f_2 = -\mu_2 \, \mathrm{SGN}(\dot{\theta}_2 - \dot{\theta}_1)$$

Summing these torques yield:

$$U_1 = \frac{K_{\gamma 1} u_1}{R_1} - \frac{K_{\gamma 1} K_{\beta 1}}{R_1} \dot{\theta}_1 - \mu_1 \, \mathrm{SGN}(\dot{\theta}_1)$$
$$U_2 = K_{\gamma 2} u_2 - \mu_2 \, \mathrm{SGN}(\dot{\theta}_2 - \dot{\theta}_1)$$

All constants for the above equations are defined in Table 1.

STATE SPACE FORMULATION

The following state variables were chosen to represent the system:

$$x_1 = \theta_1 \,, \qquad x_3 = \theta_2 - \theta_1$$
$$x_2 = \dot{\theta}_1 \,, \qquad x_4 = \dot{\theta}_2$$

The following constants are introduced to simplify the conversion process,

$$A = I_A + m_2 r_3^2 \qquad B = m_2 r_2 r_3 \qquad C = m_1 r_1 g$$

$$D = m_2 r_3 g \qquad E = I_B \qquad F = m_2 r_2 g$$

In terms of these constants, the state variable equations of motion become:

$$A\dot{x}_2 + B\cos(x_3)\dot{x}_4 - B\sin(x_3)x_4^2 + C\sin(x_1) + D\sin(x_1) = U_1 - U_2$$

$$E\dot{x}_4 + B\cos(x_3)\dot{x}_2 + B\sin(x_3)x_2^2 + F\sin(x_3+x_1) = U_2$$

Cramer's Rule is used to solve for $\dot{x}_2$ and $\dot{x}_4$ and along with the state definitions, we obtain

$$\dot{x}_1 = x_2$$

$$\dot{x}_2 = \frac{[U_1 - U_2 + B\sin(x_3)x_4^2 - (C+D)\sin(x_1)]E - [U_2 - B\sin(x_3)x_2^2 - F\sin(x_3+x_1)]B\cos(x_3)}{AE - B^2\cos^2(x_3)}$$

$$\dot{x}_3 = x_4 - x_2$$

$$\dot{x}_4 = \frac{[U_2 - B\sin(x_3)x_2^2 - F\sin(x_3+x_1)]A - [U_1 - U_2 + B\sin(x_3)x_4^2 - (C+D)\sin(x_1)]B\cos(x_3)}{AE - B^2\cos^2(x_3)}$$

We put the equations in this state space form for ease of use with standard numerical integration techniques. It can be shown that the denominator will always be positive so there is no chance of a singularity occurring during the integration. At this point we make note of an interesting quote from Richter and Scholz [8], "... books ... tend to stop short of the interesting part of the analysis. They seem to tacitly assume that deriving the equations of motion is the hard part. Nowhere except in Arnold's book have we found a hint to the bewildering complexity that this simple system exhibits."

GENERATING CHAOS

The double pendulum system will exhibit a variety of complex behaviors depending on the nature of the inputs to the two motors. Chaotic motion can be induced by driving the upper motor with a sine or cosine input voltage at a frequency near the natural frequency of the lower pendulum. If the amplitude of the input is low, the pendulums will settle into periodic motion (in phase with each other). As the amplitude is increased, the lower pendulum will be driven into chaotic motion as evidenced by its spinning around its pivot point, sometimes clockwise, sometimes counterclockwise. If the voltage amplitude is increased further, both pendulums will eventually be driven into chaotic motion as evidenced by both pendulums spinning around their axis in sometimes clockwise, sometimes counterclockwise directions. Only the first case of chaotic motion for the lower pendulum will be studied here.

To study the chaotic motion described above, input was restricted to motor 1 in the form of a cosine wave.

$$u_1 = E\cos(\omega t)$$

Where, E is the input amplitude in volts, ω is the driving frequency in rad/sec, and t is time. The second pendulum was allowed to swing freely.

By trial and error, it was found that an amplitude of 5.3 volts and a frequency of 5 rad/sec would produce the desired chaotic motion in the second pendulum. Both the model and the physical system exhibited the same behavior with this input. The lower pendulum would spin around randomly while the upper pendulum swung back and forth with quasi periodic motion.

A bifurcation diagram is useful in determining the region where chaotic motion begins with respect to a particular parameter. We are particularly interested in knowing how close our driving amplitude is to the "border" of chaos. Our bifurcation program is patterned after one written by Baker and Gollub [1] for a single pendulum although the "snapshot" algorithm is a little different. It is important for accuracy that the recording of the states occurs exactly at the same place in each drive cycle. A "snapshot" must be taken exactly when $\omega t = n2\pi + \rho$ for (n= 51,52,...,150), where ρ is a phase angle. The first fifty drive cycles are ignored to let any transient responses die out.

For any given point in time, the value n for the next "snapshot" can be determined by the following equation

$$n = \mathrm{INT}\left(\frac{\omega t - \rho}{2\pi}\right) + 1$$

where the function INT(x) keeps only the integer portion of x.

Knowing n, the time to go till the next "snapshot" can be determined by

$$\omega \Delta t = \rho + n2\pi - \omega t$$

When Δt is less than the integration time step, the correct time for a "snapshot" will be straddled by the integration routine, so an intermediate integration is carried out with a time step of Δt, the states are recorded, and normal integration is resumed.

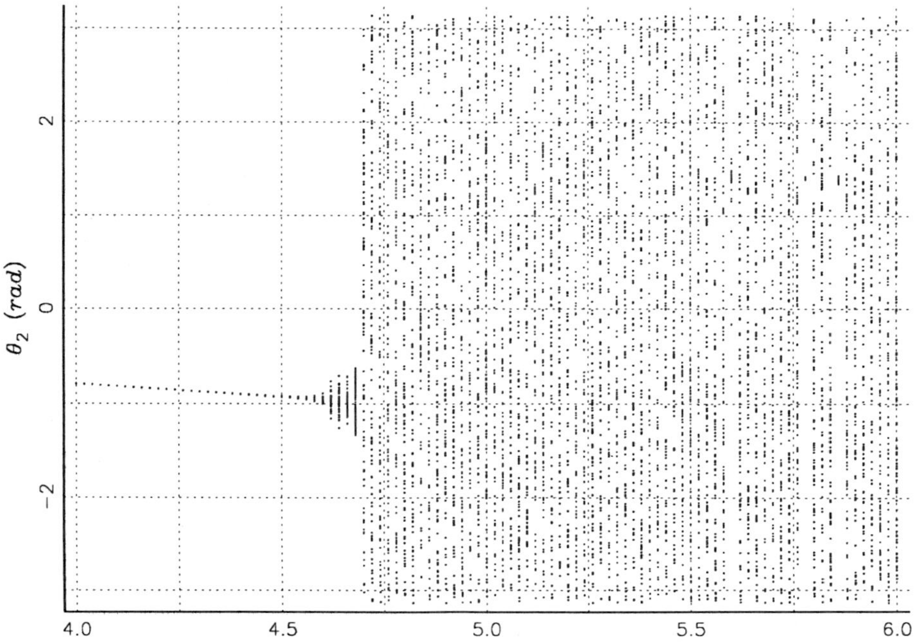

Figure 3. Bifurcation diagram for the double pendulum.

Figure 3, shows the bifurcation diagram of the double pendulum for input amplitudes between 4 and 6 volts. The phase angle ρ for this picture was zero. If one of the pendulums flipped around during integration, its angle from the vertical was set back between π and −π so that all plotted points are between these values for the angles. The picture shows a plot of E verses θ_2; diagrams of the other states are similar. Below 4.4 volts the steady state motion is clearly periodic. Above approximately 4.7 volts the behavior appears chaotic, however there are streaks in the diagram above 5.5 volts where the behavior is almost steady state again. There is a transition region between 4.4 and 4.7 volts but, because of the data resolution, it is not clear if bifurcations exist. Further refinement of the data may answer this.

The driving amplitude of 5.3 volts that was chosen to produce chaotic motion does appear to be inside the chaotic region, although it is not too far from the transition region. To study this choice further, a Poincaré map of the double pendulum model as illustrated in Figure 4 was made. Data was collected for these pictures the same way as for the bifurcation diagram with a few exceptions. First, the driving amplitude was fixed at 5.3 volts ($\rho = 0$ rad). Second, instead of collecting 100 data points, the integration was allowed to continue until 10,000 data points were collected. Third, the data was plotted in the pair (θ_2, $\dot{\theta}_2$). The map represents a two dimensional slice through the five dimensional "strange attractor" of the double pendulum. Figure 4 illustrates the fact that points on the attractor fill in most of the area between the maximum and minimum angular velocity bounds. While the picture is grainy, it is still possible to see some form to the attractor. There are denser patches where the second pendulum is more likely to be found and thinner patches where it is less likely to be found.

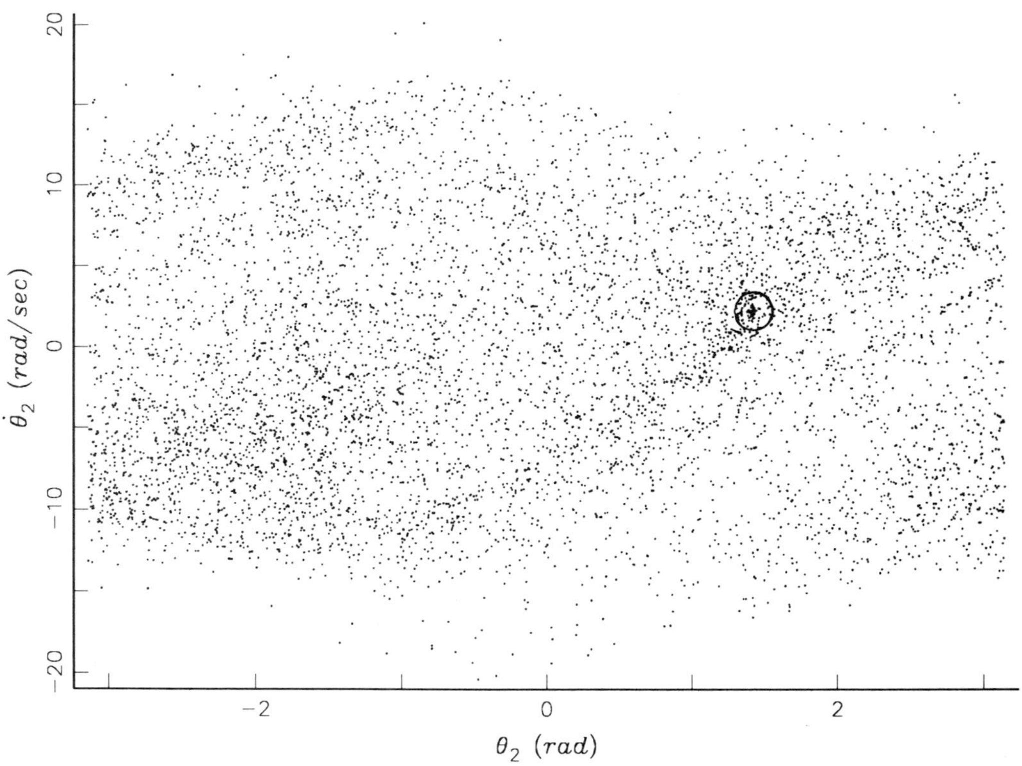

Figure 4. Poincaré map for the double pendulum.

A surprising feature of this map is that a point attractor is also obtained. The location of this stable attractor has been circled in Figure 4. Within the 10,000 drive cycles, the states of the pendulums have come close enough to this attractor to be drawn into it. Since the attractor is stable, once the system is drawn in, it cannot escape. This same behavior was also noted with the physical system. Once at the stable attractor, a light kick is necessary to push it back into chaotic motion. The values of the states at this point attractor are:

$$x_1 = \theta_1 = -0.8726$$

$$x_2 = \dot{\theta}_1 = 2.514$$

$$x_3 = \theta_2 - \theta_1 = 2.299$$

$$x_4 = \dot{\theta}_2 = 2.354$$

If the integration is restarted using these numbers for the initial conditions, the resulting motion is periodic. The pendulums are now swinging out of phase with each other by 180°. Thus the chaotic motion produced by our driving amplitude of 5.3 volts is actually a chaotic transient. Vincent and Yu [11] found the same type of behavior with the Lorenz Equations when a stable equilibrium point was generated by means of adding feedback control to the equations of motion.

For long run times, the pendulum tends to rotate clockwise more often than counterclockwise until the point attractor is encountered. Baker and Gollub [1] noticed a similar phenomenon for a sinusoidally driven single pendulum. While the motion of the single pendulum was chaotic, the average angular velocity was either positive or negative depending on the initial conditions. Two sets of initial conditions formed basins of attraction to the positive or negative average velocities and each basin had its own attractor (symmetric to each other). Further study may reveal similar behavior with the double pendulum.

Even though our open loop controller for producing chaotic motion does in fact only produce a chaotic transient, if we start the system from the stable downward position for both pendulums, the motion produced is sufficiently rich in the states it visits for a sufficiently long time period to be used as a part of the total control program.

STABILIZING THE SYSTEM

We will take as our objective the problem of stabilizing the system about the target unstable equilibrium point where the first pendulum hangs vertically downward with the second pendulum vertically upward. Since we are going to use chaotic motion to produce a favorable "starting condition" for applying feedback control, we can assume that this favorable starting condition is near the desired target. Thus any controller which will stabilize the system in the vicinity of the target will do. Toward this end we rewrite the equations of motion for the pendulums as

$$(I_A + m_2 r_3^2)\ddot{\theta}_1 + (m_1 r_1 g + m_2 r_3 g)\sin(\theta_1) = \frac{K_{\gamma 1}}{R_1} u_1 - K_{\gamma 2} u_2 + V_1$$

$$I_b \ddot{\theta}_2 + m_2 r_2 g \sin(\theta_2) = K_{\gamma 2} u_2 + V_2$$

where

$$V_1 = -\mu_1 \mathrm{SGN}(\dot{\theta}_1) - m_2 r_2 r_3 \cos(\theta_2 - \theta_1)\ddot{\theta}_2 + m_2 r_2 r_3 \sin(\theta_2 - \theta_1)\dot{\theta}_2^2$$

$$- \frac{K_{\gamma 1} K_{\beta 1}}{R_1}\dot{\theta}_1 + \mu_2 \mathrm{SGN}(\dot{\theta}_2 - \dot{\theta}_1)$$

and

$$V_2 = -\mu_2 \mathrm{SGN}(\dot{\theta}_2 - \dot{\theta}_1) - m_2 r_2 r_3 \cos(\theta_2 - \theta_1)\ddot{\theta}_1 - m_2 r_2 r_3 \sin(\theta_2 - \theta_1)\dot{\theta}_1^2$$

may be thought of as uncertain inputs to the first and second pendulums respectively. When the system is in a small neighborhood of the target ($\theta_1 = 0$, $\theta_2 = \pi$, $\dot{\theta}_1 = \dot{\theta}_2 = 0$) V_1 and V_2 will be small. If we set $V_1 = V_2 = 0$ and linearize the left hand side of the above equations of motion about the target point we obtain

$$\ddot{\theta}_1 + p_1 \theta_1 = q_1 u_1 - q_3 u_2$$

$$\ddot{\alpha}_2 + p_2 \alpha_2 = q_2 u_2$$

where $\alpha_2 = \theta_2 - \pi$ is the small angle perturbation in θ_2 about $\theta_2 = \pi$ and θ_1 is now confined to small angles about $\theta_1 = 0$. The coefficients are given by $p_1 = g(m_1 r_1 + m_2 r_3)/(I_A + m_2 r_3^2) = 25.264\mathrm{sec}^{-2}$, $q_1 = K_{\gamma 1}/R_1(I_A + m_2 r_3^2) = 0.42273\mathrm{volt}^{-1}\mathrm{sec}^{-2}$, $p_2 = -m_2 r_2 g/I_b = -31.324\mathrm{sec}^{-2}$, $q_2 = K_{\gamma 2}/I_b = 6.2286\mathrm{amp}^{-1}\mathrm{sec}^{-2}$, *and* $q_3 = K_{\gamma 2}/(I_A + m_2 r_3^2) = 0.36577\mathrm{amp}^{-1}\mathrm{sec}^{-2}$. Both systems may be stabilized by means of PD feedback controllers of the form

$$u_1 = -k_{11}\theta_1 - k_{12}\dot{\theta}_1 + q_3 u_2$$

$$u_2 = -k_{21}\alpha_2 - k_{22}\dot{\alpha}_2$$

In particular choosing $k_{11} = -.387$, $k_{12} = 14.193$, $k_{21} = 13.82$ and $k_{22} = 2.38$ yields a controlled systems of the form

$$\ddot{\theta}_1 + 2\zeta_1 \omega_{n1} \dot{\theta}_1 + \omega_{n1}^2 \theta_1 = 0$$

$$\ddot{\alpha}_2 + 2\zeta_2 \omega_{n2} \dot{\alpha}_2 + \omega_{n2}^2 \alpha_2 = 0$$

where $\zeta_1 = 0.6$, $\omega_{n1} = 5$ rad/sec, $\zeta_2 = 1$ and $\omega_{n2} = 7.4$ rad/sec. If either pendulum is held stationary, these controllers will provide reasonably fast return time with little or no overshoot for the other one. Of course neither pendulum will in general be stationary and we need to be sure that when simultaneous control action is initiated, it will be in such a way so that it will aid in stabilizing the second pendulum. Toward this end, we will assume that when both pendulums are near the equilibrium point (control torques are small) the above independent control actions are adequate to do the job. However away from the equilibrium point (control torques are large), we couple the two controllers in a fashion such that the first pendulum will tend to swing the second pendulum to the vertical position. That is, when the torque to the second pendulum is saturated, the torque to the first pendulum should also be saturated and in the same direction.

The following code fragment was used for the control algorithm suggested above, except that the q_3u_2 component of u_1 was not implemented:

```
IF ABS(ATN(TAN(θ2))) < .75 AND COS(θ2) < 0 THEN
        u2 = -k21 ATN(TAN(θ2)) - k22θ̇2
        u1 = u1max SGN(u2)
        IF ABS(u2) < u2max THEN u1 = -k11 θ1 - k12 θ̇2
ELSE
        u1 = Amplitude  COS(Frequency  t)
        u2 = 0
END IF
IF ABS(u1) > u1max THEN u1 = u1max  SGN(u1)
IF ABS(u2) > u2max THEN u2 = u2max SGN(u2)
```

The first line of the code tests to see if θ_2 is within .75 radian of the vertical. If it is, then the control for u_1 and u_2 is PD as per above unless u_2 is saturated, in which case u_1 is also saturated with the same sign. If θ_2 is greater than .75 radian from the vertical, then control torque is applied only through motor #1 in such a fashion to create chaotic motion.

DOMAIN OF ATTRACTION

The lines of code between the first IF statement and the ELSE statement given above will only be implemented provided the system is in the region of state space defined by the first IF statement. As we shall see, this condition on θ_2 is only a crude approximation to the actual domain of attraction. The actual domain of attraction, in state space, for which the above modified PD control law would drive the system to the target point, has not been determined for this problem. If we knew the actual domain of attraction, then the above modified PD controller would only have to be "turned on" once in order to obtain the desired operating point. As we will see, in what follows, since the above code only approximates the domain of attraction, the PD controller may have to be "turned on" several times before the second pendulum can be captured to the vertical upright position. While the above algorithm is not elegant, it is easy to implement and it does work quite well.

A more elegant algorithm would require the determination of when the system is in the actual domain of attraction. Since the double pendulum system is a nonlinear fourth order system, the domain of attraction is not easily calculated. One way to obtain a better approximation would be to first linearize the system about the operating point, then use the eigenvectors to convert the system to block diagonal form. The system would then be expressed in terms of second order subsystems for the complex eigenvalues and in terms of first order subsystems for the real eigenvalues. Since each of these subsystems are linear and at most second order, the method proposed by Gayek and Vincent [2-3] can be readily employed to determine the actual boundaries of the domain of attraction in the decoupled space. The inverse of the transformation matrix used to put the system in block diagonal form could then be used to numerically test any point in state space to see if it lies in the decoupled domain of attraction. Future study will be directed toward improving our estimate of the domain of attraction and toward determining control laws which will give the largest possible domain of attraction.

RESULTS

The above algorithm as given by the code fragment, was used to drive both the nonlinear equations of motion and the actual laboratory system. We will refer to these as two systems, "the simulated system" and "the laboratory system." Both systems are started initially at rest with both pendulums hanging straight down. The parameters used to generate Figures 5-10 are as follows:

k_{11} = -.387	k_{12} = 14.193
k_{21} = 13.82	k_{22} = 2.38
Amplitude = 5.3 volts	*Frequency* = 5 rad/sec
u_{1max} = 19.9 volts	u_{2max} = 1.7 amps.

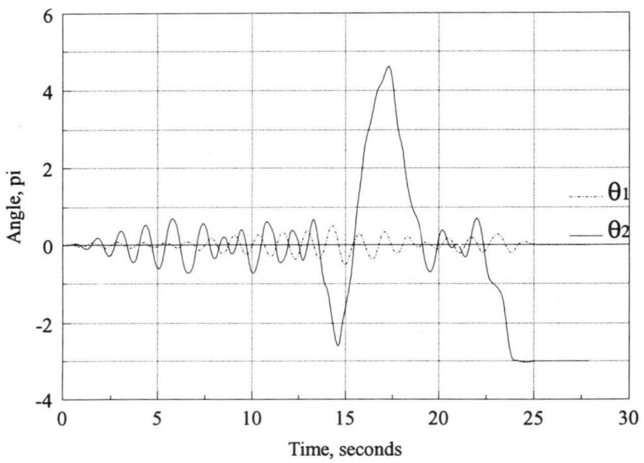

Figure 5. Response of the simulated system with *Frequency* = 5 rad/sec.

Figure 5 represents the time response of θ_1 and θ_2 for the simulated system. Note that the *y*-axis is in units of π. This allows us to easily visualize the target point ($\theta_1 = 0$, $\theta_2 = \pm 1, \pm 3, \pm 5$, etc.). The simulated system achieves the target through a complicated route. After a series of downward pointed oscillations, the second pendulum swings clockwise one full revolution, then counterclockwise three revolutions, then clockwise two revolutions where it again oscillates in the downward position a few times before finally swinging clockwise one and one half revolutions to be captured at the target point.

Figure 6 shows the voltage control input to the first motor as a function of time. The role this motor plays in producing chaotic motion with its cosine input and in switching to aid in stabilizing the second pendulum is clearly evident. There were seven times that the system satisfied the condition given by the first IF statement of the control algorithm, and seven times the control algorithm switched strategy in an attempt to capture the second pendulum. It was finally successful on the eight try. Note that the PD control was unsaturated only at the very end of the control effort. Figure 7 illustrates the current control input to the second motor.

Figure 8 illustrates that the laboratory system achieves the desired operating point in a more direct way. This large difference between Figure 5 and Figure 8 does not indicate that the simulation model is necessarily inaccurate. Rather it is reflective of the sensitivity of the solution to initial conditions (and/or parameters) whenever $|\theta_2|$ becomes large. Attempting to "fine tune" any given system parameter in an effort to have the simulated system track the real system more closely (once $|\theta_2|$ becomes large) is futile.

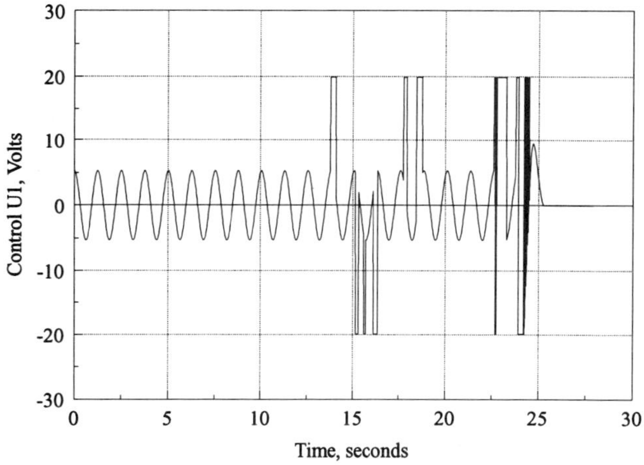

Figure 6. Control voltage input u_1 used in the simulation of Figure 5.

We have shown, incidently, that time plots for the simulated system and the real system match closely when the inputs are such that $|\theta_2|$ remains small. Figures 9 and 10 illustrate the control inputs to the two motors used to generate Figure 8. Due to friction, the final upright position of the second pendulum was not exactly upright and hence a small control signal to both motors is still present at the end.

Figures 11 and 12 were generated using the same conditions as in Figures 5-10, except that Frequency = 4.8. Theses figures illustrate the fact that the actual system can be in good agreement with the simulated system. However any small change in a system parameter or initial condition will, in general, result in a different path to the target point. Figures 1, 5, 8, 11, and 12 are typical for both systems. All figures are interchangeable in the sense that they are all representative of what the results of this controller could look like under small changes in initial conditions or system parameters.

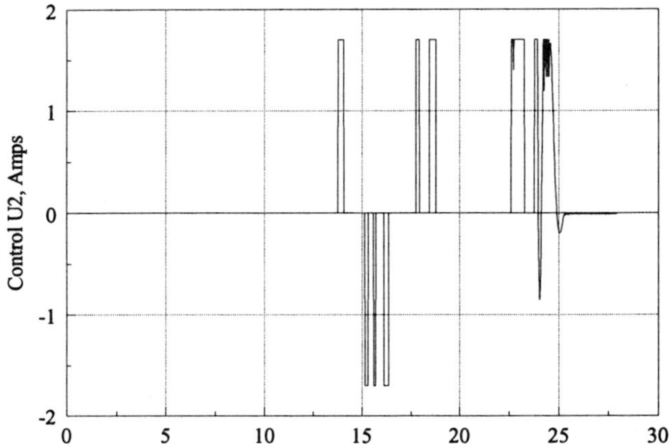

Figure 7. Control current input u_2 used in the simulation of Figure 5.

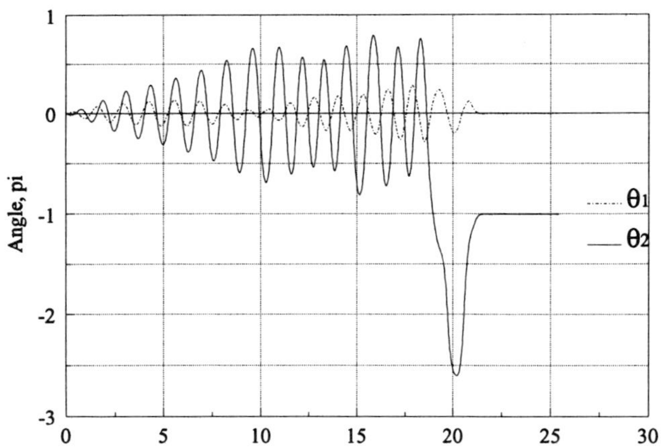

Figure 8. Response of the laboratory system with *Frequency* = 5 rad/sec

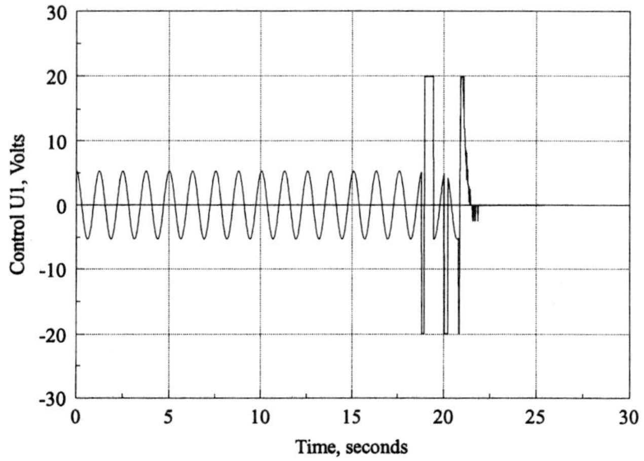

Figure 9. Control voltage input u_1 used to generate the data of Figure 8.

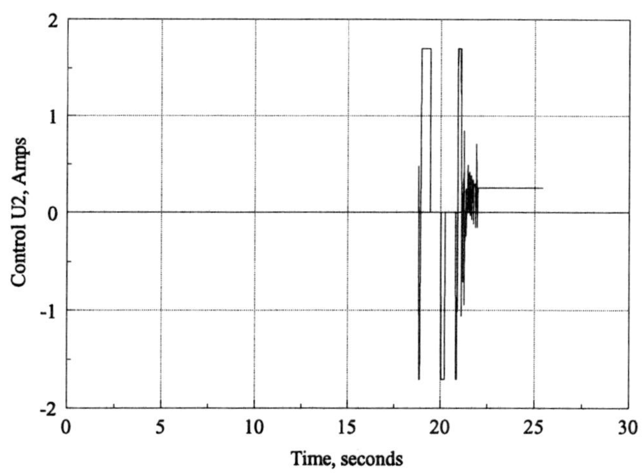

Figure 10. Control current input u_2 used to generate the data of Figure 8.

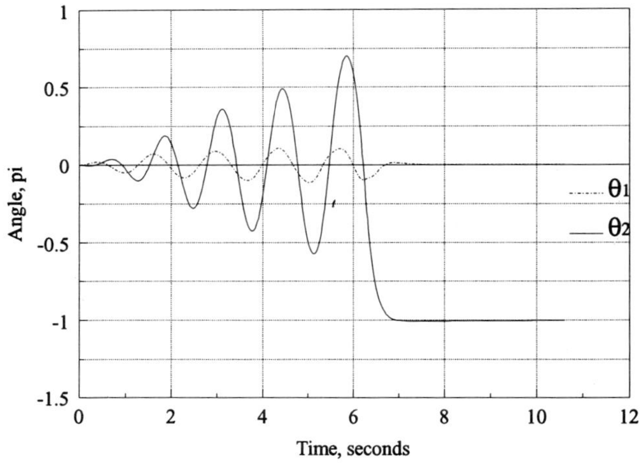

Figure 11. Response of the simulated system with *Frequency* = 4.8 rad/sec.

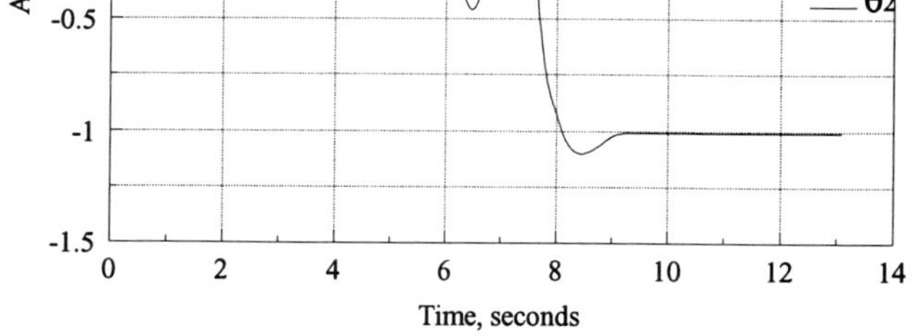

Figure 12. Response of the laboratory system with *Frequency* = 4.8 rad/sec

DISCUSSION

Considering the usual deterministic methods used in control design, one might question an approach for which one can not accurately predict when the system will achieve the desired control objective. Perhaps the best way to answer this question is to look at the alternatives. Here we have presented a combination of open loop control and closed loop control that is simple to implement and ultimately achieves the objective. The alternative is to apply a nonlinear technique based on other methods such as optimal control theory. Due to the complexity of the system equations, numerical methods would have to be used (e.g. MIZER [4]) which will result in an open loop control law for the two controllers u_1 and u_2. Since the system we are dealing with has sensitivity with respect to initial conditions as we move away from the downward stable position, it is not clear that any optimal open loop control law could achieve stabilization about the desired operating point. An alternative could be to teach a neural network closed loop control based on optimal open loop solutions [5] or perhaps some other training tool. Clearly, if we are willing to give up certainty in the time required to achieve our objective, we can avoid such complications with a chaotic controller such as the one described here.

REFERENCES

1. G.L. Baker and J.P. Gollub, *Chaotic Dynamics an Introduction.* Cambridge University Press: Cambridge, 1990.

2. J. E. Gayek and T. L. Vincent, "On the asymptotic stability of boundary trajectories," *International Journal of Control,* vol. 41, pp.1077-1086, 1985.

3. J.E. Gayek and T.L. Vincent, "An existence theorem for boundary trajectories," *Journal of Mathematical Analysis and Applications*, vol. 132, pp. 290-299, 1988.

4. C.J. Goh and K.L. Teo, *MISER: An Optimal Control Software.* University of Singapore: Singapore, 1987.

5. C.J. Goh, N.J. Edwards and A.Y. Zomaya, "Feedback control of minimum-time optimal control problems using neural networks," *Optimal Control Applications and Methods*, In Press, 1993.

6. Z. Meier, H. Farwig, and H. Unbehauen, "Discrete computer control of a triple-inverted pendulum," *Optimal Control Applications & Methods*, vol. 11, pp. 157-171, 1990.

7. E. Ott, C. Grebogi, and J.A. Yorke, "Chaos," in *Phys. Rev. Lett.* 64 1196, (D.K. Campbell, ed.) pp. 153-72, American Institute of Physics: New York, 1990.

8. P.H. Richter and H.J. Scholz, "Chaos in classical mechanics: The double pendulum," *Stochastic Phenomena and Chaotic Behavior in Complex Systems.* Springer: Berlin, pp. 86-97, 1984.

9. T.J. Schmitt, "Double pendulum simulation," *Masters report*, Aerospace and Mechanical Engineering, University of Arizona, Tucson, 1992.

10. T. Tél, "Controlling transient chaos," *J. Phys. Math. Gen.*, vol. 24, pp.L1359-L1368, 1991.

11. T.L. Vincent and J. Yu, "Control of a chaotic system," *Dynamics and Control* , vol. 1, pp. 35-54, 1991.

ADAPTIVE CONTROL OF REDUNDANT MULTIPLE ROBOTS IN COOPERATIVE MOTION

Mohamed Zribi, Shaheen Ahmad and, Shengwu Luo

Real-time Robot Control Laboratory
School of Electrical Engineering
Purdue University
West Lafayette, IN 47907-1285
U.S.A

ABSTRACT

A redundant robot has more degrees of freedom than what is needed to uniquely position the robot end-effector. In practical applications the extra degrees of freedom increase the orientation and reach of the robot. The load carrying capacity of a single robot can be increased by cooperative manipulation of the load by two or more robots. In this paper we develop an adaptive control scheme for kinematically redundant multiple robots in cooperative motion.

In a usual robotic task, only the end-effector position trajectory is specified. The joint position trajectory will therefore be unknown, for a redundant multirobot system and it must be selected from a self-motion manifold for a specified end-effector or load motion. We show that the adaptive control of cooperative multiple redundant robots can be addressed as a reference velocity tracking problem in the joint space. A stable adaptive velocity control law is derived it ensures bounded parameter convergence, exponential convergence to zero of the load position error, the internal force error and the reference velocity error. The individual robot joint motions is shown to be stable by decomposing the joint coordinates into two variables one which is homeomorphic to the load coordinated, the other to the coordinates of the self-motion manifold. The dynamics on the self-motion manifold is directly shown to be related to the concept of *zero-dynamics*. It is shown that if the reference joint trajectory is selected to optimize a certain type of objective functions, then stable dynamics on the self-motion manifold results. The overall stability of the joint angle is established from the stability of two cascaded dynamic systems involving the two decomposed coordinates.

Mechanics and Control, Edited by R.S. Guttalu
Plenum Press, New York, 1994

INTRODUCTION

Recently considerable amount of research has focused on the problem of cooperative control and coordination of multiple robots. Interest in multi-robot systems has arisen because several tasks require the use of two or more robots. Examples of such tasks include the joining and securing of large pipes for the construction of space structures, picking up and carrying heavy loads, and grasping odd shaped loads. Cooperative robots may be used in hazardous or unsafe environments such as in space, in deep waters and in radioactive environments. By using more than one robot the manipulation capability and the workspace of the system may be further increased. However cooperative multiple-robot systems are more difficult to control than single robots. Additional problems may arise in the control if the parameters of the robots and the manipulated load may not be known exactly.

Several control schemes, adaptive and non adaptive schemes have been proposed for cooperative multiple robots with rigid joints manipulating a common load. Zheng and Luh [32] considered the kinematic and dynamic model of the multi-robot system and developed an inverse dynamics schemes for load position control. Hsu et al [23] developed a control algorithm for the coordinated manipulation of multi-fingered robot hands. Tarn, et al [28] developed a robust nonlinear control scheme using nonlinear transformation techniques. Yun, et al [29] also used exact linearization and output decoupling techniques to control multiple robots. Yoshikawa and Zheng [31] also developed linearizing tracking control laws for multiple robots, experiments were also reported. Few adaptive control schemes for cooperative robots manipulating a common load have been proposed. Walker et al. [30] developed an adaptive algorithm for the control of two robots handling a common load of an unknown mass. Zribi and Ahmad [24] proposed a robust adaptive controller for the multi-robot system manipulating a rigid object cooperatively when subject to bounded disturbances. The problem of manipulating a load using multiple robots when the load makes contact with an environment was addressed by Hyati [33], Cole [34] and, Hu and Goldenberg [27]. Ahmad and Guo [26] addressed the problem of controlling multiple flexible-joint robots with linear dynamics. Ahmad [35] developed a feedback linearizing controller for multiple flexible joint robots.

There are a very few papers in the area of control of multiple redundant robots, other than the recent paper by Tarn et al. [36] and that by Tao and Luh [37]. This is primarily because non-redundant robot control schemes cannot be easily extended to control redundant robot systems. This is because a redundant robot has more joints than what is required to position the end-effector. Usually the end-effector trajectory is known and thus the joint trajectory cannot be found uniquely. In fact, for a fixed end effector position there is a self-motion manifold on which joint motions could occur without effecting the end-effector position. Any arbitrary joint trajectory which ensures end effector position cannot be used as this may not result in stable joint motions on the self motion manifold and would therefore effect overall stability. These two problems have prevented the simple extension of non-redundant strategies being adopted for multiple redundant robots. We should note here that the extra joints are extremely useful in real applications as they can be used to configure the manipulator posture, to avoid obstacles in the workspace or to avoid joint singularities.

Initial interest in the control of redundant robots started with the work of Whitney [21] who devised a kinematical resolved motion rate control strategy. Since then a number of researchers have addressed the the joint coordination and control of redundant robots (see Nenchev [13] for a review of those developments). The tutorial review by Siciliano [17] and the tutorial workshop report on the theory and application of redundant robots at the 1989 IEEE robotics and automation conference [1] covered some more recent developments (see also [1-3,7,11-14,16,17]). In the area of redundant robot adaptive control, Seraji [16] presented an approach based on the model reference adaptive control theory. He resolved the redundancy problem by adding additional task dependent kinematic constraints to the end-effector kinematics. This effectively ensured the joint solutions were unique. Niemeyer and Slotine [14] applied sliding mode adaptive control to redundant manipulators. They used the passivity principle to prove the stability of the adaptive system. Niemeyer and Slotine also performed some experiments to demonstrate their control law. Colbaugh et al. [3] proposed an adaptive inverse kinematics algorithm that did not require the knowledge of the kinematics of the robots. However their algorithm required persistent excitation conditions; also their algorithm did not consider the dynamics of the robot. Luo, Ahmad and Zribi [38] developed an adaptive control law for redundant robots making use of weighted scaling functions and the concept of zero dynamics to show both the joint motions on the self motion manifold and the end-effector motions would be stable for their control law.

In this paper, we address the problem of controlling redundant multiple robots manipulating a load cooperatively. We assume the load mass/inertial parameters and the robot joints mass/inertial parameters are unknown. We first state the dynamic models of the robots and the load and give a few properties of the multi-robot system. Next the redundancy resolution problem is discussed, and a model for adaptive resolution of the redundancy is established. A controller that leads to the exponential tracking of the load position and the convergence of the internal forces to their desired values is then derived. The boundedness of the joint motion and control torques are proved next. The conclusions can be found in the final section of the paper.

MULTI-ROBOT SYSTEM MODEL

Dynamics Model

The general dynamic model for a cooperative multi-robot system has been investigated thoroughly in the literature, and is also described in the below for completeness. We first state a few assumptions related to the robots grasp of the load and the reachability of the trajectory that will be used in the subsequent derivation.

Assumptions

(A1) The manipulators are rigidly grasping the load, (i.e., there is no motion between the contact point of the load and the robots end-effectors).
(A2) The desired trajectory is reachable and the end effector can be positioned at those workspace (of dimension six) positions without exceeding any joint motion limits.

The dynamic equation of the ith manipulator in cooperative manipulation can be written as,

$$D_i(q_i)\ddot{q}_i + C_i(q_i,\dot{q}_i)\dot{q}_i + G_i(q_i) + J_{e_i}(q_i)^T F_{e_i} = \tau_i \qquad i=1,...,k. \tag{1}$$

where, $q_i \in R^{n_i}$ is the vector of joint displacements, and $n_i>6$ is the number of joints of the ith robot. The inertia matrix of the ith robot is $D_i(q_i) \in R^{n_i \times n_i}$, this is a positive definite and symmetric matrix. The matrix of centrifugal and Coriolis forces is $C_i(q_i,\dot{q}_i) \in R^{n_i \times n_i}$; the vector of gravity forces is $G_i(q_i) \in R^{n_i}$, and the manipulator Jacobian is $J_{e_i}(q_i) \in R^{6\times n_i}$. The control input torque for the ith robot is $\tau_i \in R^{n_i}$. We will define the total number of joints of the k robots as n, $n = \sum_{i=1}^{i=k} n_i$.

The forces/moments applied by the ith manipulator on to the object at the point of contact are F_{e_i}. The contact forces/moments $F_{e_i} \in R^6$ can be written in terms of the contact forces $f_{e_i} \in R^3$ and contact moments $\eta_{e_i} \in R^3$, (where 6 represents the dimension of the Cartesian work space), such that,

$$F_{e_i} = \begin{bmatrix} f_{e_i}{}^T & \eta_{e_i}{}^T \end{bmatrix}^T \qquad i = 1,...,k\ . \tag{2}$$

Now we will group the dynamics of the k-robots system to get,

$$D(q)\ddot{q} + C(q,\dot{q})\dot{q} + G_r(q) + J_e(q)^T F_e = \tau\ , \tag{3}$$

where $D \in R^{n\times n}$ is a block diagonal matrix whose diagonal elements are $D_i \in R^{n_i\times n_i}$. $C \in R^{n\times n}$ is a block diagonal matrix whose diagonal elements are $C_i \in R^{n_i\times n_i}$ and $J \in R^{6k\times n}$ is a block diagonal matrix whose diagonal elements are $J_{e_i} \in R^{6\times n_i}$. Also we will define the following vectors as,

$$q=[q_1{}^T ... q_k{}^T]^T, G_r=[G_1{}^T ... G_k{}^T]^T, \tau=[\tau_1{}^T ... \tau_k{}^T]^T \ and,\ F_e=[F_{e_1}^T ... F_{e_k}{}^T]^T. \tag{4}$$

If we assume that the object is rigidly grasped, then the equations of motion of the object are obtained from the Newton-Euler mechanics as,

$$M_1\ddot{x}_p + M_1 g_l = \sum_{i=1}^{i=k} f_{e_i}\ , \tag{5}$$

$$and,\ \ I\dot{\omega} + \omega\times(I\omega) = \sum_{i=1}^{i=k}(\eta_{e_i} + r_i\times f_{e_i})\ , \tag{6}$$

where the position of the center of mass of the object expressed in the world coordinate frame is $x_p \in R^3$. The rotational velocity of the object expressed in the world coordinate frame is $\omega \in R^3$, and the gravity force vector of the object, expressed in the world coordinate reference frame is $g_l \in R^3$. The mass matrix $M_1 \in R^{3\times3}$ is a diagonal matrix whose diagonal elements are the mass of the load; the matrix $I \in R^{3\times3}$ is the inertia matrix of the load. The vector $r_i = [\ r_{ix}, r_{iy}, r_{iz}]^T \in R^3$ represents the translational displacements from the center of mass of the object to the contact point of the object and the ith manipulator.

If we let $\dot{x} = [\ \dot{x}_p^T \quad \omega^T\ ]^T$, then the motion of the object expressed by equation (5) and (6) can be written as,

$$M\ddot{x} + N\dot{x} + G_l = GF_e = F_o\ , \tag{7}$$

where $G \in R^{6\times 6k}$ is the grasp matrix; G is defined as,

$$G = \begin{bmatrix} T_1 & T_2 & \dots & T_k \end{bmatrix} . \tag{8}$$

The matrix $T_i \in R^{6\times6}$ is such,

$$T_i = \begin{bmatrix} I_{3\times3} & 0 \\ \Omega_i(r_i) & I_{3\times3} \end{bmatrix} \quad \text{and ,} \quad \Omega_i(r_i) = \begin{bmatrix} 0 & -r_{iz} & r_{iy} \\ r_{iz} & 0 & -r_{ix} \\ -r_{iy} & r_{ix} & 0 \end{bmatrix} .$$

The matrix $I_{3\times3} \in R^{3\times3}$ is the identity matrix. Also we have,

$$M = \begin{bmatrix} M_1 & 0 \\ 0 & I \end{bmatrix}, \quad N\dot{x} = \begin{bmatrix} 0 \\ \omega\times(I\omega) \end{bmatrix} \quad \text{and} \quad G_l = \begin{bmatrix} M_1 g_l \\ 0 \end{bmatrix} . \tag{9}$$

Kinematic Model

We are interested in controlling the manipulators in some predefined Cartesian task space such that,

$$x_{e_i} = K_{e_i}(q_i) \qquad i = 1,\dots,k , \tag{10}$$

where $K_{e_i}(.) : R^{n_i} \rightarrow R^6$ is the transformation from the joint angle space of q_i to the task space containing x_{e_i}, and $x_{e_i} \in R^6$ is the position and orientation of the point of contact of the ith manipulator, with the load, expressed in the world coordinate frame. If we differentiate equation (10) with respect to time, and if we define $J_{e_i}(q_i)$ to be the differential map from the q_i space to x_{e_i} space, then we can write,

$$\dot{x}_{e_i} = J_{e_i}(q_i)\dot{q}_i \qquad i = 1,\dots,k . \tag{11}$$

If these equations are stacked into a single vector by forming the J_{e_i} into a block diagonal matrix, and concatenating the $\dot{q}_i$'s into one vector $\dot{q}$, we get,

$$v_c = J_e \dot{q} , \tag{12}$$

where $v_c = \begin{bmatrix} \dot{x}_{e_1}^T & \dot{x}_{e_2}^T & \dots & \dot{x}_{e_k}^T \end{bmatrix}^T$ is the velocity vector at the contact points and $J_e = diag(J_{e_1}, \dots, J_{e_n})$.

Using equation (7), we can write,

$$F_o = G F_e . \tag{13}$$

Now from the duality between the forces and the velocities, we can write,

$$G^T \dot{x} = v_c , \tag{14}$$

where $\dot{x}$ is the velocity of the object. Thus for the k robots system, we can combine equations (12) and (14) to get,

$$G^T \dot{x} = J_e \dot{q} . \tag{15}$$

where G is the grasp matrix defined earlier.

Definition of Internal Forces and Internal Force Errors

The end-effector force of the ith manipulator, F_{e_i}, can be decomposed into two forces, the motion force and the internal grasping force. The internal grasping forces $F_I = [F_{Ie1},..,F_{Iek}]^T \in R^{6k}$ do not cause any motion of the load. However we must control these end-effector internal forces, $F_{Iei} \in R^6$ *with* $i=1,..,k$, in order to prevent excessive compressive or expansive forces being applied to the load. We can calculate the internal force F_I from equation (7) if F_o is known and rank (G)= 6, then

$$F_e = G^+ F_o + F_I \,. \tag{16}$$

Here, $G^+ = G^T(GG^T)^{-1}$ and $GG^+ = I_6$, given I_6 is an 6×6 identity matrix. (For a discussion related to other choices of the inverse of the G^+ matrix see [36] and [39]. Notice that other choices of the inverse of G does not effect the derivations presented in this paper.) Therefore we see that $GF_I=0$ and $GF_e=F_o$, i.e, the internal forces do not contribute to the motion of the load. The desired internal forces $F_{I,d} \in R^{6k}$ also satisfy $GF_{I,d}=0$. The internal force error, $e_f = F_{I,d} - F_I$, also satisfies

$$G\ e_f = 0.$$

These internal force properties will be used to derive the control law.

A Few Properties of the Multi-Robot System

In the following we will state several properties which will be used in the derivations of the controller.

System Dynamics Properties

(P1) D and M are symmetric positive definite matrices.
(P2) $\dot{D} - 2C$ is skew symmetric matrix or $\frac{1}{2}\ \dot{q}^T(\dot{D} - 2C)\dot{q} = 0$. The proof of property P2 can be found in [19] and [40].
(P3) $\dot{M} - 2N_2$ is skew symmetric matrix or, $\frac{1}{2}\ \dot{x}^T(\dot{M} - 2N)\dot{x} = 0$. This property can be seen from energy considerations. In general if M is not expressed in the object center of mass coordinate frame, $M=M(x)$ and $N\dot{x}=N(x,\dot{x})\dot{x}$. The total energy of the load is given by $E_L = \frac{1}{2}\ \dot{x}^T M\dot{x} + h(x)$ where $h(x)$ is the potential energy and $G_l = \frac{\partial}{\partial x}\ h(x)$. As the power input to the load is given by, $\frac{d}{dt}E_L = \dot{x}^T F_o = \dot{x}^T(M\ddot{x} + \frac{1}{2}\ \dot{M}\dot{x} + G_l) = \dot{x}^T(M\ddot{x} + N\dot{x} + G_l)$, thus we have the property, $\frac{1}{2}\ \dot{x}^T(\dot{M} - 2N)\dot{x} = 0$.

Two important properties of the inertia matrix, the centrifugal/Coriolis matrix and the gravity vector which will be used in the developments are now given.

(P4) Linear Parameterization of the Robot Dynamics

The linearity of D, C and G_r with respect to the manipulators dynamic parameters $P_r \in R^{s_r}$ is now stated, these parameters will be estimated by the proposed adaptive scheme. The robot dynamics can be linearly parameterized [19], [40] and,

$$Da_r + Cv_r + G_r = Y_r P_r \ , \tag{17}$$

where $a_r \in R^n$ and $v_r \in R^n$ are vectors and we denote a_r as the "reference acceleration of the robots" and also, v_r is the " reference velocity of the robots." The regressor matrix $Y_r(q, \dot{q}, v_r, a_r) \in R^{n \times s_r}$ represents the structure of the robots dynamics, hence its elements are combinations of the nonlinear functions present in the inertia matrix, centrifugal/Coriolis matrix and the gravity vector.

(P5) Linear Parameterization of the Object Dynamics

The second property deals with the linearity of M, N_2 and G_l with respect to the load parameter vector P_o,

$$Ma_o + N_2 v_o + G_l = Y_o P_o \tag{18}$$

where $a_o \in R^6$ and $v_o \in R^6$ are the "reference acceleration of the load" and the "reference velocity of the load," respectively. We will denote by $P_o \in R^{s_o}$ the vector of s_o load parameters which are constants for a given load. These parameters will be estimated by the proposed adaptive scheme. The regressor matrix $Y_o(x, \dot{x}, v_o, a_o) \in R^{6 \times s_o}$ represents the structure of the load dynamics.

Remark 1

Let $\hat{P}_r$ be the vector of estimates of the parameters of the robots, then the error vector in the estimates of the robots parameters is $\tilde{P}_r = \hat{P}_r - P_r$. Similarly, we can write the parameter estimation error vector for the load as $\tilde{P}_o = \hat{P}_o - P_o$. Notice that we can write,

$$\hat{D}a_r + \hat{C}v_r + \hat{G}_r = Y_r \hat{P}_r \ , \tag{19}$$

where $\hat{D}$ is the estimate of the inertia matrix D, $\hat{C}$ is the estimate of the Coriolis/centrifugal matrix and $\hat{G}_r$ is the estimate of the the gravity vector. Also notice that, $\tilde{D}a_r + \tilde{C}v_r + \tilde{G}_r = Y_r \tilde{P}_r = Y_r(\hat{P}_r - P_r)$, where $\tilde{D}$ is the error in the inertia matrix, $\tilde{C}$ is the error in the Coriolis/centrifugal matrix and $\tilde{G}_r$ is the error in the gravity vector.

Similarly we can write,

$$\hat{M}a_o + \hat{N}_2 v_o + \hat{G}_l = Y_o \hat{P}_o \ , \tag{20}$$

where $\hat{M}$ is the estimate of M, $\hat{N}_2$ is the estimate of N_2 and $\hat{G}_l$ is the estimate of G_l. Also notice that $\tilde{M}a_o + \tilde{N}_2 v_o + \tilde{G}_l = Y_o \tilde{P}_o$, where $\tilde{M}$, $\tilde{N}_2$ and $\tilde{G}_l$ are the differences between the estimates and the true values.

REDUNDANCY RESOLUTION PROBLEM

Preliminaries

Consider a kinematically redundant manipulator with the carried load center of mass positioned at point x, and the joint position q_i. Then the differentiable kinematic mapping is K_i such that,

$$x = K_i(q_i) \qquad i = 1, \ldots, k \tag{21}$$

where $x \in R^6$ is the position of the load and $q_i \in R^{n_i}$ is the vector of joint positions of the ith manipulators and as $n_i > 6$, the degree of redundancy of the ith robot is $r_i = n_i - 6$. As a result of the joint redundancy at the end effector point $x = x_d$, there will exist a set of joint angles, the self motion manifold such that $Q^q_{N_i} = \{q_i \mid x = x_d = K_i(q)\}$. Thus, in order to find a unique joint angle q, additional requirements are necessary; these will be stated later. We will denote the Jacobian of the kinematic map (21) by, $J_i = T_i^{-T} J_{e_i} \in R^{6 \times n_i}$. This relates differential map between the load position kinematics $K_i(\bullet)$ and the end effector kinematics $K_{e_i}(\bullet)$. The projection operator onto the null space of J_i is denoted by $P_{J_i}(q_i)$ (i=1,...,k). Also let all the columns of matrix N_{J_i} be the basis of $ker\ (J_i)$, which is the null space of J_i. Hence we have,

$$J_i P_{J_i} = 0, \qquad and \qquad ker\ (J_i) = span\ (N_{J_i}). \tag{22}$$

The matrix $N_{J_i} \in R^{n_i \times r_i}$ has the following properties that will be used in the text,

$$J_i N_{J_i} = 0 \in R^{6 \times r_i}\ ,\quad N_{J_i}^T J_i^T = 0 \in R^{r_i \times 6}\ ,\quad N_{J_i}^T J_i^+ = 0 \in R^{r_i \times 6}\ , \tag{23}$$

$$N_{J_i}^T P_{J_i} = N_{J_i}^T \in R^{r_i \times n_i}\ ,\quad N_{J_i}^T N_{J_i} = I_{r_i \times r_i}\ ,\quad N_{J_i} N_{J_i}^T = P_{J_i} \in R^{n_i \times n_i}\ ; \tag{24}$$

$$for\ any\ vector\quad \dot{q}_i \in R^{n_i} \qquad \text{if}\ \ N_{J_i}^T \dot{q}_i = 0 \in R^{r_i}\ \ then\ P_{J_i} \dot{q}_i = 0 \in R^{n_i}\ . \tag{25}$$

Notice also that $\begin{bmatrix} J_i \\ N_{J_i}^T \end{bmatrix}$ is a square matrix of full rank, thus we have,

$$\begin{bmatrix} J_i \\ N_{J_i}^T \end{bmatrix}^{-1} = [J_i^+ \quad N_{J_i}]\ . \tag{26}$$

These properties show that the pairs $(J_i, N_{J_i}^T)$ and (J_i^+, N_{J_i}) are orthogonal complement operator pairs.

Statement of the Problem of the Redundancy Resolution

The redundancy is usually resolved by the constrained optimization of a performance index H_i (i=1,...,k), this function can be used to avoid joint limits, obstacles and singularities (see the review papers listed in the reference). The problem can be formulated as follows: given a desired position x_d, find the joint position q_i (i=1,..,k) such that,

$$\min_{q_i} \qquad H_i(q_i) \qquad with \quad x_d = K_i(q_i) \qquad\quad i=1,...,k\ . \tag{27}$$

We can conclude from the Lagrange multiplier method that the solution of the constrained optimization problem (27) necessarily satisfies the following set of constrained differential equations:

$$P_{J_i} \nabla H_i(q_i) = 0 \qquad and \qquad x_d = K_i(q_i) \qquad\quad i=1,...,k\ . \tag{28}$$

We will define the end-effector path tracking error e as,

$$e = K_i(q_i) - x_d \qquad i=1,...,k. \tag{29}$$

Our goal is to resolve the "asymptotic resolution of the redundancy problem" such that as $t \longrightarrow \infty$, we have,

$$\dot{e} \longrightarrow 0, \qquad\quad e \longrightarrow 0\ , \quad and, \qquad P_{J_i} \nabla H_i(q_i) \longrightarrow 0 \qquad i=1,...,k. \tag{30}$$

We want to optimize H_i (i=1,...,k) by appropriate joint motion on the self-motion manifold, $Q^q_{N_i}$ (i=1,...,k). At the optimal point, we do not desire further motion on the self motion manifold. Therefore the projection of the joint velocity on the self-motion manifold must be zero, and $N_{J_i}^T \dot{q}_i \rightarrow 0$ as, $t \rightarrow \infty$. Thus it is sufficient (not necessary) to write the asymptotic redundancy resolution as $t \rightarrow \infty$,

$$(\dot{e} + \gamma e \rightarrow 0) \text{ and, } N_{J_i}^T(\dot{q}_i - \mu_i \nabla H_i) \rightarrow 0 \text{ , with } N_{J_i}^T \dot{q}_i \rightarrow 0 \quad i=1,...,k. \quad (31)$$

Here, $\gamma > 0$ and $\mu_i \neq 0$. The first equation can be written as $J_i\dot{q}_i - \dot{x}_d + \gamma e \rightarrow 0$. After grouping terms and using the matrix inversion expressed by (26), we find $\dot{q}_i$, and, as $t \rightarrow \infty$,

$$\dot{q}_i - \begin{bmatrix} J_i^+ & N_{J_i} \end{bmatrix} \begin{bmatrix} \dot{x}_d - \gamma e \\ \mu_i N_{J_i}^T \nabla H_i \end{bmatrix} \rightarrow 0 \text{ with } q_i \rightarrow \{ q_i \mid N_{J_i}^T \nabla H_i = 0 \text{ and } x_d = K_i(q_i) \}. \quad (32)$$

Therefore the "asymptotic resolution of redundancy problem" can be expressed by the conditions given by (32). These conditions result in the joint velocities approaching their desired values, while the joint positions satisfy a set of constraint equations. Notice that the redundancy resolution problem is characterized by the fact that the desired joint positions are not known in advance. This fact prevents us from directly using the existing adaptive schemes that achieves joint position tracking.

We will denote by v_{r_i} $(i=1,...,k)$ the joint reference velocity for the ith robot. We also will denote by v_o the load reference velocity. We will choose v_{r_i} $(i=1,...,k)$ such that,

$$v_{r_i} = \begin{bmatrix} J_i^+ & N_{J_i} \end{bmatrix} \begin{bmatrix} \dot{x}_d - \gamma e \\ \mu_i N_{J_i}^T \nabla H_i \end{bmatrix} = J_i^+(\dot{x}_d - \gamma e) + \mu_i P_{J_i} \nabla H_i \ . \quad (33)$$

We will group the v_{r_i} $(i=1,...,k)$ into one vector v_r such that, $v_r = \begin{bmatrix} v_{r_1}^T & v_{r_2}^T & \dots & v_{r_k}^T \end{bmatrix}^T$ is the joint reference velocity of the robots. We will choose v_o such that,

$$v_o = \dot{x}_d - \gamma e \ . \quad (34)$$

It should be noted that the choice of v_o guarantees that,

$$v_o = J_i v_{r_i} = T_i^{-T} J_{e_i} v_{r_i} \qquad i=1,...,k. \quad (35)$$

The asymptotic resolution of redundancy problem can be solved by a mechanism that ensures $\dot{q}_i - v_{r_i} \rightarrow 0$, $(for\ i=1,...,k)$, as $t \rightarrow \infty$.

In order to proceed further we will state a few more assumptions these will be needed in the control law development.

Assumptions - Continued

(A3) The desired paths $x_d(t)$, $\dot{x}_d(t)$ and $\ddot{x}_d(t)$ are bounded for all time t.

(A4) The Jacobian $J_i(q_i)$ $(i=1,...,k)$ is a full rank continuously differentiable function matrix, that is, $J_i(q_i)$ is of class C^r, $r \geq 2$. (i.e., at least twice differentiable).

(A5) The cost function $H_i(q_i)$ $(i=1,...,k)$ given in (27) is a twice differentiable real valued function.

In assumption (A4) the full rank restriction on $J_i(q_i)$ (i=1,...,k) requires that all possible joint motions q_i, do not pass through any singularity configuration of $J_i(q_i)$, this will be shown to be possible with the control law derived in this paper, this will be addressed in the final section of the paper (see also [38]). If $J_i(q_i)$ is continuous and full rank in some subset $G_{J_i} \subset R^{n_i}$, then $J_i^+ = J_i^T(J_iJ_i^T)^{-1}$, $P_{J_i} = I_{n_i} - J_i^+ J_i$ and N_{J_i} are continuous in G_{J_i}. The matrices J_i, J_i^+, P_{J_i} and N_{J_i} are shift varying linear operators. It is easy to show that any continuous linear operator is bounded, hence J_i, J_i^+, P_{J_i} and N_{J_i} are bounded in G_{J_i}, (i.e. the induced norm of J_i, J_i^+, P_{J_i} and N_{J_i} are finite in G_{J_i}). Furthermore, if $\dot{J}_i$ is continuous in G_{J_i}, then $\frac{dJ_i^+}{dt}$ and $\dot{P}_{J_i}$ are continuous on any path with continuous $\dot{q}_i$ in G_{J_i}.

DESIGN OF THE CONTROL AND UPDATE LAWS

Design of the Control Law

Our goal is to design an adaptive controller that guarantees the asymptotic convergence of the load tracking error to zero, the convergence of the internal forces to their desired values and the redundancy resolution. We will start by defining a few variables needed for the development. The weighted reference velocity error for the ith robot is defined as,

$$\rho_{r_i} = w_t(\dot{q}_i - v_{r_i}) \qquad i=1,...,k \, . \tag{36}$$

The scalar weighting function w_t will be chosen as, $w_t = e^{\lambda t}$, where λ is a positive constant (see [18] for the use of weighting functions in the adaptive control of single rigid robots). We will group the ρ_{r_i} (i=1,...,k) into one vector ρ_r such that, $\rho_r = \left[\rho_{r_1}^T \; \rho_{r_2}^T \; \cdots \; \rho_{r_k}^T\right]^T$. Also the weighted reference velocity error for the load is defined as,

$$\rho_o = w_t(\dot{x} - v_o) \, . \tag{37}$$

It is easy to show that,

$$\rho_o = J_i \rho_{r_i} \qquad i=1,...,k \, . \tag{38}$$

We will choose $\dot{\rho}_r$ and $\dot{\rho}_o$ such that,

$$\dot{\rho}_{r_i} = w_t(\ddot{q}_i - a_{r_i}) \qquad i=1,...,k \, , \tag{39}$$

$$and, \; \dot{\rho}_o = w_t(\ddot{x} - a_o) \, . \tag{40}$$

The choice of ρ_{r_i} given by equation (36), and the choice of $\dot{\rho}_{r_i}$ given by equation (39) will result in the following value for a_{r_i},

$$a_{r_i} = \dot{v}_{r_i} + \lambda(v_{r_i} - \dot{q}_i) \qquad i=1,...,k \, . \tag{41}$$

We can group the a_{r_i} (i=1,...,k) into one vector a_r such that, $a_r = \left[a_{r_1}^T \; a_{r_2}^T \; \cdots \; a_{r_k}^T\right]^T$. The choice of ρ_o given by equation (37), and the choice of $\dot{\rho}_o$ given by equation (40) will result in the following value for a_o,

$$a_o = \dot{v}_o + \lambda(v_o - \dot{x}) \ . \tag{42}$$

Notice that v_r and v_o are independent of $\dot{q}$ and $\dot{x}$, hence a_r and a_o are not functions of $\ddot{q}$ and $\ddot{x}$. Therefore the proposed adaptive scheme does not require the measurements of the accelerations $\ddot{q}$ and $\ddot{x}$.

Theorem 1

Given that the matrices K_o, K_r, Γ_r and Γ_o are positive definite matrices, K_f is a positive semidefinite diagonal matrix, the control law given by (43) and the parameter update laws given by (45) and (46) ensure that ρ_r , $\rho_o \in L_2 \cap L_\infty$ and that P_r , $P_o \in L_\infty$.

$$\begin{aligned}\tau &= \hat{D}a_r + \hat{C}v_r + \hat{G}_r - K_r(\dot{q}-v_r) + J_e^T G^+[\hat{M}a_o + \hat{N}_2 v_o + \hat{G}_l - K_o(\dot{x}-v_o)] + J_e^T \tau_f \\ &= Y_r \hat{P}_r - K_r(\dot{q}-v_r) + J_e^T G^+[Y_o \hat{P}_o - K_o(\dot{x}-v_o)] + J_e^T \tau_f \ ,\end{aligned} \tag{43}$$

The force torque τ_f is given by,

$$\tau_f = F_{eI_d} - K_f \textstyle\int e_f \ , \tag{44}$$

The parameters update laws are such,

$$\dot{\hat{P}}_r = -\,\Gamma_r^{-1} Y_r^T \rho_r w_t \ , \tag{45}$$

$$\textit{and,} \quad \dot{\hat{P}}_o = -\,\Gamma_o^{-1} Y_o^T \rho_o w_t \ . \tag{46}$$

Preliminaries to the Proof

Before proving theorem 1, we will derive the equation of the closed loop system. We can solve for the force from equation (7), thus we get,

$$F_e = G^+(M\ddot{x} + N_2\dot{x} + G_l) + F_{eI} \ . \tag{47}$$

If we combine equations (3) and (47), we get,

$$D\ddot{q} + C\dot{q} + G_r + J_e^T G^+(M\ddot{x} + N_2\dot{x} + G_l) + J_e^T F_{eI} = \tau \ . \tag{48}$$

Now we will multiply both sides of equation (48) by $G(J_e^T)^+$, we get,

$$G(J_e^T)^+[\, D\ddot{q} + C\dot{q} + G_r \,] + M\ddot{x} + N_2\dot{x} + G_l = G(J_e^T)^+\tau \ . \tag{49}$$

Here we used the fact that $GF_{eI} = 0$.
Replacing τ by its value from equation (43), and using the fact that $G\tau_f = 0$, we get,

$$\begin{aligned}&G(J_e^T)^+[\, D\ddot{q} + C\dot{q} + G_r \,] + M\ddot{x} + N_2\dot{x} + G_l \\ &\quad = G(J_e^T)^+[\, Y_r\hat{P}_r - K_r(\dot{q}-v_r) \,] + Y_o\hat{P}_o - K_o(\dot{x}-v_o) \\ &\quad = G(J_e^T)^+[\, Y_r\tilde{P}_r + Y_r P_r - K_r(\dot{q}-v_r) \,] + Y_o\tilde{P}_o + Y_o P_o - K_o(\dot{x}-v_o) \ .\end{aligned} \tag{50}$$

Replacing $Y_r P_r$ and $Y_o P_o$ by their values from equations (17) and (6), we obtain,

$$\begin{aligned}&G(J_e^T)^+[\, D(\ddot{q}-a_r) + C(\dot{q}-v_r) + K_r(\dot{q}-v_r) \,] + M(\ddot{x} - a_o) + N_2(\dot{x}-v_o) + K_o(\dot{x}-v_o) \\ &\quad = G(J_e^T)^+ Y_r\tilde{P}_r + Y_o\tilde{P}_o \ .\end{aligned} \tag{51}$$

Thus the composite system can be written as,

$$G(J_e^T)^+[D\dot{\rho}_r + C\rho_r + K_r\rho_r] + M\dot{\rho}_o + N_2\rho_o + K_o\rho_o = G(J_e^T)^+ Y_r\tilde{P}_r w_t + Y_o\tilde{P}_o w_t. \quad (52)$$

Proof of Theorem 1:

Consider the following Lyapunov function candidate

$$V = \frac{1}{2}\rho_r^T D\rho_r + \frac{1}{2}\tilde{P}_r^T\Gamma_r\tilde{P}_r + \frac{1}{2}\rho_o^T M\rho_o + \frac{1}{2}\tilde{P}_o^T\Gamma_o\tilde{P}_o\,. \quad (53)$$

Now if we differentiate V with respect to time and use properties P1 - P3, we get,

$$\dot{V} = \rho_r^T[\,D\dot{\rho}_r + C\rho_r\,] + \tilde{P}_r^T\Gamma_r\dot{\tilde{P}}_r + \rho_o^T[\,M\dot{\rho}_o + N_2\rho_o\,] + \tilde{P}_o^T\Gamma_o\dot{\tilde{P}}_o\,, \quad (54)$$

using the fact that $\rho_r = J_e^+ G^T\rho_o$, $\dot{V}$ becomes,

$$\dot{V} = \rho_o^T[\,G(J_e^T)^+(D\dot{\rho}_r + C\rho_r) + M\dot{\rho}_o + N_2\rho_o\,] + \tilde{P}_r^T\Gamma_r\dot{\tilde{P}}_r + \tilde{P}_o^T\Gamma_o\dot{\tilde{P}}_o\,. \quad (55)$$

Using equation (52), we get,

$$\begin{aligned}\dot{V} &= \rho_o^T[\,-G(J_e^T)^+K_r\rho_r - K_o\rho_o + G(J_e^T)^+ Y_r\tilde{P}_r w_t + Y_o\tilde{P}_o w_t\,] + \tilde{P}_r^T\Gamma_r\dot{\tilde{P}}_r + \tilde{P}_o^T\Gamma_o\dot{\tilde{P}}_o \\ &= -\,\rho_o^T G(J_e^T)^+K_r J_e^+ G^T\rho_o - \rho_o^T K_o\rho_o + \rho_o^T G(J_e^T)^+ Y_r\tilde{P}_r w_t + \rho_o^T Y_o\tilde{P}_o w_t \\ &\quad + \tilde{P}_r^T\Gamma_r\dot{\tilde{P}}_r + \tilde{P}_o^T\Gamma_o\dot{\tilde{P}}_o\,. \end{aligned} \quad (56)$$

Using the update laws given by equations (45) and (46), we get,

$$\begin{aligned}\dot{V} &= -\,\rho_o^T G(J_e^T)^+K_r J_e^+ G^T\rho_o - \rho_o^T K_o\rho_o + \rho_o^T G(J_e^T)^+ Y_r\tilde{P}_r w_t - \tilde{P}_r^T Y_r^T\rho_r w_t \\ &\quad + \rho_o^T Y_o\tilde{P}_o w_t \;\; - \tilde{P}_o^T Y_o^T\rho_o w_t\,. \end{aligned} \quad (57)$$

Thus,

$$\dot{V} = -\,\rho_o^T G(J_e^T)^+K_r J_e^+ G^T\rho_o - \rho_o^T K_o\rho_o = -\,\rho_r^T K_r\rho_r - \rho_o^T K_o\rho_o\,. \quad (58)$$

Hence $V > 0$ and $\dot{V} \leq 0$. Thus we can conclude that $\rho_r \in L_2 \cap L_\infty$, $\rho_o \in L_2 \cap L_\infty$, $\tilde{P}_r \in L_\infty$ and $\tilde{P}_o \in L_\infty$.

□

Corollary 1

$\dot{q}_i - v_{r_i} \longrightarrow 0$ (i=1,...,k) and $\dot{x} - v_o \longrightarrow 0$ at the rate of $e^{-\lambda t}$.

Proof:

Using equation (36), we can write $\dot{q}_i - v_{r_i} = \rho_{r_i} e^{-\lambda t}$. Hence $\dot{q}_i - v_{r_i} \longrightarrow 0$ (i=1,...,k) at the rate of $e^{-\lambda t}$. Similarly, from equation (37), we can write, $\dot{x} - v_o = \rho_o e^{-\lambda t}$. Hence, $\dot{x} - v_o \longrightarrow 0$, at the rate of $e^{-\lambda t}$.

□

Hence we can conclude from equation (51) that,

$$G(J_e^T)^+ D\dot{\rho}_r e^{-\lambda t} + M\dot{\rho}_o e^{-\lambda t} - G(J_e^T)^+ Y_r \tilde{P}_r - Y_o \tilde{P}_o \longrightarrow 0 \qquad as\ t \longrightarrow \infty. \quad (59)$$

provided that the joint angles q are bounded and $v_r(q)$ is bounded. We can show through the analysis of the perturbed dynamical systems $\dot{q} - v_r(q) = w_t^{-1}\rho_r \longrightarrow 0$, as, $t \longrightarrow \infty$ that q for an appropriate choice of $v_r(q)$ will be bounded and stable. This will be shown next (see also [38]). In fact the boundedness q and the boundedness $J_e^+ G^T$ (i.e, robot trajectories do not pass through singular configuration) both depend on the stability of $\dot{q} = v_r(q)$ and therefore on the choice of $v_r(q)$, this will be seen in the final sections (see also [38]). If $J_e^+ G^T$ is bounded (J_e nonsingular and q is bounded), we can write,

$$D\dot{\rho}_r e^{-\lambda t} + J_e^T G^+ M\dot{\rho}_o e^{-\lambda t} - Y_r \tilde{P}_r - J_e^T G^+ Y_o \tilde{P}_o \longrightarrow 0 \qquad as\ t \longrightarrow \infty. \quad (60)$$

Boundedness of the Internal Forces

Theorem 2

The control law given by (43) and the parameter update laws given by (45) and (46) ensure the convergence of the internal forces to their desired trajectories. (i.e., $e_f \longrightarrow 0\ as\ t \longrightarrow \infty$) .

Proof:

If we combine equations (48) and (43), we get,

$$\begin{aligned} D\ddot{q} + C\dot{q} + G_r + J_e^T G^+(M\ddot{x}+N_2\dot{x}+G_l) + J_e^T F_{eI} &= Y_r\hat{P}_r - K_r(\dot{q}-v_r) \\ &+ J_e^T G^+[Y_o\hat{P}_o - K_o(\dot{x}-v_o)] + J_e^T \tau_f. \end{aligned} \quad (61)$$

Using the facts that $\hat{P}_r = \tilde{P}_r + P_r$, and $\hat{P}_o = \tilde{P}_o + P_o$, and replacing $Y_r P_r$ and $Y_o P_o$ by their values from equations (17) and (18), we obtain,

$$\begin{aligned} D(\ddot{q}-a_r) + C(\dot{q}-v_r) + K_r(\dot{q}-v_r) + J_e^T G^+[\, M(\ddot{x} - a_o) + N_2(\dot{x}-v_o) + K_o(\dot{x}-v_o)\,] \\ = Y_r\tilde{P}_r + J_e^T G^+ Y_o \tilde{P}_o + J_e^T(\tau_f - F_{eI}) \,. \end{aligned} \quad (62)$$

or,

$$\begin{aligned} D\dot{\rho}_r e^{-\lambda t} + C\rho_r e^{-\lambda t} + K_r\rho_r e^{-\lambda t} + J_e^T G^+[\, M\dot{\rho}_o e^{-\lambda t} + N_2\rho_o e^{-\lambda t} + K_o\rho_o e^{-\lambda t}\,] \\ = Y_r\tilde{P}_r + J_e^T G^+ Y_o\tilde{P}_o - J_e^T(e_f + K_f\textstyle\int e_f). \end{aligned} \quad (63)$$

Using corollary 1 and equation (60) (assuming $J_e^+ G^T$ and q is bounded), we can conclude that,

$$J_e^T(e_f + K_f\textstyle\int e_f) \longrightarrow 0 \quad as \quad t \longrightarrow \infty\,. \quad (64)$$

The matrix J_e^T is not a singular matrix, and it is a full rank matrix, thus we can conclude from equation (64) and with appropriate choice of K_f that,

$e_f \longrightarrow 0$, *as*, $t \longrightarrow \infty$. Notice that K_f can be set to zero if the internal forces are not measurable.

□

BOUNDEDNESS OF THE JOINT MOTIONS AND CONTROL TORQUES

In this section we will show the boundedness of q, $\dot{q}$, and the control torque τ based on a perturbation model. We notice that equation (32) can be written as a decayed perturbation system,

$$\dot{q}_i = v_{r_i}(q_i) + \delta_{v_i}(q_i, t) \qquad i=1,...,k. \tag{65}$$

Recall from Corollary 1 that $||\delta_{v_i}(q_i,t)|| \longrightarrow 0$, *as*, $t \longrightarrow \infty$, thus the perturbation $\delta_{v_i} = w_t^{-1} \rho_{ri}$ $(i=1,...,k)$ is bounded and tends to zero as $t \longrightarrow \infty$.

The boundedness of q_i in the perturbed system, described by equation (65), can be proved by ensuring the boundedness of q_i in the unperturbed system $\dot{q}_i = v_{r_i}(q_i)$. This extends from the result of Markus and Opial (see [5] pp. 282) (see also [38] for details). We note from the assumption (A3)-(A5) boundedness of $\dot{q}_i$ ensures boundedness of q_i and therefore boundedness of the control torque; for further details see [38].

THE STABILITY OF THE UNPERTURBED SYSTEM

The trajectories q_i $(i=1,...,k)$ of the unperturbed system are bounded if $\dot{q}_i$ $(i=1,...,k)$ of the self motion manifold is bounded. The dynamics of q_i on the self motion manifold have to be shown to result into joint angle q_i which is bounded. We will show that the quadratic form cost function $H_i(q_i)$ $(i=1,...,k)$ is a special choice which guarantees the boundedness of q_i $(i=1,...,k)$.

Below, we will examine the boundedness of the unperturbed system by using a homeomorphic transformation of the coordinates. A homeomorphism is a continuous mapping between two topological spaces if its inverse mapping is also continuous. A homeomorphism also maps a continuous function to another continuous function. A homeomorphism preserves the topological properties such as the openness, connectedness, and the convergence of a set. We will find a homeomorphism which transforms the coordinates of the configuration q_i $(i=1,...,k)$ into a decomposable coordinates ξ_i and ς_i $(i=1,...,k)$, where ξ_i is homeomorphic to the workspace coordinates x. The variable ς_i will be used to represent the dynamics on the self motion manifold. Hence the unperturbed system $\dot{q}_i = v_{r_i}(q_i)$ $(i=1,...,k)$ is transformed into a cascaded system,

$$\dot{\varsigma}_i = v_{\varsigma_i}(\varsigma_i, \xi_i) \, , \qquad \dot{\xi}_i = v_{\xi_i}(\xi_i) \qquad i=1,...,k \, . \tag{66}$$

The boundedness of q_i $(i=1,...,k)$ will be deduced from the boundedness of ξ_i and ς_i. We will adopt the method used to prove the sufficiency of the Frobenius' theorem [9], to find the homeomorphism. We will construct the diffeomorphism based on the self-motion manifold. For any given x, all the points q_i such that $x = K_i(q_i)$ lie on the leaf of the self-motion manifold $Q_N^{q_i^0}$. The leaf of the self-motion manifold will be denoted by $Q_N^{q_i^0}$. This manifold is a connected region. By assumption $N_{J_i}(q_i)$ is nonsingular, then the distribution $\Delta_i = ker\ (J_i) = span\ (N_{J_i})$ is nonsingular. The null space of a Jacobian matrix is always completely integrable, hence Δ_i is involutive. The distribution $\Delta_i = ker\ (J_i)$ has an annihilator Δ_i which is spanned by J_i which is the exact differential of the kinematic map K_i. The integrability of Δ_i allows us to construct the integral manifold by piecewise integrating every column of N_{J_i}.

Let $\Phi_t^{f_i}$ denote the flow of the vector field f_i, such that $q_i(t) = \Phi_t^{f_i}(q_i^0)$ solves the ordinary differential equation $\dot{q}_i = f_i(q_i)$ with initial condition q_i^0. The transition mapping $\Phi_t^{f_i}$ which maps q_i^0 to $q_i(t)$ is a diffeomorphism, and has the property $\frac{\partial \Phi_t^{f_i}(q_i^0)}{\partial t} = f_i(q_i(t))$ [6,9]. The flow of each vector field represented by a column of N_{J_i} is the solution of the following differential equations,

$$\dot{q}_i = N_{J_i^l}(q_i) \quad with \quad q_i(0) = q_i^0, \qquad \text{for} \quad l{=}1,...,r_i; \tag{67}$$

and can be written as,

$$q_i(t{=}\varsigma_i^l) = \Phi_{\varsigma_i^l}^{N_{J_i^l}}(q_i^0) \qquad l{=}1,...,r_i; \tag{68}$$

Thus we have,

$$\frac{\partial \Phi_{\varsigma_i^l}^{N_{J_i^l}}(q_i^0)}{\partial \varsigma_i} = N_{J_i^l}(q_i) \qquad l{=}1,...,r_i. \tag{69}$$

Lemma 4 (The parameterized equation of the self-motion manifold)

Given a kinematic mapping $x = K_i(q_i)$. The composite mapping $Q_{\varsigma_i} : R^{r_i} \longrightarrow Q_{N_i}^{q}$ such that,

$$(\varsigma_i^1, ..., \varsigma_i^{r_i}) \longrightarrow q_i(t) = \Phi_{\varsigma_i^{r_i}}^{N_{J_i^{r_i}}} o ... o \Phi_{\varsigma_i^1}^{N_{J_i^1}}(q_i^0) \quad and \quad t = \varsigma_i^1 + ... + \varsigma_i^{r_i}\ . \tag{70}$$

is a locally parametrized equation of the manifold $Q_N^{q_i^0} = \{q_i \in C(q_i^0) \quad such\ that \quad x_0 = K_i(q_i) = K_i(q_i^0)\}$, which passes through q_i^0. Here $C(q_i^0)$ is used to denote the connected regions of the self-motion manifold and $C(q_i^0)$ passes through the initial joint configuration q_i^0.

Proof:

We shall show that for $t = \varsigma_i^1 + ... + \varsigma_i^{r_i}$, we have $K_i(q_i(t)) = K_i(q_i^0)$. Since $x = K_i(q_i)$, it suffices to show that x is unchanged whenever ς_i varies locally, i.e. $\frac{\partial x}{\partial \varsigma_i^l} = 0$ for $l{=}1,...,r_i$.

First, consider the rightmost integral $\Phi_{\varsigma_i^1}^{N_{J_i^1}}$ in (70). Let $q_{\varsigma_i^1} = \Phi_{\varsigma_i^1}^{N_{J_i^1}}(q_i^0)$. Then

$$\frac{\partial x}{\partial \varsigma_i^1} = \frac{\partial K_i}{\partial \varsigma_i^1}\Big|_{q_{\varsigma_i^1}} = \frac{\partial K_i}{\partial q_i}\Big|_{q_{\varsigma_i^1}} \frac{\partial q_{\varsigma_i^1}}{\partial \varsigma_i^1} = \frac{\partial K_i}{\partial q_i}\Big|_{q_{\varsigma_i^1}} \frac{\partial \Phi_{\varsigma_i^1}^{N_{J_i^1}}(q_i^0)}{\partial \varsigma_i^1}$$

$$= J_i(q_{\varsigma_i^1})\, N_{J_{\varsigma_i^1}}(q_{\varsigma_i^1}) = 0\ . \tag{71}$$

Hence $q_{\varsigma_i^1} \in Q_N^{q_i^0}$ when $q_i^0 \in Q_N^{q_i^0}$. Similarly, we have

$$\frac{\partial x}{\partial \varsigma_i^l} = 0 \qquad for \;\; l = 2, ..., r_i \, . \tag{72}$$

Then for the lth transition, we have, $q_i(t=\varsigma_i^1+...+\varsigma_i^l) = \Phi_{\varsigma_i^l}^{N_{J_i^l}}(q_i(t=\varsigma_i^l+...+\varsigma_i^{l-1}))$ for l=1,...,r_i. Hence $q(t=\varsigma_i^1+...+\varsigma_i^l) \in Q_N^{q_i^0}$. Moreover these $q_i's$ are connected since $\Phi_{\varsigma_i^l}^{N_{J_i^l}}$ for l=1,...,r_i. are continuous mapping with respect to ς_i^l (l=1,...,r_i). Therefore (70) maps ς_i to $q_i(t) \in Q_N^{q_i^0}$. This mapping is a diffeomorphism because it is a composition of the diffeomorphisms $\Phi_{\varsigma_i^l}^{N_{J_i^l}}$.

Lemma 5 (Decomposition of the coordinates)
Given a kinematic mapping $x = K_i(q_i)$ (i=1,...,k), and let U_i be the image of the joint space Q_i. At any point $q_i \in Q_i$, there exists a diffeomorphism F_i^{-1}, which decomposes q_i into $\varsigma_i \in R^{r_i}$ and $\xi_i \in R^{m_i}$ (here m_i=6 for all i=1,...,k), such that $\begin{bmatrix}\varsigma_i \\ \xi_i\end{bmatrix} = F_i^{-1}(q_i)$. The mapping $\varsigma_i(q_i)$ maps a point q_i on the corresponding self-motion manifold $Q_{N_i}^q$ into ς_i.

Proof:

We will construct the desired diffeomorphism on the given leaf of the self-motion manifold. Recall that N_{J_i} is the orthogonal complement of J_i^+. The matrix J_i is assumed to be full rank and has the right inverse J_i^+, $J_i^+ = J_i^T(J_iJ_i^T)^{-1}$. Then the range space of J_i^+ and the range space of J_i^T are equal. The domain space of any matrix is the direct-sum of its row space and its null space, hence the domain of J_i is R^{n_i}. Thus we have,

$$rank([N_{J_i} \; , \; J_i^+]) = n_i \, . \tag{73}$$

Consider the composite mapping $F_i : R^{n_i} \longrightarrow Q_i$

$$(\varsigma_i^1, ..., \varsigma_i^{r_i}, \xi_i^1, ..., \xi_i^{m_i}) \longrightarrow q_i(t) = \Phi_{\xi_i^m}^{J_{i_m}^+} o ... o \Phi_{\xi_i^1}^{J_{i_1}^+} o \Phi_{\varsigma_i^{r_i}}^{N_{J_i^{r_i}}} o ... o \Phi_{\varsigma_i^1}^{N_{J_i^1}}(q_i^0). \tag{74}$$

The mapping F_i is a diffeomorphism, since the composition of diffeomorphisms is a diffeomorphism. Hence the inverse of F_i , F_i^{-1}, exists and it is a smooth mapping. Thus,

$$\begin{bmatrix}\varsigma_i \\ \xi_i\end{bmatrix} = F_i^{-1}(q_i) \tag{75}$$

where $\varsigma_i = (\varsigma_i^1, ..., \varsigma_i^{r_i})^T$ and $\xi_i = (\xi_i^1, ..., \xi_i^{m_i})^T$ are real functions.
We have,

$$(\varsigma_i \; , \; \xi_i) = F_i^{-1} \circ F_i(\varsigma_i, \xi_i) \, , \tag{76}$$

then the Jacobian matrices F_i^{-1} and F_i should satisfy the following equation,

$$\begin{bmatrix}\dfrac{\partial \varsigma_i}{\partial q_i} \\ \dfrac{\partial \xi_i}{\partial q_i}\end{bmatrix} \begin{bmatrix}\dfrac{\partial F_i}{\partial \varsigma_i} & \dfrac{\partial F_i}{\partial \xi_i}\end{bmatrix} = I_{n_i} \, . \tag{77}$$

In the next lemma we can find the relationships between the derivatives of (ς_i, ξ_i) and q_i.

As the distribution $\Delta_i = ker\ (J_i)$ is involutive, the diffeomorphism F_i has the property, ([4] pp. 27) that for every q_i, the r_i columns of the Jacobian matrix $\frac{\partial F_i}{\partial \varsigma_i}$ are linearly independent vectors in the distribution Δ_i.

Lemma 6 (The time derivatives of the transformed coordinates)

The transformation F_i given in Lemma 5 allows us to write,

$$\dot{\xi}_i = M_{J_i}\ J_i\ \dot{q}_i \tag{78}$$

$$\dot{\varsigma}_i = M_{N_i}^{-1}\ N_{J_i}^T\ \dot{q}_i\ . \tag{79}$$

Proof:

We can always find a nonsingular $r_i \times r_i$ matrix M_{N_i}, which expresses $\frac{\partial F_i}{\partial \varsigma_i}$ as a linear combination of the columns in N_{J_i},

$$\frac{\partial F_i}{\partial \varsigma_i} = N_{J_i}\ M_{N_i}\ . \tag{80}$$

From (77) we have $\frac{\partial \xi_i}{\partial q_i}\ \frac{\partial F_i}{\partial \varsigma_i} = 0$, thus,

$$\frac{\partial \xi_i}{\partial q_i}\ N_{J_i}\ M_{N_i} = \ 0 \in R^{m_i \times r_i}\ . \tag{81}$$

Hence N_{J_i} annihilates $\frac{\partial \xi_i}{\partial q_i}$. Recall that $J_i N_{J_i} = 0$, thus each row of $\frac{\partial \xi_i}{\partial q_i}$ must be a linear combination of the rows of J_i. Hence,

$$\frac{\partial \xi_i}{\partial q_i} = M_{J_i}\ J_i \tag{82}$$

where M_{J_i} is a nonsingular $m_i \times m_i$ matrix. Therefore $\dot{\xi}_i = \frac{\partial \xi_i}{\partial q_i}\dot{q}_i = M_{J_i}\ J_i\ \dot{q}_i$ yields (78). From (77) we have, $\frac{\partial \xi_i}{\partial q_i}\ \frac{\partial F_i}{\partial \xi_i} = I_{m_i}$; combining this equation with (82) yields,

$$\frac{\partial F_i}{\partial \xi_i} = J_i^{+} M_{J_i}^{-1}\ , \tag{83}$$

because the nonsingular matrix J_i has a unique pseudo-inverse J_i^{+} such that $J_i J_i^{+} = I_{m_i}$. We can write,

$$\dot{q}_i = \frac{\partial F_i}{\partial \varsigma_i}\ \dot{\varsigma}_i + \frac{\partial F_i}{\partial \xi_i}\ \dot{\xi}_i\ . \tag{84}$$

Thus we have,

$$\frac{\partial F_i}{\partial \varsigma_i}\ \dot{\varsigma}_i = (I_{n_i} - \frac{\partial F_i}{\partial \xi_i} M_{J_i}\ J_i\ )\dot{q}_i = (I_{n_i} - J_i^{+}\ J_i)\dot{q}_i = P_{J_i}\ \dot{q}_i\ . \tag{85}$$

To obtain (79), we substitute (80) into the above equation and premultiply both sides by $N_{J_i}^T$. Notice that $N_{J_i}^T N_{J_i} = I_{r_i}$, since each column of N_{J_i} is a normalized basis vector.

Remark 2

Equation (78) implies that $\dot{\xi}_i = M_{J_i}\dot{x}$ and $\frac{\partial \xi_i}{\partial x} = M_{J_i}$. From the implicit mapping theorem, the non singularity of M_{J_i} ensures that ξ_i is homeomorphic to x.

Lemma 7 (The decomposition of the unperturbed system)

Using the transformation F_i given by lemma 5, we can write the unperturbed system $\dot{q}_i = v_{r_i}(q)$, (v_{r_i} is expressed by (33)) as a cascaded system in the following form,

$$\dot{\varsigma}_i = \mu_i M_{N_i}^{-1} (N_{J_i}^T \ \nabla H_i) \ (q(\varsigma_i, e)), \tag{86}$$

$$\dot{e} + \gamma e = 0. \tag{87}$$

The notation used in (86) means that $N_{J_i}^T$ and ∇H_i are functions of (ς_i, e) through dependency on the joint variable q_i.

Proof:

The unperturbed system is now given by,

$$\dot{q}_i = J_i^+(\dot{x}_d - \gamma e) + \mu_i P_{J_i} \nabla H_i \qquad i{=}1,...,k\ . \tag{88}$$

Equation (87) is obtained by premultiplying both sides of (88) by J_i and recalling that $J_i P_{J_i} = 0$. Similarly, equation (86) is obtained by premultiplying both sides of (88) by $M_{N_i}^{-1} N_{J_i}^T$ and recalling that $N_{J_i}^T J_i^+ = 0$. Notice that q_i can be decomposed into (ς_i, ξ_i) by F_i^{-1} given by (75). Also notice that ξ_i is homeomorphic to x. Thus ξ_i is homeomorphic to e because there is a one to one mapping between x and e. Then e is independent of ς_i, so q_i can be decomposed into (ς_i, e).

Lemma 8 (The stability of a cascaded system)

Consider the system (86) and (88) in hierarchical form,

$$\dot{\varsigma}_i = f_i(\varsigma_i, \xi_i) \qquad and, \qquad \dot{\xi}_i = g_i(\xi_i)\ . \tag{89}$$

If the functions f_i and g_i are continuously differentiable, then $(\varsigma_i, \xi_i) = (0,0)$ is a locally asymptotically stable equilibrium of the system, if and only if $\xi_i = 0$ is a locally asymptotically stable equilibrium of $g_i(\xi_i)$ and $\varsigma_i = 0$ is a locally asymptotically stable equilibrium of $f_i(\varsigma_i, 0)$.

The proof of this lemma can be found in Vidyasagar [9].

The equilibrium point of the cascaded system given in lemma 7 is $e = 0$, $\varsigma_i = \varsigma_i^*$ (for i=1,...,k). Here ς_i^* is the coordinates such that,

$$(N_{J_i}^T \nabla H_i)(q_i(\varsigma_i^*, 0)) = 0 \qquad i{=}1,...,k. \tag{90}$$

The equilibrium joint position q_i^* is then,

$$q_i^* = F_i(\varsigma_i^*, 0) \qquad i{=}1,...,k. \tag{91}$$

Remark 3

Setting $e = 0$ in (86) gives us the zero-dynamics [4],

$$\dot{\varsigma}_i = \mu_i M_{N_i}^{-1} (N_{J_i}^T \nabla H_i)\,(q_i(\varsigma_i,0)) \qquad i{=}1,...,k, \tag{92}$$

of the unperturbed system. The zero dynamics is defined on the manifold R^{r_i}. Equations (79) and (92) lead to,

$$N_{J_i}^T \dot{q}_i = \mu_i(N_{J_i}^T \nabla H_i)(q_i(\varsigma_i,0)) \;\; or \;\; P_{J_i}\dot{q}_i = \mu_i(P_{J_i} \nabla H_i)(q_i(\varsigma_i,0)) \quad i{=}1,...,k. \tag{93}$$

Notice that $q_i(\varsigma_i,0) \in Q_{N_i}^q$. Equation (93) is defined on the manifold of $\{q_i = F_i(\varsigma_i,\xi_i)$ *such that* $\varsigma_i \in R^{r_i}$and $e = 0\,\}$. This manifold is also expressed by,

$$Q_{N_i}^q = \{q_i \in Q_i \;\; such\ that \;\; x_d = K_i(q_i) \;\text{ and }\; J_i\dot{q}_i = 0\} \;\text{ for } i{=}1,...,k, \tag{94}$$

and it is indeed the self-motion manifold over x_d. We observe that the identity $\dot{q}_i = (J_i^+ J_i + P_{J_i})\dot{q}_i$ is satisfied on any $q_i \in Q_i$. However for motions on the self-motion manifold $\dot{x} = J_i(q_i)\dot{q}_i = 0$, and thus for motions on the self motion manifold we also have $\dot{q}_i = P_{J_i}\dot{q}_i$. Equation (93) can be rewritten as,

$$q_i = \mu_i(P_{J_i} \nabla H_i)(q_i) \qquad for\ all\ q_i \in Q_{N_i}^q \;. \tag{95}$$

Equation (95) will be called the "equivalent zero dynamics" expressed in the joint space and defined on the manifold $Q_{N_i}^q$.

Proposition 4 (Boundedness of the unperturbed system)

The equilibrium points q_i^* $(i{=}1,...,k)$ of the unperturbed system is asymptotically stable if the equilibrium point $(\varsigma_i^*,0)$ of the zero-dynamics (95) is asymptotically stable. The trajectory $q_i(t)$ of the unperturbed system starting from any finite initial configuration q_i^0 is bounded if the solution trajectory of the zero dynamics defined on the self-motion manifold $Q_N^{q_i^0} = \{q_i \in C(q_i^0)$ *such that* $K_i(q_i) = K_i(q_i^0) = x_0\}$ is bounded.

Proof:

Lemma 7 asserts that the unperturbed system given by (95)) can be decomposed into a cascaded system, then the asymptotic stability results are obtained immediately from lemma 8.

Proposition 5 (Boundedness is guaranteed with the choice of H_i)

Let the cost function $H_i(q_i)$ be a quadratic of the form :

$$H_i(q_i) = \frac{1}{2}(q_i - q_{C_i})^T M_{h_i}(q_i - q_{C_i}) \qquad i{=}1,...,k \tag{96}$$

where q_{C_i} $(i{=}1,...,k)$ is fixed, and M_{h_i} is a symmetric positive definite matrix. Further let q_{C_i} be given in a set of isolated points. Consider the zero-dynamics,

$$\dot{q}_i = \mu_i P_{J_i}\nabla H_i(q_i) = \mu_i P_{J_i} M_{h_i}(q_i - q_{C_i}) \quad with \quad q_i \in Q_{N_i}{}^q \;. \tag{97}$$

The vector q_i is bounded and $q_i \to q_i^*$ as $t \to \infty$ for every fixed q_{C_i}. Where q_i^* $(i{=}1,...,k)$ is the optimal solution of the problem given by (27).

Proof:

Let the Lyapunov function candidate V_i be,

$$V_i = \frac{1}{2}(q_i - q_{C_i})^T M_{h_i}(q_i - q_{C_i}) \qquad q_i \in Q^q_{N_i}\ , \tag{98}$$

The derivative of V_i is,

$$\dot{V}_i = \mu_i(q_i - q_{C_i})^T M_{h_i} P_{J_i} M_{h_i}(q_i - q_{C_i}) = \mu_i \| P_{J_i} M_{h_i}(q_i - q_{C_i})\|^2 \leq 0\ ,\ for\ \mu_i < 0. \tag{99}$$

Here the fact that P_{J_i} is a projector was used. Hence $q_i - q_{C_i} \in L_\infty$, in addition, because of the boundedness of q_{C_i} we have $q_i \in L_\infty$. Notice that the set $E_i = \{q_i \mid \dot{V}_i = 0\}$ is the the set of equilibrium points of (97), and is therefore an invariant set. From LaSalle's extension of Lyapunov direct method [5], $q_i(t) \longrightarrow q_i^*$ (i=1,...,k) as $t \longrightarrow \infty$ because q_i is in a bounded set.

Remark 4

Thus we see from the last proposition the choice of q_{C_i} and M_{h_i} for i=1,...,k can be used to ensure that point q_i^* is far from singular configurations. Thus ensuring that the robot joints do not go through the singular configuration this was assumed in A3 for the purpose of the development of the control law at the beginning of the paper. We should note the exact value of the joint angles $q_i^* \in Q^q_{N_i}$ for all i=1,...,k can be obtained by simulation of the equation (93).

Remark 5

The quadratic performance function defined in (90) ensures that the function v_{r_i} (i=1,...,k) is locally Lipschitz.

$$v_{r_i} = J_{i^+}(\dot{x}_d - \gamma e) + \mu_i P_{J_i} M_{h_i}(q_i - q_{C_i}) \qquad i=1,...,k. \tag{100}$$

The matrices J_i^+ and P_{J_i} are differentiable because of assumption (A4). A continuously differentiable function is locally Lipschitz. Also notice that M_{h_i} is a constant matrix. Hence the function given in (100) is differentiable with respect to q_i, and is therefore Lipschitz.

CONCLUSION

In this paper, we addressed the problem of controlling redundant multiple robots manipulating a load cooperatively. We proposed an adaptive controller that ensures the exponential tracking of the load position to its desired value and the convergence of the internal forces to their desired values. The controller also guaranteed that the parameters errors remained bounded, and that the redundancy resolution error was asymptotically stable. Measurements of the joint or load accelerations were not required. The concepts of zero dynamics and stability of perturbed nonlinear dynamical systems were used to prove the stability of the adaptive system, particularly the stability of the joint motions on the self motion manifold. The overall stability of the adaptive system was established for a certain class of optimization functions used for redundancy resolution.

REFERENCES

[1] S. Ahmad and Y. Nakamure, Workshop Report on "Theory and Application of Redundant Robots," *International Conference on Robotics and Automation,* Scottsdale, Arizona (1989).

[2] J. Baillieul, J.M. Hollerbach, and R.W. Brockett, "Programming and Control of Kinematically Redundant Manipulators," Proceedings of the 1984 IEEE Decision and Control Conference pp. 768-774 (1984).

[3] R. Colbaugh, K. Glass, and H. Seraji, "An Adaptive Kinematics Algorithm for Robot Manipulators," *Proceedings of the 1990 American Control Conference,* San Diego, pp. 2281-2286 (1990).

[4] A. DeLuca, "Zero Dynamics in Robotics Systems Nonlinear Synthesis," Progress in Systems and Control Series, Birkhauser Boston pp. 68-87 (1991).

[5] J.K. Hale and J.P. LaSalle, "Differential Equations and Dynamical Systems," New York: Academic Press.

[6] M.W. Hirsch and S. Smale, "Differential Equations, Dynamical Systems, and Linear Algebra," Academic Press, New York (1974).

[7] P. Hsu, J. Hauser, and S. Sastry, "Dynamic Control of Redundant Manipulators," *Journal of Robotic Systems,* Vol. 6, No. 2, pp. 133-148 (1989).

[8] A. Ilchmann and D.H. Owens, "Adaptive Stability with Exponential Decay," *Systems and Control Letters,* Vol. 14, pp. 437-443 (1990).

[9] A. Isidori, "Nonlinear Control Systems," 2nd Ed. New York: Springer-Verlag (1989).

[10] J.P. LaSalle, "Some Extensions of Liapunov's Second Method," *IRE Transactions on Circuit Theory,* Vol. , pp. 520-527 (1960).

[11] A. Liegeois, "Automatic Supervisory Control of Configuration and Behavior of Multibody Mechanisms," *IEEE Transaction on Systems, Man and Cybernetics,* Vol. SMC-7, pp. 868-871 (1977).

[12] Y. Nakamura and H. Hanafusa, "Optimal Redundancy Control of Robot Manipulators," *International Journal of Robotics Research,* Vol. 6, No. 1, pp. 32-42 (1987).

[13] D.N. Nenchev, "Redundancy Resolution through Local optimization: A Review," *Journal of Robotic Systems,* Vol. 6, No. 6, pp. 769-798 (1989).

[14] G. Niemeyer and J-J.E. Slotine, "Adaptive Cartesian Control of Redundant Manipulators," *Proceedings of the American Control Conference,* San Diego, pp. 234-241 (1990).

[15] H.L. Royden, "Real Analysis," MacMillan, New York (1963).

[16] H. Seraji, "Configuration Control of Redundant Manipulators: Theory and Implementation," *IEEE Transactions on Robotics and Automation,* Vol. 5, No. 4, pp. 472-490 (1989).

[17] B. Siciliano, "Kinematic Control of Redundant Robot Manipulators: A Tutorial," *Journal of Intelligent and Robotic Systems,* Vol. 3, pp. 201-212 (1990).

[18] Y.D. Song, R.H. Middleton, and J.N. Anderson, "Study on the Exponential Path Tracking Control of Robot Manipulators via Direct Adaptive Methods," *1991 IEEE International Conference on Robotics and Automation,* pp. 22-27 (1991).

[19] M.W. Spong and M. Vidyasagar, "Robot Dynamics and Control," John Wiley & Sons, New York, NY (1989).

[20] M. Vidyasagar, "Decomposition Techniques for Large-Scale Systems with

Nonadditive Interactions: Stability and Stabilizability," *IEEE Transactions on Automatic Control,* Vol. AC-25, pp. 773-779 (1980).

[21] D.E. Whitney, "Resolved Motion Rate Control of Manipulators and Human Prostheses," *IEEE Transactions on Man-Machine Systems,* Vol. MMS-10, pp. 47-53 (1969).

[22] C. Carignan and D. Akin, "Cooperative Control of Two Arms in the Transport of an Inertial Load in Zero Gravity," *IEEE Journal of Robotics and Automation,* Vol. 4, No 4, pp. 414-419 (1988).

[23] P. Hsu, Z. Li, and S. Sastry, "On Grasping and Coordinated Manipulation by a Multifingered Robot Hand," *Proceedings of the 1988 IEEE International Conference on Robotics and Automation,* Philadelphia, PA, pp. 384-389 (1988).

[24] M. Zribi and S. Ahmad, "Adaptive Control for Multiple Coordinated Robots Cooperative Arms," *Proceedings of the 1992 IEEE CDC Conference,* Tuscon, AZ, pp. 1392-1398 (1992).

[25] M. Zribi and S. Ahmad, "Lyapunov Based Control of Multiple Flexible Joint Robots," *Proceedings of the 1992 American Control Conference,* Chicago, IL, pp. 3324-3329 (1992).

[26] S. Ahmad and H. Guo, "Dynamic Coordination of Dual-Arm Robotic Systems With Joint Flexibility," *Proceedings of the 1988 IEEE International Conference on Robotics and Automation,* Philadelphia, Pennsylvania, pp. 332-337 (1988).

[27] Y. Hu and A. Goldenberg, "An Adaptive Approach to Motion and Force Control of Multiple Coordinated Robot Arms," *Proceedings of the 1989 IEEE International Conference on Robotics and Automation,* Scottsdale, Arizona, pp. 1091-1096 (1989).

[28] H. Seraji, "Decentralized Adaptive Control of Manipulators: Theory, Simulation, and Experimentation," *IEEE Trans. on Robotics and Automation,* Vol. 5, No. 2, pp. 183-201 (1989).

[29] T. Tarn, A. Bejczy, and X. Yun, "Design of Dynamic Control of Two Cooperating Robot Arms: Closed chain Formulation," *Proceedings of the 1987 IEEE International Conference on Robotics and Automation,* Raleigh, NC, pp. 7-13 (1987).

[30] X. Yun, T. Tarn, and A. Bejczy, "Dynamic Coordinated Control of Two Robot Manipulators," *Proceedings of 28th Conference on Decision and Control,* Tampa, FL, pp. 2476-2481 (1989).

[31] M. Walker, D. Kim and J. Dionise, "Adaptive Coordinated Motion Control of Two Manipulator Arms," *Proceedings of the 1989 IEEE International Conference on Robotics and Automation,* Scottsdale, Arizona, pp. 1084-1090 (1989).

[32] T. Yoshikawa and X. Zheng, "Coordinated Dynamic Hybrid Position/Force Control for Multiple Robot Manipulators Handling One Constraint Object," *Proceedings of the 1990 IEEE International Conference on Robotics and Automation,* Cincinnati, Ohio, pp. 1178-1183 (1990).

[33] Y. Zheng and J. Luh, "Joint Torques Control of Two Coordinated Moving Robots," *Proceedings of the 1986 IEEE International Conference on Robotics and Automation,* San Francisco, CA, pp. 1375-1380 (1986).

[34] S. Hayati, "Hybrid Position/Force Control of Multi-Arm Cooperating Robots," *Proceedings of the 1986 IEEE International Conference on Robotics and Automation,* San Francisco, CA., pp. 82-89 (1986).

[35] A. Cole, J. E. Hauser and S. S. Sastry, "Kinematics and Control of Multifingered Hands with Rolling Contacts," IEEE Transactions on Automatic Control, Vol. 34, No. 4, pp. 398-404 (1989).

[36] S. Ahmad, "Control of Cooperative Multiple Flexible Joint Robots," to appear in IEEE Transactions on Systems, Man, Cybernetics, May-April 1993. (Also in IEEE CDC Conference Brighton, pp. 1413-1418 (1991)).

[36] T.J. Tarn, A.K. Bejczy, "Coordinated Control of Multiple Redundant Robots," Proceedings of the 1992 IEEE International Conference on Robotics and Automation, Nice, France (1992).

[37] J. Tao and J.Y.S. Luh, "Coordination of Two Redundent Robots," Proceedings of the 1989 IEEE International Conference on Robotics and Automation, Scottsdale, AZ, pp. 425-430 (1989).

[38] S. Luo, S. Ahmad, M. Zribi, "Adaptive Control of Kinematically Redundant Robots," TR-EE-92-22, June 1992, Electrical Engineering, Purdue University, West Lafayette, IN 47906 - USA. (Also in the Proceedings of the 1992 IEEE CDC Conference, Tuscon, AZ, pp. 602-607) (1992).

[39] I.D. Walker, S.I. Marcus, and R.A. Freeman, "Internal Object Loading for Multiple Cooperating Robot Manipulators," Proceedings of the 1989 IEEE International Conference on Robotics and Automation, Scottsdale, AZ, pp. 606-611 (1989).

[40] J.J.E. Slotine and W. Li, "Applied Nonlinear Control," Prentice Hall, N.J. (1991).

COMPUTATIONAL TECHNIQUES

JACOBIAN MOTION: A DIRECT APPROACH

Wolfram Stadler

San Francisco State University
Engineering Division
1600 Holloway Avenue
San Francisco, California 94132

ABSTRACT

We present a direct approach to the calculation of the Jacobian matrix of a manipulator. The approach is based on the use of the expression for the forward kinematics of the manipulator in terms of the usual product of Denavit-Hartenberg matrices, combined with the usual velocity calculations in the presence of rotating reference frames. The method is illustrated for the AdeptOne SCARA robot.

INTRODUCTION

The robotics literature generally adheres to two methods for the calculation of the Jacobian matrix of a manipulator: that of Whitney [1] and that of Paul [2], the latter approach having become most prevalent. We offer a third possibility for the calculation of the manipulator Jacobian matrix.

Within the present context, the terms *Jacobian* and *Jacobian matrix* have often been used interchangeably. The Jacobian is the determinant of the Jacobian matrix. With this laxity in mind, we thus briefly introduce some of the standard mathematical terminology.

For functions of several variables, the Jacobian matrix plays much the same role as does the differential quotient of a single variable: it provides a notion of tangency in n-space. Thus, if a transformation $f(\cdot)\colon \mathbb{R}^2 \to \mathbb{R}^2$ is defined by

$$f(x,y) = \begin{pmatrix} g(x,y) \\ h(x,y) \end{pmatrix},$$

then the differential of $f(\cdot)$ at (x_0, y_0) is

Mechanics and Control, Edited by R.S. Guttalu
Plenum Press, New York, 1994

$$df(x_0,y_0) = \begin{bmatrix} \frac{\partial g}{\partial x}(x_0,y_0) & \frac{\partial g}{\partial y}(x_0,y_0) \\ \frac{\partial h}{\partial x}(x_0,y_0) & \frac{\partial h}{\partial y}(x_0,y_0) \end{bmatrix} \begin{pmatrix} dx \\ dy \end{pmatrix}$$

where the transformation matrix is the Jacobian matrix for the transformation. More formally, for the transformation

$$f(\cdot): \mathbb{R}^n \rightarrow \mathbb{R}^m$$

defined by

$$f(z) = \left(u_1(z)\ u_2(z) \dots u_m(z)\right)^{\mathrm{T}}$$

with

$$z = (x_1 x_2 \dots x_n)^{\mathrm{T}}$$

we take

$$J(z) = \frac{\partial(u_1,\dots,u_m)}{\partial(x_1,\dots,x_n)}(z) = \begin{bmatrix} \frac{\partial u_1}{\partial x_1}(z) & \dots & \frac{\partial u_1}{\partial x_n}(z) \\ \frac{\partial u_m}{\partial x_1}(z) & \dots & \frac{\partial u_m}{\partial x_n}(z) \end{bmatrix}$$

to be the Jacobian matrix of the transformation $f(\cdot)$ at z.

Our calculation of the Jacobian matrix of a manipulator is based on velocity considerations. Occasionally, this approach is mentioned in the literature, but it appears not to have been carried out for specific examples. We precede our calculation with some essential kinematic results and notations involving the rotation matrix.

KINEMATIC FUNDAMENTALS

Our results are based on three views of the kinematics of rigid body motion: the vector description and matrix descriptions based on the configuration depicted in Figure 1.

We suppose the origin O_n of the nth reference frame to be located at a judicious point of the end-effector, with an imbedded reference frame whose orientation is that of the end-effector. Of the three descriptions of the motion, only the vectorial description is independent of the choice of components. For the vector description, we have

$$\boldsymbol{r}_0 = \boldsymbol{c}_n + \boldsymbol{r}_n \tag{1}$$

for the motion of a point P fixed in the body $\mathcal{B}$. In matrix form, we have the rigid body motion

$$\boldsymbol{r}_0 = \boldsymbol{c}_n + \boldsymbol{A}_n \boldsymbol{r}_n \tag{2}$$

where the components of $\boldsymbol{r}_0$ and $\boldsymbol{c}_n$ are taken with respect to $(0xyz)_0$ and those of $\boldsymbol{r}_n$ with

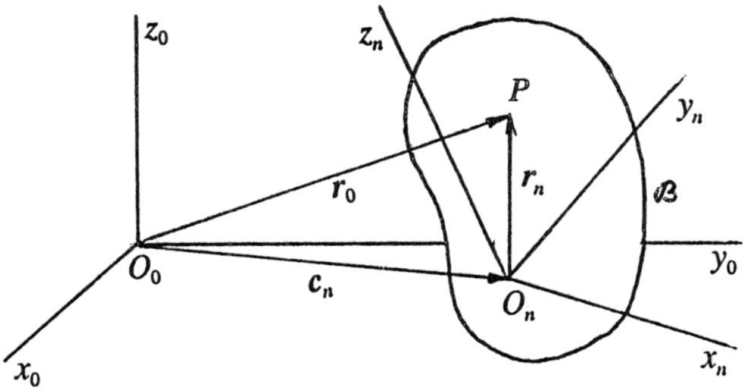

Figure 1. Motion of a Manipulator Point

respect to $(0xyz)_n$; $\boldsymbol{A}_n$ is the rotation matrix describing the orientation of $(0xyz)_n$ with respect to the frame $(0xyz)_0$. Finally, we may also represent the motion as a product of homogeneous Denavit-Hartenberg (DH) matrices,

$$\boldsymbol{H}_n = \sum_{i=1}^{n} \boldsymbol{H}_{i-1,i} = \boldsymbol{H} = \begin{bmatrix} a_x & n_x & o_x & p_x \\ a_y & n_y & o_y & p_y \\ a_z & n_z & 0_z & p_z \\ 0 & 0 & 0 & 1 \end{bmatrix}, \tag{3}$$

with

$$\boldsymbol{r}_0 = \boldsymbol{H}_n \boldsymbol{r}_n ,$$

where $\boldsymbol{H}$ is termed the *hand matrix* used to describe the usually specified motion and orientation of the end-effector. Here, p_x, p_y and p_z locate the origin O_n, and $\boldsymbol{a}$, $\boldsymbol{n}$ and $\boldsymbol{o}$ are unit vectors along the coordinate axes of $(0xyz)_n$. The identification of $\boldsymbol{a}$, $\boldsymbol{n}$ and $\boldsymbol{o}$ with specific coordinate axes may vary.

The differential displacement of the end-effector and the resultant expression for the Jacobian matrix are based on velocity calculations. In particular, we make use of the expressions for the velocity as derived from Equations (1) and (2). We have

$$\dot{\boldsymbol{r}}_0 = \dot{\boldsymbol{c}}_n + \boldsymbol{\omega}_n \times \boldsymbol{r}_n \tag{4}$$

$$\dot{\boldsymbol{r}}_0 = \dot{\boldsymbol{c}}_n + \dot{\boldsymbol{A}}_n \boldsymbol{r}_n , \tag{5}$$

and

$$\dot{\boldsymbol{r}}_0 = \dot{\boldsymbol{H}} \boldsymbol{r}_n , \tag{6}$$

where ω_n is the absolute angular velocity of $(0xyz)_n$. Expressions (4) and (5) are related in the following manner. We use Equation (2) in the form

$$r_n = A_n^{\mathrm{T}}(r_0 - c_n)$$

to write Equation (5) as

$$\dot{r}_0 - \dot{c}_n = \dot{A}_n r_n = \dot{A}_n A_n^{\mathrm{T}}(r_0 - c_n) = W(r_0 - c_n) \ .$$

The skew-symmetric matrix

$$W = \dot{A}_n A_n^{\mathrm{T}} = \begin{bmatrix} 0 & -\omega_{z_0} & \omega_{y_0} \\ \omega_{z_0} & 0 & -\omega_{x_0} \\ -\omega_{y_0} & \omega_{x_0} & 0 \end{bmatrix} \tag{7}$$

is related to the absolute angular velocity ω_n with

$$\omega_n = \omega_{x_0}\hat{\imath}_0 + \omega_{y_0}\hat{\jmath}_0 + \omega_{z_0}\hat{k}_0 \ , \tag{8}$$

in terms of its components with respect to the base reference frame; ω_n is also termed the axial vector of the skew-symmetric matrix W.

THE JACOBIAN MATRIX FOR THE AdeptOne

We now turn to the definition of the Jacobian matrix of an n-axis manipulator. Its calculation is based on the differential displacement of the end-effector or, equivalently, the differential of H. Such a differential may be expressed by the homogeneous matrix

$$\Delta = \begin{bmatrix} 0 & -d\theta & d\phi & da \\ d\theta & 0 & -d\alpha & db \\ -d\phi & d\alpha & 0 & dc \\ 0 & 0 & 0 & 0 \end{bmatrix},$$

whose skew symmetric components represent the differential rotation, and the remainder, the differential displacement of O_n. The variables θ, ϕ, α, a, b and c are taken to be functions of the robot variables $z = (\xi_1,\ldots,\xi_n)$, the n actively controlled axes of the robot. We define the column matrices

$$dD = (da\ db\ dc\ d\alpha\ d\phi\ d\theta)^{\mathrm{T}}$$

and

$$dz = (d\xi_1\ d\xi_2 \ldots d\xi_n)^{\mathrm{T}} \ .$$

Definition: *Jacobian Matrix of a Manipulator*. The Jacobian matrix $J(z)$ such that

$$dD = J(z)\,dz \tag{9}$$

is termed the Jacobian matrix of the manipulator.

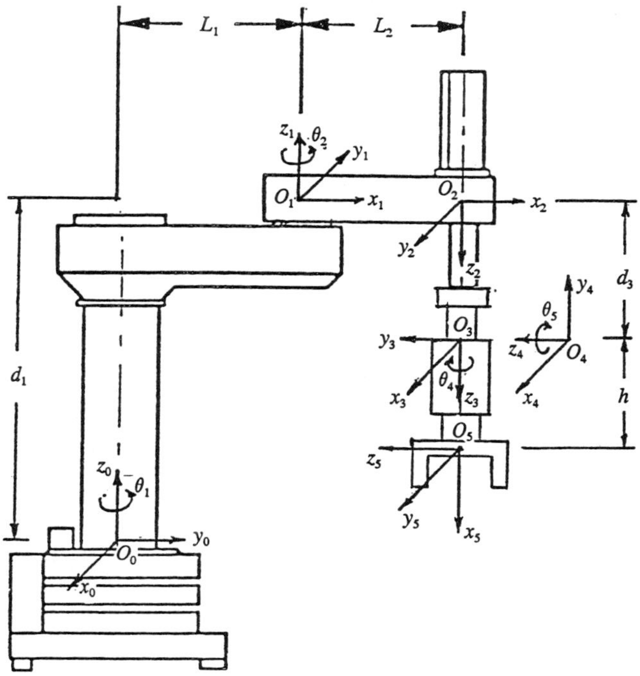

Figure 2. The AdeptOne Robot

This definition is now applied to the calculation of the Jacobian matrix of the AdeptOne manipulator.

The AdeptOne, a SCARA robot, is one of the earlier production models of Adept, Inc. It was selected as an example because of its relatively simple kinematics. It thus helps to illustrate the method rather than obscuring it with algebraic detail. The AdeptOne is a five-axes robot whose axes and their relative location are shown in Figure 2. The DH-parameters for the robot are presented in Table 1. Based on these parameters, the forward kinematics of the robot are given by the DH-matrix

$$\boldsymbol{H}_5 = \prod_{i=1}^{5} \boldsymbol{H}_{i-1,i}$$

$$= \begin{bmatrix} C\theta_5 C\psi & -S\theta_5 C\psi & S\psi & hC\theta_5 C\psi + L_2 C\beta + L_1 C\theta_1 \\ C\theta_5 S\psi & -S\theta_5 S\psi & -C\psi & hC\theta_5 S\psi + L_2 S\beta + L_1 S\theta_1 \\ S\theta_5 & C\theta_5 & 0 & hS\theta_5 - d_3 + d_1 \\ 0 & 0 & 0 & 1 \end{bmatrix} \quad \textbf{(10)}$$

where $\psi = \theta_1 + \theta_2 - \theta_4$ and $\beta = \theta_1 + \theta_2$.

By virtue of the definition of H_5 as a homogeneous rigid motion, we have

$$\boldsymbol{r}_0 = \boldsymbol{H}_5 \boldsymbol{r}_5$$

and

$$dr_0 = dH_5 r_5$$

since r_5 denotes a fixed location in $(0xyz)_5$. The calculation of dH_5 and the elimination of constraint equations eventually yield the desired Jacobian matrix.

Table 1

link	variable	θ_i	α_i	a_i	d_i
1	θ_1	θ_i	0	L_1	d_1
2	θ_2	θ_2	$-180°$	L_2	0
3	d_3	0	0	0	d_3
4	θ_4	θ_4	$-90°$	0	0
5	θ_5	θ_5	0	h	0

We begin with the identity

$$H_5 = H$$

or, equivalently,

$$\begin{array}{llll} a_x = C\theta_5 C\psi & o_x = -S\theta_5 C\psi & n_x = S\psi & p_x = hC\theta_5 C\psi + L_2 C\beta + L_1 C\theta_1 \\ a_y = C\theta_5 S\psi & o_y = -S\theta_5 S\psi & n_y = -C\psi & p_y = hC\theta_5 S\psi + L_2 S\beta + L_1 S\theta_1 \\ a_z = S\theta_5 & o_z = C\theta_5 & n_z = 0 & p_z = hS\theta_5 - d_3 + d_1 \end{array} .$$

With the robot variables $z = (\theta_1\ \theta_2\ d_3\ \theta_4\ \theta_5)^T$ we now obtain the differential displacements and rotations contained in the matrix Δ. The differential displacement of O_5 is immediate with

$$\begin{aligned} da = dp_x &= (-hC\theta_5 S\psi - L_2 S\beta - L_1 S\theta_1)\, d\theta_1 + (-hC\theta_5 S\psi - L_2 S\beta)\, d\theta_2 \\ &\quad + hC\theta_5 S\psi\, d\theta_4 - hS\theta_5 C\psi\, d\theta_5 \\ db = dp_y &= (hC\theta_5 C\psi + L_2 C\beta + L_1 C\theta_1)\, d\theta_1 + (hC\theta_5 C\psi + L_2 C\beta)\, d\theta_2 \\ &\quad - hC\theta_5 C\psi\, d\theta_4 - hS\theta_5 S\psi\, d\theta_5 \\ dc = dp_z &= -d d_3 + hC\theta_5\, d\theta_5 \ . \end{aligned}$$

The differentials of the rotational part of the hand matrix are

$$\begin{array}{lll} da_x = -C\theta_5 S\psi\, d\psi - S\theta_5 C\psi\, d\theta_5 & do_x = S\theta_5 S\psi\, d\psi - C\theta_5 C\psi\, d\theta_5 & dn_x = C\psi\, d\psi \\ da_y = C\theta_5 C\psi\, d\psi - S\theta_5 S\psi\, d\theta_5 & do_y = -S\theta_5 C\psi\, d\psi - C\theta_5 S\psi\, d\theta_5 & dn_y = S\psi\, d\psi \\ da_z = C\theta_5\, d\theta_5 & do_z = -S\theta_5\, d\theta_5 & dn_z = 0 \end{array} .$$

The postmultiplication of the resultant matrix of differentials by the transpose of the rotational partition of the matrix H_5 yields the product

$$\begin{bmatrix} da_x & do_x & dn_x \\ da_y & do_y & dn_y \\ da_z & do_z & dn_z \end{bmatrix} \begin{bmatrix} C\theta_5 C\psi & C\theta_5 S\psi & S\theta_5 \\ -S\theta_5 C\psi & -S\theta_5 S\psi & C\theta_5 \\ S\psi & -C\psi & 0 \end{bmatrix}$$

$$= \begin{bmatrix} 0 & -d\psi & -C\psi\, d\theta_5 \\ d\psi & 0 & -S\psi\, d\theta_5 \\ C\psi\, d\theta_5 & S\psi\, d\theta_5 & 0 \end{bmatrix} = \begin{bmatrix} 0 & -d\theta & d\phi \\ d\theta & 0 & -d\alpha \\ -d\phi & d\alpha & 0 \end{bmatrix}$$

from which we conclude

$$\begin{pmatrix} d\alpha \\ d\phi \\ d\theta \end{pmatrix} = \begin{pmatrix} S\psi\, d\theta_5 \\ -C\psi\, d\theta_5 \\ d\theta_1 + d\theta_2 - d\theta_4 \end{pmatrix} .$$

The Jacobian matrix for the manipulator thus is given by

$J(z)$ =

$$\begin{bmatrix} -hC\theta_5 S\psi - L_2 S\beta - L_1 S\theta_1 & -hC\theta_5 S\psi - L_2 S\beta & 0 & hC\theta_5 S\psi & -hS\theta_5 C\psi \\ hC\theta_5 C\psi + L_2 C\beta + L_1 C\theta_1 & hC\theta_5 C\psi + L_2 C\beta & 0 & -hC\theta_5 C\psi & -hS\theta_5 S\psi \\ 0 & 0 & -1 & 0 & hC\theta_5 \\ 0 & 0 & 0 & 0 & S\psi \\ 0 & 0 & 0 & 0 & -C\psi \\ 1 & 1 & 0 & -1 & 0 \end{bmatrix}$$

This calculation coincides with that of Whitney [1]. A more detailed discussion of this approach, including additional examples, may be found in Stadler [3].

REFERENCES

1. Whitney, D. E., "The Mathematics of Coordinated Control of Prosthetic Arms and Manipulators," *Transactions of the ASME, Journal of Dynamic Systems, Measurement and Control*, Vol. 122, pp. 303-309, 1972.

2. Paul, R. P., *Robot Manipulators (Mathematics, Programming and Control)*, The MIT Press, Cambridge, Massachusetts, 1981.

3. Stadler, W., *Analytical Robotics*, McGraw-Hill, New York, to appear Spring 1994.

NEURAL MODELING AND IDENTIFICATION OF NONLINEAR SYSTEMS IN AN ABSTRACT SPACE SETTING

Rui J.P. de Figueiredo

Department of Electrical
and Computer Engineering
and Department of Mathematics
University of California, Irvine
Irvine, CA 92717

ABSTRACT

A rigorous framework is available for the modeling and identification of nonlinear dynamical systems by artificial neural networks. The system model is obtained as a best approximation of the operator(s) representing the system in a "neural space", under interpolating or smoothing constraints imposed by the input-output training data. This optimal *modeling* results in one of four types of neural networks proposed and discussed by the author elsewhere, namely the OI, OS, OMNI and OSMAN nets. The *identification* of a system so modeled can take place instantaneously by batch processing of the training data, or sequentially by *adaptation*, *learning*, and/or *evolution*. These concepts are briefly discussed in the present article, and illustrated in terms of a self-tuning nonlinear regulator system.

INTRODUCTION

Control systems of today often involve the control of large nonlinear systems, called *plants*, for which a description in terms of detailed physical equations is not available. In such cases, one has to obtain a "mathematical representation" for the plant behavior, called the plant *model*, based on input-output test data and on the available à priori knowledge of the class of systems to which the plant belongs. The problem of obtaining such a mathematical representation is called "system modeling", and that of acquiring the model parameters from the data, "system identification". In this chapter, we formulate both these problems as a best approximation of the underlying nonlinear plant mappings in an appropriate abstract space under input/output data interpolating or smoothing constraints. The surprising result is that these strictly mathematical arguments lead to a solution in the form of an artificial neural network.

Mechanics and Control, Edited by R.S. Guttalu
Plenum Press, New York, 1994

It may be pointed out that for the linear case, a similar approach to the solution of the system modeling and identification problems, in the form of an abstract operator spline, had been presented by the author and associates in Angelo Miele's journal in 1973 [1].

The results presented here are based on the theory of nonlinear system modeling and identification in a Generalized Fock Space (GFS) [2-5] setting developed by us in the late 1970's and early 1980's. Those results provide a rigorous basis for the current thrusts of the control systems community in the area of artificial neural networks, and for this reason they are interpreted for this community in the present article. For simplicity, we limit our attention to discrete-time parameter systems, the results being clearly generalizable to the continuous time-parameter case.

The GFS $F(E^N)$ is a Reproducing Kernel Hilbert Space (RKHS) of power series (Volterra series) $\tilde{f}(x)$ in the components of the vector variable $x \in E^N$, where E^N denotes the N-dimensional real Euclidean space. The values of $\tilde{f}$ are of course real. The neural space $\eta(E^N, E^M)$ is defined as the vector space of M-tuples of elements of $F(E^N)$. In other words, if f belongs to $\eta(E^N, E^M)$, we may write $f(x) = (f_1(x),...,f_m(x))^T$, where T denotes the transpose and each f_j, $j=1,...,M$, belongs to $F(E^N)$.

We have derived elsewhere [6-14] four types of generic artificial neural networks as a result of best approximation in $\eta(E^N, E^M)$ of the input-output map of a system subject to input-output training data constraints. If these constraints are *interpolating* constraints we obtain the **OI** (Optimal Interpolating) net, shown in Fig. 1, or the **OMNI** (Optimal Multilayer Neural Interpolating) net, consisting of a cascade of n OI nets as depicted in Fig. 2. If the constraints are *smoothing* rather than interpolating, we are led to the **OS** (Optimal Smoothing) net, or the **OSMAN** (Optimal Smoothing Multilayer Artificial Neural) net. The OS net has the same structure as the OI net except that the number of neurons in the first hidden layer may be smaller and the synaptic weights are computed using smoothing (least squares or Gauss-Markov) formulas. The OSMAN net, in a way analogous to the OMNI net, consists of a cascade of n OS nets.

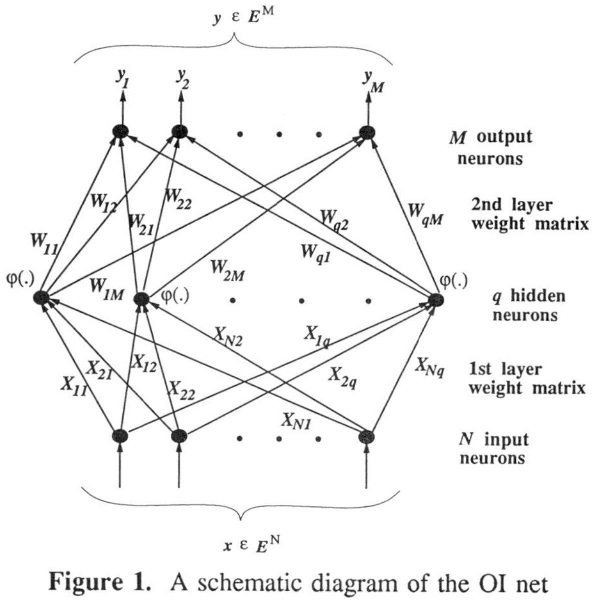

Figure 1. A schematic diagram of the OI net

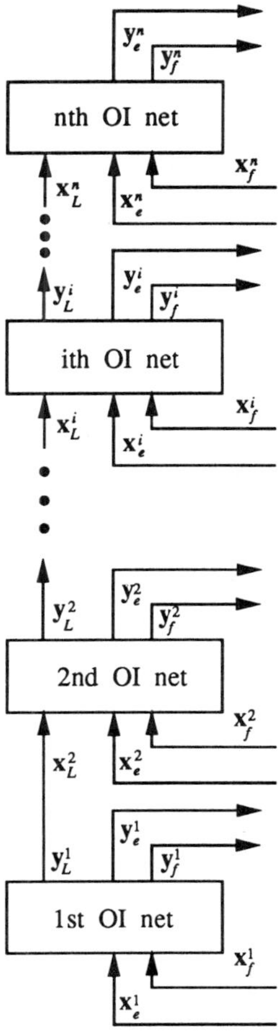

Figure 2. Block diagram of an OMNI net

In what follows, we consider three generic nonlinear dynamical system models, namely the *input-output* (or nonlinear covolutional) model, the *output-input* (or nonlinear difference equation) model, and the *state-space* model. The desired neural net realizations for these models can then be obtained by implementing the nonlinear operators appearing in them by the above neural networks.

NONLINEAR SYSTEM MODELS

Let $\{u(k') : k' = k_o, k_o + 1, \ldots, k\}$, where k_o may be finite or $-\infty$, denote the sequence of system inputs up to time k, and y(k) the system output at k. We will assume that u(k′) and y(k) belong respectively to n- and m-dimensional real Euclidean spaces E^n and E^m.

In the absence of noise, the three models referred to in the preceding section are described by the following sets of equations, and illustrated by the block diagrams of Figs. 3-5.

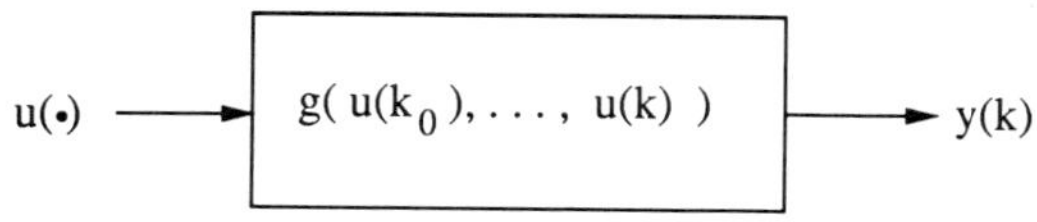

Figure 3. Input-Output (nonlinear convolutional) model

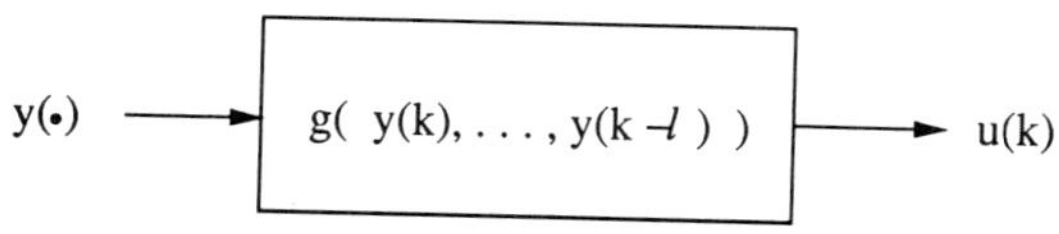

Figure 4. Output-Input (nonlinear difference equation) model

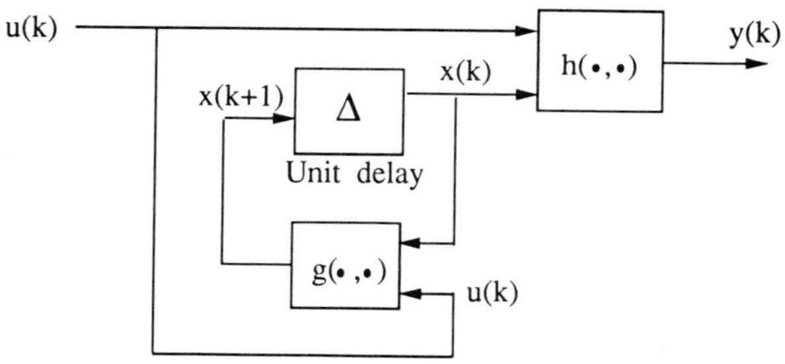

Figure 5. State-space model

The *Input-Output* model maps the set of past inputs $u(k_o)$, $u(k_o + 1)$, $u(k_o + p-1)$ and present input $u(k_o + p) = u(k)$ into the present output y(k); i.e.

$$y(k) = g(u(k_o), u(k_o+1), \ldots, u(k)) \qquad (1)$$

where the operator $g : E^{(p+1)n} \rightarrow E^m$ is a nonlinear function of its arguments belonging to the neural space $\eta(E^{(p+1)m}, E^m)$. This model may be viewed as a *nonlinear convolutional* or *moving-average* model.

The *Output-Input* model expresses the current input u(k) as a nonlinear function of the current output y(k) and *l* preceding outputs y(k-1),...,y(k-*l*). Symbolically, this is written as

$$g(y(k), y(k-1), \ldots, y(k-)) = u(k) \qquad (2)$$

where $\eta(E^{(l+1)m}, E^n)$. This model is also called the *nonlinear autoregressive* or *nonlinear difference-equation* model.

Finally, the *state-space* model is described by the set of equations

$$x(k+1) = g(x(k), u(k)) \qquad (3a)$$

$$y(k) = h(x(k), u(k)) \qquad (3b)$$

where $x(k) \in E^s$ represents the state of the system at time k, u(k) and y(k) are as before, and g and h are nonlinear operators belonging to $\eta(E^{s+n}, E^s)$ and $\eta(E^{s+n}, E^m)$.

Each of the above three models can be converted into *neural models* by realizing the functions g and h in the form of any one of the neural networks introduced in the preceding section [14].

ADAPTATION, LEARNING, AND EVOLUTION

If the input-output training data appear sequentially, the process whereby the preceding neural networks acquire their realizations from the data may be classified as follows:

• *Adaptation* occurs when the network parameters are adjusted so as to enable it to best approximate (along the lines described in the preceding section) a desired function, under the input-output training data constraints taking into account the new data.

• *Learning* takes place when the network is preforming a cognitive function, such as detection, classification, or interpretation. In this case the network parameters are adjusted so as to enable it to improve its performance of the cognitive function taking into account the new data.

• *Evolution* occurs when neurons are added to or subtracted from the networks structure so as to enable the network to achieve one or both of the preceding objectives with the least effort.

Algorithms for implementing all the above three processes have been proposed and extensively discussed by us in the literature [7-14].

Some of these algorithms are based on the propagation of the inverse of an appropriate Gram matrix using the Matrix Inversion Lemma.

NEURAL IDENTIFICATION AND CONTROL

Once one of the three neural models is obtained for a nonlinear plant, the parameters are acquired: either instantaneously if the training data is available in a batch form; or by *adaptation*, *learning*, and/or *evolution* if the training data appears sequentially. A control system in which a nonlinear plant is modeled and identified as above can be called *adaptive*, or *learning* or *evolutionary* depending on the procedure selected in using the criteria just mentioned.

Fig. 6 gives a block diagram of a self-tuning nonlinear regulator [12]. From the input-output data, the plant model parameters are estimated as indicated in the present article and the results used in the design and adjustment of the regulator parameters.

In order to illustrate the capabilities of an OI net to learn to discriminate between two different types of data sets, we illustrate in Figs 7 and 8, a version of the 2-Spirals example studied by Lang and Witbrock [15]. These figures have been borrowed from [11]. The set of training samples consists of 65 input-output pairs (x_k, y_k), where $x_k = (x_{1k}, x_{2k})^T$ corresponds to the coordinates of the kth input sample shown as circle or cross in Fig. 7, and y_k is a binary scalar, with zero corresponding to a circle, and one to a cross. The curve in Fig. 7 shows how the net separates the two classes after training using only the fraction of the training samples (prototypes) shown in Fig. 8.

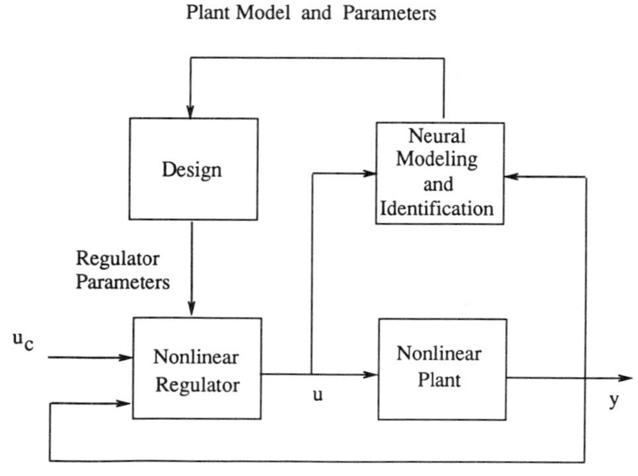

Figure 6. Block Diagram of a Self-tuning Nonlinear Regulator

CONCLUSION

Artificial neural networks constitute generic structures which naturally lend themselves to the representation of nonlinear dynamical systems. In this article, we have provided a framework, in an abstract space setting, for neural modeling and neural identification of such systems. The modeling takes place via OI, OS, OMNI, and OSMAN nets introduced by us elsewhere, while the identification occurs by processes germane to adaptation, learning, and evolution. The concepts have been illustrated in terms of a self-tuning nonlinear regulator.

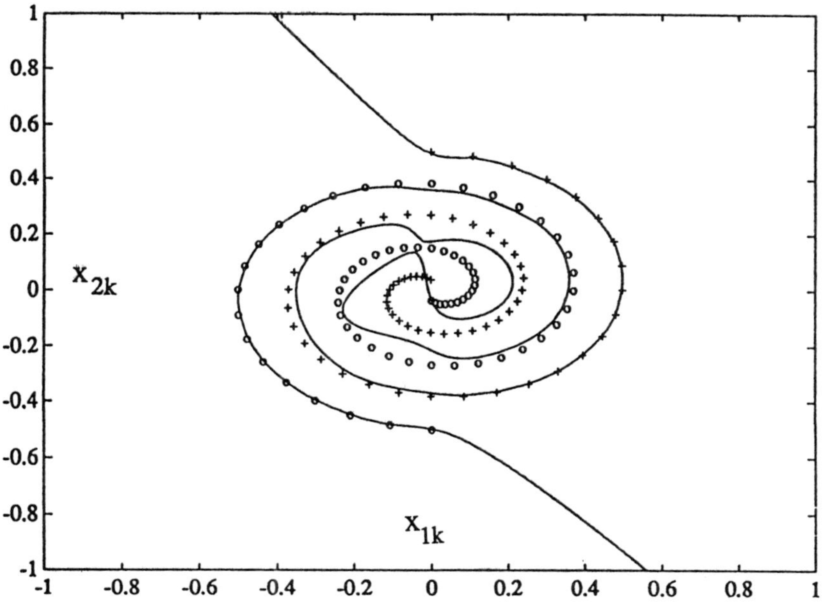

Figure 7. Training data and decision boundary for a version of the 2-Spirals studied by Lang and Witbrock [15]

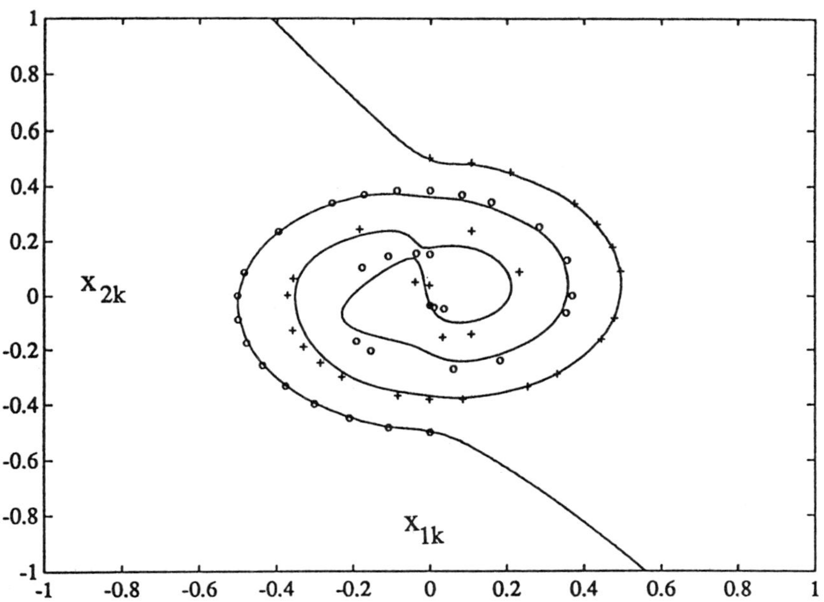

Figure 8. Prototypes selected in Example of Fig. 7 (2-Spirals). Also shown here for reference is the decision boundary as defined by the trained network obtained in this example.

ACKNOWLEDGEMENT

This article is dedicated to Professor Angelo Miele in honor of his 70th birthday. The work reported here was supported in part by the ONR Contract N00014-91-J-1072.

REFERENCES

[1] R.J.P. de Figueiredo, A. Caprihan, and A.N. Netravadi, "On Optimal modeling of Systems", *J. of Optimization Theory and Applications*, Vol.11, 1973, pp. 68-83.

[2] R.J.P. de Figueiredo and L.V. Zyla, "Nonlinear system identifications by an abstract Volterra operator space approach, *Proc. the 20th Midwest Symposium or Circuits and Systems*, Lubbock, Texas, August 1977, p. 16.

[3] R.J.P. de Figueiredo and L.V. Zyla, "An interpolative and smoothing approach for modeling large scale nonlinear multivariable systems", *Proc. of the 12th Asilomar Conference on Cicruits, Systems, and Computers*, (C.C. Hsieh, edition), IEEE Publication 78 CH1369-8C/CA/CS, 1978, pp. 455-459.

[4] R.J.P. de Figueiredo and T.A.W. Dwyer, III, "A best approximation framework and implementation for simulation of large-scale nonlinear systems", *IEEE Transactions on Circuits and Systems*, Vol. CAS-27, No. 11, November 1980, pp. 1005-1014

[5] R.J.P. deFigueiredo, "A generalized Fock Space framework for nonlinear system and signal analysis", *IEEE Transactions on Circuits and Systems*, Vol. CAS-30, No. 9, September 1983, pp. 637-647.

[6] R.J.P. deFigueiredo, "An optimal matching score net for pattern classification", *Proceedings of the 1990 International Joint Conference on Neural Networks (IJCNN)*, San Diego, CA, June 17-21, 1990, Vol. 2, pp. 909-916.

[7] S.K. Sin and R.J.P. deFigueiredo, "An incremental adjustment algorithm for the design of optimal interpolating neural networks", *International Journal of Pattern Recognition and Artificial Intelligence*, Vol. 5, No. 4, Oct. 1991, pp. 563-579.

[8] R.J.P. deFigueiredo "Mathematical foundations of optimal interpolative neural networks", in *Artificial Intelligence, Expert Systems, and Symbolic Computing*, E. Houstis and J.R. Rice, Editors, Elsevier, Amsterdam, 1992, pp. 303-319.

[9] R.J.P. deFigueiredo, "An optimal multilayer neural interpolating (OMNI) net in a Generalized Fock Space setting" *Proc. of the 1992 International Joint Conference on Neural Networks (IJCNN-92)*, Baltimore, MD. Vol. 1, 1992, pp. 111-118.

[10] S.K. Sin and R.J.P. deFigueiredo, "An evolution-oriented learning algorithm for the optimal interpolative net", *IEEE Transactions on Neural Networks*, Vol. 3, No. 2, March 1992, pp. 315-323.

[11] S.K. Sin and R.J.P. deFigueiredo, "Efficient learning procedures for optimal interpolative nets", *Neural Networks*, Vol. 6, 1993, pp. 99-113.

[12] R.J.P. de Figueiredo and G. Chen, *Nonlinear Feedback Control Systems: An Operator Theory Approach.* Cambridge, Mass.: Academic Press, 1993.

[13] S.K. Sin and R.J.P. de Figueiredo, "A method for the design of evolutionary multilayer neural networks", *Proc. of the 1993 International Conference on Neural Networks (ICNN-93)*, San Francisco, CA, 1993.

[14] R.J.P. de Figueiredo, *Artificial Neural Networks: Theory and Applications.* Lecture Notes and Manuscript of a book (to be submitted for publication).

[15] K. Lang and M.J. Witbrock, "Learning to tell two spirals apart", *Proc. of the 1988 Connectionist Models Summer School (pp. 52-59)*, San Mateo, CA, M. Kaufmann, 1988.

GLOBAL BEHAVIOR OF BRANIN'S METHOD

Pedro J. Zufiria[1] and Ramesh S. Guttalu[2]

[1]Departamento de Matemática Aplicada
a las Tecnologías de la Información, and GTI
ETSI de Telecomunicación, UPM
28040 Madrid, Spain
[2]Department of Mechanical Engineering
University of Southern California
Los Angeles, CA 90089-1453

ABSTRACT

An analysis of Branin's method for finding all the real zeros of a vector function is carried out. It is based on a global study of this method perceived of as a dynamical system. Since Branin's algorithm is closely related to homotopy methods, this paper sheds some light on the global performance of these methods when employed for locating *all* the zeros of a vector function. We provide two examples for which the global behavior of Branin's algorithm is carried out using the cell-to-cell mapping computational technique.

INTRODUCTION

This paper is concerned with an analysis of the method developed by Branin [5] for the purpose of determining *all* the zeros of a system of nonlinear algebraic equations

$$f(x) = 0, \quad f : \mathbf{R}^{\mathbf{N}} \to \mathbf{R}^{\mathbf{N}}, \quad f \in C^1 \tag{1}$$

The method was originally proposed by Davidenko [7,8] and is based on the construction of a certain dynamical system (whose form is further discussed in the next section) to aid in finding the zeros of a nonlinear vector function. The problem of finding all the zeros has direct applications in global optimization, determination of fixed points of a map and bifurcation curves, solution of inverse kinematics in robotics, location of multiple equilibria of nonlinear systems, etc.

There exist a vast literature which addresses the problem of locating a zero of f, see Ortega and Rheinboldt [19], Smale [20], and Hirsch and Smale [15]. Besides the approaches mentioned so far, several other investigations have been performed for locating *all* the zeros of f, see Branin [5], Hsu and Guttalu [16], Hsu and Zhu [17], Hsu [18], and Ying and Katz [23]. We will not digress into the details of any of these investigations except the one concerning Branin's algorithm [5] which is strongly related to the homotopy idea and is based on the theory of dynamical systems.

Branin's algorithm consists of computing trajectories of a certain dynamical system associated with f (see equations (2)-(3)). The computation is carried out in an alternate manner by switching the sign of the vector field F of the dynamical system. The switch of sign is performed either when a zero of f is discovered or when the determinant of the gradient of f changes sign. Branin's algorithm cannot always provide all the zeros of f; a well-known deficiency of the global convergence of this algorithm has been documented by Brent [6] and Treccani [21]. Some effort has been made in analyzing and improving the performance of the algorithm since its inception, see Gomulka [10,11], Hardy [14], and Treccani [21]. The results provided in these references are not based on a dynamical system treatment of the algorithm which is essential in order to assess its global performance.

The main objective here is only to examine the behavior of Branin's method from the point of view of local and global dynamics to provide both a justification and an evaluation of the algorithm. For this purpose, a theory developed by Zufiria and Guttalu [24] for locating all the zeros of f will be employed.

The rest of the paper is organized as follows. We state the algorithm of Branin in the next section and carry out a global analysis of Branin's dynamical system. After motivating the performance of Branin's algorithm, its numerical implementation issues are discussed. Finally, two examples are provided to illustrate how one can characterize the global behavior of Branin's method.

BRANIN'S ALGORITHM

For the purpose of locating the zeros of f, a parameter t (for convenience taken here to mean time) is introduced so that the state vector $x(t)$ satisfies one of the following two systems of differential equations

$$\dot{x} = +F(x) = -J^{-1}(x)f(x), \quad J = \nabla_x f \tag{2}$$

$$\dot{x} = -F(x) = +J^{-1}(x)f(x) \tag{3}$$

where $\dot{x}$ is the time derivative of the vector $x(t)$ and the Jacobian matrix J is the gradient of the vector function f. The algorithm provided by Branin [5] for finding zeros of f is based on numerically integrating the differential equations (2) and (3) alternately as follows. One begins to integrate with either (2) or (3). The switch from one equation to the other is performed whenever *det* $J(x(t))$ changes sign for some t or when a new zero of f is found.

Before constructing an algorithm based on the combined system (2)-(3), it is essential to understand the relationship between the solutions of the algebraic system (1) and those of the differential system (2)-(3). Zufiria and Guttalu [24] have elaborated on the relationship between the solutions of (1) and the trajectories of the dynamical system (2) (referred to as continuous Newton method) as well as on its typical global behavior. They also have shown that the system (2) can be integrated to obtain

$$f(x(t)) = e^{t_0 - t} f(x(t_0)) \tag{4}$$

where $x(t_0)$ is a starting point.

An advantage of studying methods based on the properties of (2)-(3) when compared to homotopy formulations is that the dynamical system point of view permits a characterization of their global performance. For such a characterization, the next section brings together the global analysis developed by Zufiria and Guttalu [24] for the system (2) with the switch of sign algorithm of Branin.

GLOBAL DYNAMICS OF SYSTEM (2)-(3)

The purpose of switching the sign of the vector field $F(x)$ in the dynamical system (2)-(3) is not rigorously justified by Branin [5]. The role of this procedure in the algorithm becomes evident when a detailed study of the global behavior of the dynamical system composed of both the systems (2) and (3) is performed. A brief review of the global behavior of the system (2), taken from Zufiria and Guttalu [24], is first provided which is then accompanied by results for the system (3). The results for the combined dynamical system (2)-(3) follow later.

A Brief Review of Global Results for Dynamical System (2)

The relationship between the zeros of f and the equilibria of (2) (that is, zeros of F) is explored by Zufiria and Guttalu [24]. It is shown that when a zero of F (that is, an equilibrium point of (2)) exists, then it always corresponds to a zero of f. On the other hand, a zero of f need not always imply that a corresponding zero of F exists. Whenever $J(\cdot)$ is regular at a zero of f, then this zero becomes an asymptotically stable equilibrium point of (2). When J is singular, F may not be defined at a zero of f. This case has been further examined by Guttalu and Zufiria [12].

Some results are also presented by Zufiria and Guttalu [24] on the global behavior of (2) based on the Lyapunov approach. It is shown that $V(x) = f(x)^T f(x)$ is a strict Lyapunov function of (2) as stated in the following theorem:

Theorem 1 *Given f, consider the scalar function*

$$V(x) = f^T(x)f(x) = f(x) \cdot f(x), \;\; V : \mathbf{R^N} \to \mathbf{R} \tag{5}$$

where $f^T(x)$ is the transpose of $f(x)$. Suppose that $F(x^*) = 0$, $x^* \in \mathbf{R^N}$, that is, x^* is an equilibrium point of (2). Then V is a strict Lyapunov function of (2) for x^* and x^* is asymptotically stable.

The Lyapunov function V is applicable to every equilibrium point of the system (2). The global structure of this Lyapunov function is provided in the following theorem.

Theorem 2 *Assume that J is regular in a region $\Phi \subset \mathbf{R^N}$. Then, the Lyapunov function V given by (5) can only have minima (not a maximum or a saddle point). Besides, a point $x_m \in \Phi$ can only be a minimum of $V(x)$ if and only if x_m is a zero of f.*

In Zufiria and Guttalu [24], the concepts of *singular manifolds, barrier manifolds,* and *isolated regions* are developed and are shown to play a crucial role in determining the global behavior of the system. Singular manifolds are manifolds of dimension $(N-1)$ for which $det\ J(x) = 0$. These manifolds partition the state space into different regions such that in each region $det\ J(x)$ does not switch sign. The barrier manifolds are singular manifolds with an additional property that the trajectories of (2) cannot cross them. In a neighborhood of the barrier manifolds, the vector field F usually experiences a change of direction as we evaluate it on each side of the manifold. Barrier manifolds, in general, are composed of points for which any trajectory that reaches these points is not defined any more for future times. Isolated regions are regions whose boundaries are made up of barrier manifolds. The importance of these new definitions in delineating the global behavior of the dynamical system (2) have been explored by Zufiria and Guttalu [24].

The different possible trajectory behavior for the system (2) can be summarized in the following theorem, which is a modified version of Theorem 6 in Zufiria and Guttalu [24].

Theorem 3 *Suppose that f has only isolated zeros. Then, every trajectory of the dynamical system (2) can only have one of the following behaviors:*

I. $\| x(t) \| \to \infty$ *as* $t \to T \in \mathbf{R}^+$
(If T is finite we have a finite escape time*).*

II. $x(t) \to x_S$ *as* $t \to T$ *where det* $J(x_S) = 0$.

III. $x(t) \to x^*$ *as* $t \to \infty$ *where* $F(x^*) = 0$ *(that is x^* is a zero of f)*

One can conclude from Theorem 6 of Zufiria and Guttalu [24] that no periodic solutions or chaotic behavior is possible. It is clear that Branin's algorithm, when using the system (2), can fail if case I of the above theorem is encountered.

Extension of results to dynamical system (2)

The trajectories of (3) are equivalent to backward-time integration of the trajectories of (2). Hence, both (3) and (2) have the same equilibria. In addition, the same function V defined by (5) can be used to prove instability of the equilibria of (3). We state this result in the following corollary and omit the proof.

Corollary 1 *Suppose that $F(x^*) = 0$, $x^* \in \mathbf{R}^\mathbf{N}$, that is, x^* is an equilibrium point of (3). Then V is a Lyapunov function of (3) for x^* and x^* is unstable.*

The possible trajectory behavior of (3) can be readily established in a similar manner analogous to Theorem 3 for (2). The result is stated below with proof omitted.

Corollary 2 *Suppose that f has only isolated zeros and $x(t_0) \neq x^*$. Then, the trajectory of the dynamical system (3) can only have one of the following behaviors:*

I. $\| x(t) \| \to \infty$ *as* $t \to T \in \mathbf{R}^+$
(If T is finite we have a finite escape time*).*

II. $x(t) \to x_S$ *as* $t \to T$ *where det* $J(x_S) = 0$.

Note that if $x(t_0) = x^*$, then the trajectory of (3) remains there forever.

From a numerical stand point, integration of the system (3) will in general lead to computational difficulties either because the trajectories blow up or because $\|F(x(t))\|$ may become very large when approaching a singular manifold. In the latter case, "numerical jumping" of trajectories over the singular manifolds can be expected, see Zufiria and Guttalu [25].

Computational difficulties in the neighborhood of the singular point can be avoided by using one of the several available computer codes for continuation techniques, for instance see Allgower and Georg [1], Decker and Keller [9], and Watson et al [22]. Hence, although the dynamical system formulation is very helpful from an analytical point of view, one does not need to use standard ODE integrators for numerical implementations.

Alternation of systems (2) and (3)

In order to appraise Branin's algorithm, we now analyze the combined behavior of (2) and (3) from a dynamical system point of view:

(a) Without loss of generality, one can begin by considering the system (2). The solutions of the algebraic system (1) are usually asymptotically stable equilibria of (2). The different possible trajectory behavior is stated in Theorem 3. When case II is met, a switch to system (3) is to be considered for the trajectory to continue to evolve.

(b) The behavior of the trajectories of (3) is stated in Corollary 2. When case II of Corollary 2 is encountered, the trajectory can proceed further if *det* $J(x(t))$ will have changed its sign and, accordingly, going back to system (2) with which we started.

A new variety of dynamical systems. As mentioned before, dynamical systems (2) and (3) are applied alternately, the switch from one to the other being performed when *det* $J(x)$ changes its sign (or when a root of f is located). Therefore, once an initial condition x_0 is selected, two associated trajectories are possible (until a root of f is reached) depending on which of the system (2) or (3) is first applied. Each one of these trajectories will then correspond to the following dynamical systems:

$$\dot{x} = +F_1(x) \quad = \quad -sign(det\ J(x))\, J^{-1}(x)f(x) \tag{6}$$

$$\dot{x} = -F_1(x) \quad = \quad +sign(det\ J(x))\, J^{-1}(x)f(x) \tag{7}$$

The procedure for selecting a particular dynamical system can be interpreted as follows. If x_0 is such that *det* $J(x_0) > 0$ and system (2) is chosen first, it means that system (6) is being considered for generating the trajectory. The same system (6) is also considered when *det* $J(x_0) < 0$ and (3) is chosen first. The system (6) is considered whenever either the order of the system selection or the sign of *det* $J(x_0)$ is the reverse.

The study of the global behavior of Branin's system can be addressed in many different ways, depending on which compounding systems one focuses on. On one hand, a separate dynamic analysis of systems (2) and (3) is relatively easy to perform. However, in order to obtain a complete picture of the behavior of the combined system, one must assemble the information so gathered in a non-trivial manner. On the other hand, a global analysis of systems (6) and (7) provides a better picture of the alternate system behavior. Two examples are provided later where we will study global behavior using the cell mapping computational technique.

Numerical implementation as an algorithm. Based on the above analysis, when the algorithm is successful, the behavior of a trajectory of the system defined by (2) and (3) can be roughly summarized as follows. Suppose that the trajectory is computed using (2) and that it converges to x^*, one of the zeros of f. Once this zero is detected, the algorithm uses (3) for which x^* becomes an unstable equilibrium point. Numerical computation of a trajectory starting from $x(t_0)$ close to x^* (but different from x^*) leaves x^*. Hopefully this trajectory would reach a point at some time t when $det\ J(x(t))$ changes its sign. If it is so, then the algorithm switches back to (2). In general, this point (where $det\ J = 0$) will belong to a barrier manifold and the switching guarantees that the trajectory can cross it. Hence, the trajectory passes from one isolated region to another. Then, it may be expected that a continuation of this trajectory would approach a new equilibrium point x^* of (2) (which now again is asymptotically stable), implying the discovery of a new zero of f. The procedure can be repeated, as long as the trajectories are bounded, until, hopefully, all the zeros are located.

Figure 1 illustrates a way in which trajectories of the combined system (2)-(3) may evolve. The points x_a^*, x_b^* and x_c^* are the three zeros of f, and S_1, S_2 and S_3 represent

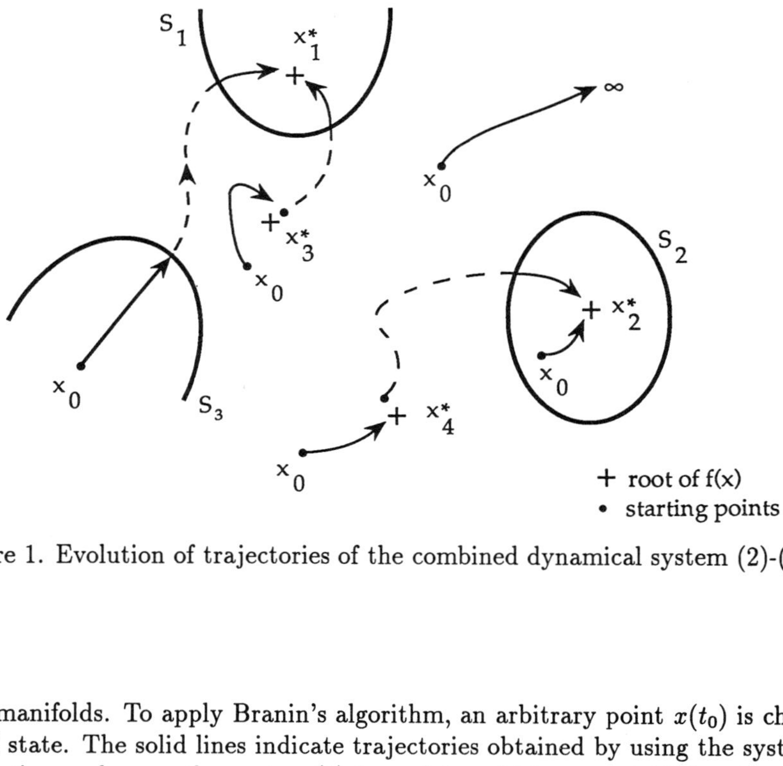

Figure 1. Evolution of trajectories of the combined dynamical system (2)-(3).

singular manifolds. To apply Branin's algorithm, an arbitrary point $x(t_0)$ is chosen as an initial state. The solid lines indicate trajectories obtained by using the system (2). After locating each zero, the system (3) is used for which a new initial state is chosen close to the zero (trajectories are represented by dashed lines).

So far we have considered mainly analytical aspects of locating zeros of f by means of Branin's algorithm. A study of numerical procedures to follow the trajectories of Branin's method should also be considered. For instance, a regular ODE integrator applied to the system (2)-(3) does not guarantee that a given path can be accurately followed in the neighborhood of a singular point. Fortunately, the main goal of Branin's method is not to follow a concrete path but to remain within some neighboring paths with an equivalent long term behavior so that eventually (as $t \to \infty$) a given solution is reached. Some of the practical numerical issues related to Branin's method will be considered later.

PERFORMANCE OF BRANIN'S ALGORITHM

From the discussion given in the previous section, it can be noted that new zeros can only be found when the system (2) is being integrated. Accordingly, Branin's computational scheme relies heavily on the fact that computed trajectories will alternate between crossing the barrier manifolds and locating new zeros. To further support the validity of this scheme, Branin [5] states the following: based on the Inverse Function Theorem, we are bound to encounter at least one Jacobian singularity on each trajectory (in this case meaning a line) which joins two adjacent solution points of (1). This need not be true as shown in the following theorem taken from Zufiria and Guttalu [24] which can be stated as follows.

Theorem 4 *Suppose that J is regular in an isolated region Ω.*

(a) Consider the case $N = 1$, $x \in \mathbf{R}$. Then, Ω contains at most *one equilibrium point x_k^* of the systems (2) and (3).*

(b) Consider the case $N \geq 2$, $x \in \mathbf{R^N}$. Then, it is possible that Ω may contain more than one equilibrium points x_k^* of (2) and (3), $k = 1, 2, \cdots$.

This result is crucial in characterizing the cases when Branin's algorithm fails.

ON PRACTICAL IMPLEMENTATION OF BRANIN'S METHOD

The approach presented in the paper for analyzing Branin's method has been analytical in nature. A different issue which should also be considered is concerned with numerical implementation of Branin's algorithm. An important aspect is the specification of a condition indicating when a zero has been detected. The second aspect is the correspondence between numerically computed trajectories and theoretical trajectories, especially in the neighborhood of singular manifolds. It is known that when a trajectory approaches a singular manifold, numerical difficulties can occur because the vector field F may be unbounded (see Zufiria and Guttalu [25]). Usually, when the theoretical trajectory approaches a singular manifold, the numerically computed trajectory obtained from an ODE integrator will jump over the manifold. The switch of sign procedure is enforced only after this jump has occurred. If the numerical integration is accurate enough, the jump should be short in order to emulate the continuous theoretical trajectory. It may so happen that numerical errors when dealing with a huge vector field may lead to a very long jump over a singular manifold implying that the behavior of the numerical system can be very different from the theoretical one. Nevertheless, the main goal of the whole procedure is not to follow a given trajectory accurately but to reach the equilibria of the system (2)-(3). Therefore, although the errors propagated by an ODE integrator may lead to a different trajectory, as long as this is close to the original (theoretical) trajectory, we may obtain a similar long time behavior.

Some numerical simulations carried out by the authors indicate that the performance of Branin's method can be highly sensitive to initial conditions selected for starting the integration, the criterion used for determining the location of a zero, and the numerical integration scheme that is employed.

The problems arising from the numerical integration via ODE solvers are partially avoided in the continuation procedures where suitable codes have been developed to deal with singularities, see Allgower and Georg [1,4]. On the other hand, as discussed in this paper, there are certain global behavior issues (for example those related with the proper selection of initial conditions) that question the overall performance of these approaches.

EXAMPLES

We provide two examples in this section for which the global behavior of Branin's method is studied using the *cell-to-cell mapping* approach (see Hsu [18] for details regarding this computational technique). In the cell mapping technique, a compact state-space region is discretized to contain a finite number of states (referred to as *cell states* or simply as *cells*) and a mapping is generated among these states by considering the evolution of a trajectory of the system initiating from each cell. The *cell mapping* C, conceived of as a mapping from a set of integer to a set of integer (or mapping in a space of integers), may be described by $z(n+1) = C(z(n))$, $n = 1, 2, \cdots$, where z is a cell state. The mapping C contains dynamic information possessed by the original dynamical system and an algorithm is available for unraveling this information. Since the cell mapping unraveling algorithm provides not only the equilibrium and periodic solutions but also attraction regions associated with them, this technique is very effective for revealing the global behavior of dynamical systems studied here.

Since the dynamical systems we want to study are "nonstandard" (that is, they possess rapidly varying vector fields and singular manifolds where the trajectory may not be defined), we will employ a modified version of the cell mapping method, referred to as *adjoining cell mapping*, see [13,26] for further details. The adjoining cell mapping is described by $z(n+1) = C_A(z(n))$ and preserves the adjoining property, that is, every cell in cell state space maps always to one of its adjoining cells (immediate neighbors) or to itself. The traditional numerical methods encounter computational difficulties for analyzing systems possessing rapidly changing vector fields and are unable to delineate global behavior. The adjoining cell mapping technique, on the other hand, provides a meaningful description of global dynamics since it can easily adapt to rapidly changing dynamics.

Example 1

Let us consider the function $f : \mathbf{R}^2 \to \mathbf{R}^2$:

$$\begin{aligned} f_1 &= x_1 - 4x_2^2 \\ f_2 &= \cos x_1 - x_2 \end{aligned} \tag{8}$$

The vector field F used in systems (2), (3), (6), and (7) is given by

$$F(x) = \frac{-1}{det\ J(x)} \begin{bmatrix} -x_1 - 4x_2^2 + 8x_2 \cos x_1 \\ \\ x_1 \sin x_1 - 4x_2^2 \sin x_1 + \cos x_1 - x_2 \end{bmatrix},$$

$$det\ J(x) = -(1 + 8x_2 \sin x_1)$$

The function f possesses three zeros (or roots) given by

$$\begin{aligned} x_a^* &= (3.5021474,\ -0.9357013) \\ x_b^* &= (1.0366739,\ 0.5090859) \\ x_c^* &= (2.4764680,\ -0.7868399) \end{aligned} \tag{9}$$

These are all regular zeros of f (they are not located on the singular manifolds).

We apply the adjoining cell mapping technique to this problem. The region of state space considered for analysis is $-2.0 \leq x_1 \leq 8.0$ and $-1.5 \leq x_2 \leq 1.5$. Each of the x_1 and x_2 axes is divided into 200 uniform intervals of size $h_1 = 0.05$ and $h_2 = 0.015$, respectively, and the region of analysis contains 40,000 cell states. An adjoining cell mapping C_A is obtained by integrating a trajectory initiating from the center point of each cell at $t = 0$. The algorithm employed by this technique adaptively determines an appropriate integration duration T for each cell to obtain its adjoining map. The initial integration duration specified is $T = 0.1$ with integration step size $\Delta t = T/10$.

System (2). Note that the equilibrium points of the system (2) are the zeros x_a^*, x_b^*, and x_c^* which are all asymptotically stable. The results of the cell mapping analysis reveal that these are replaced by equilibrium cells indicated by the symbol "×" as shown in Figure 2. In addition, periodic cells (indicated by the symbol "+")

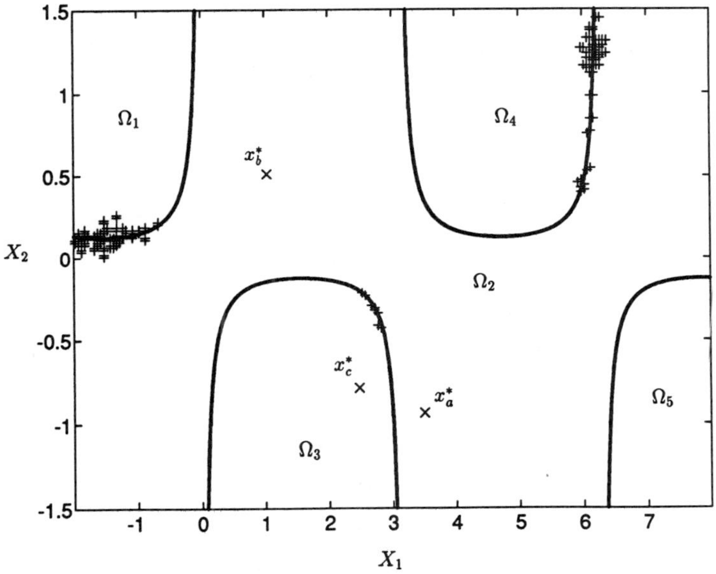

Figure 2. Location of equilibrium cells (×) and fake periodic cells (+) for system (2) corresponding to Example 1. Singular manifolds are shown by solid lines. Isolated regions are indicated by Ω_1, Ω_2, Ω_3, Ω_4, and Ω_5.

are also found in the vicinity of some of the singular manifolds as seen in this figure. These periodic cells, referred to as "fake periodic cells" exist due to the discretization of state space and do not correspond to any periodic solutions. It turns out that there are 73 of them and all are cells of period-2, that is, they are equilibrium cells for the map $C_A^2(\cdot)$. It has been computationally verified that the trajectories generated by (2) initiating from these cells do not lead to any additional solution possessed by (1). As a consequence of Theorem 3, no periodic solutions of (2) can exist for this example. Our numerical findings complements this theoretical result.

Figure 3 displays the domains of attraction associated with the equilibrium and periodic cells determined by the adjoining mapping analysis. The three dominant portions in the Ω_2 and Ω_3 isolated regions correspond to the attraction regions associated with each of the equilibria x_a^*, x_b^*, and x_c^*. Other regions indicated by different gray

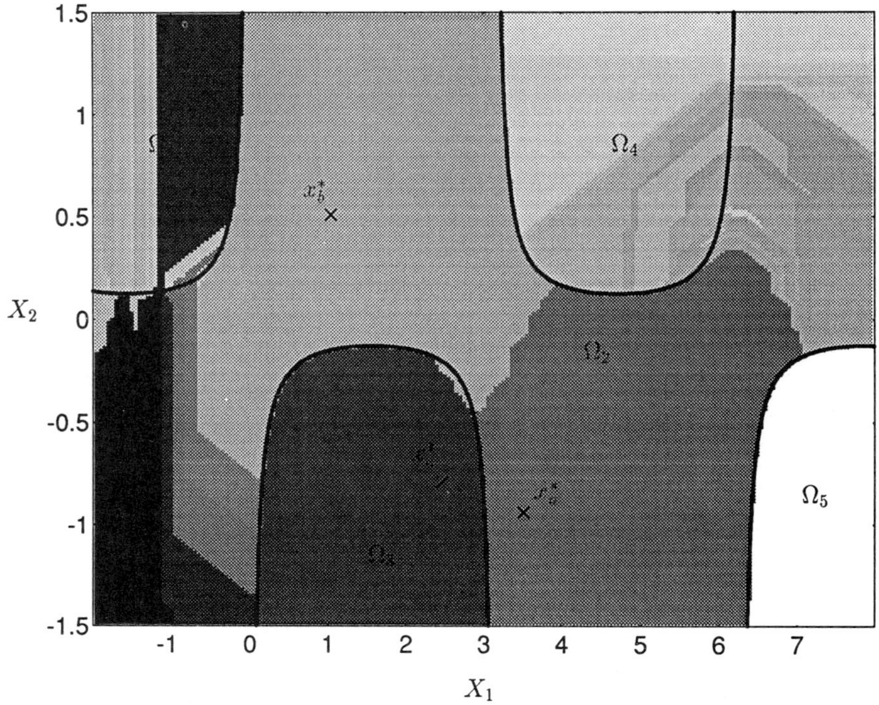

Figure 3. Domains of attraction of equilibrium points for system (2). Zeros of f in Example 1 are indicated by "×".

scales correspond to those cells in state space which map to fake periodic groups of cells located near the singular manifolds. These regions indicate the pattern of evolution of trajectories in the immediate neighborhood of the singular manifolds. The collection of cells attracted to fake periodic cells indicate also the geometric nature of the flow of trajectories starting far away from the singular manifolds. For a detailed explanation, the reader is referred to [13]. A blank space in the figure indicates that the trajectory evolving from the corresponding cell maps outside the state space region under consideration and that its long term behavior is unknown.

System (3). The three zeros of f in (9) are unstable equilibria for the dynamical system (3). The adjoining cell mapping technique for this system finds no equilibrium or periodic cells representing the unstable equilibria. However, fake periodic cells (462 each of period-2 and 3 each of period-4) are detected in the vicinity of some of the singular manifolds (indicated by "+" in Figure 4). Figure 5 shows the domains of attraction associated with these fake periodic cells which again describe the nature of the flow of trajectories before they approach the singular manifolds.

System (6). For the system (6), the zero x_c^* is asymptotically stable while the zeros x_a^* and x_b^* are unstable. The adjoining cell mapping analysis finds an equilibrium cell representing the stable point x_c^* and no equilibrium or periodic cells are found to represent the unstable points x_a^* and x_b^*. Figure 6 shows the domain of attraction associated with the equilibrium point x_c^* of the system (6).

System (7). For the system (7), zeros x_a^* and x_b^* are asymptotically stable while the zero x_c^* is unstable. Two equilibrium cells, one each representing the zeros x_a^* and

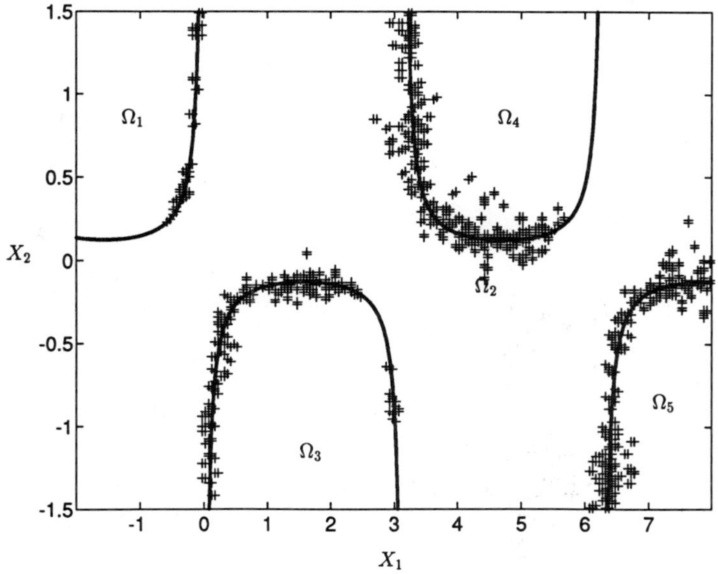

Figure 4. Location of fake periodic cells (+) for system (3). No equilibrium cells to represent the zeros of f are found.

x_b^* are found by the adjoining cell mapping analysis. No other periodic cells are found. Figure 7 shows the domains of attraction associated with the equilibrium points x_a^* and x_b^*.

The behavior of Branin's method for this example may be delineated as follows. The method starts by integrating with either system (6) or (7) from a given initial condition x_0. The success of the trajectory $x(t; x_0)$ in converging to a zero of (8) is illustrated by the domains of attraction in Figures 6 and 7. Among all the points in Figure 6 belonging to the domain of attraction of x_c^*, the points contained in Ω_1 and Ω_3 (Ω_4 and Ω_5 do not contain any such points) are successful values for x_0 if the integration procedure were to start with (2). The rest of the points in the domain (belonging to Ω_2) are successful values for x_0 if the integration procedure were to start with (3). Similarly, among all the points in Figure 7 belonging to the domains of attraction of x_a^* and x_b^*, all the points included in Ω_2 are successful values for x_0 if the integration procedure were to start with (2), the rest of the domain points are successful for x_0 with the integration of (3).

In order to indicate the alternation procedure, let us suppose that the zero x_a^* is to be reached in Figure 7. Then we start integrating system of Figure 6 from a given initial condition x_1 belonging to a small neighborhood of x_a^*. As seen in Figure 7, there are some points in this neighborhood which either lead to x_c^* or to outside the region of analysis. In the latter case, the algorithm fails to locate a new zero of the system (8). The adjoining cell mapping results provide an estimate of the percentage of points ($\sim 76\%$ in this case) in a small neighborhood which lead to a failure. Assume now that a starting point belonging to the set of successful points ($\sim 24\%$) is chosen and that x_c^* is to be located. Then, we return to Figure 7 where again, depending on which point in the neighborhood of x_c^* is selected, the trajectory may come back to x_a^*, may go to x_b^*, or may diverge. In order to reach x_b^*, initial conditions nearby x_c^* must be chosen, which belong to the domain of attraction of x_b^* in Figure 7 ($\sim 50\%$).

On the other hand, if the initial integration scheme leads to x_c^* via trajectories in Figure 6, the application of Branin's method would not provide the entire set of zeros

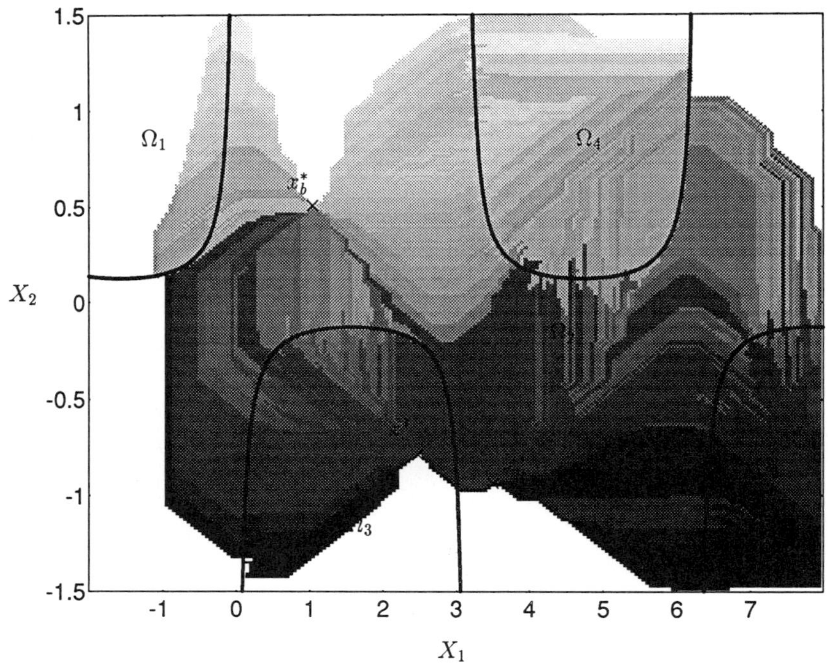

Figure 5. Domains of attraction of corresponding to fake periodic groups for system (3). Location of zeros of f are indicated by "×".

in a single attempt. In the best case, once x_a^* or x_b^* is located, one has to pass through x_c^* in order to locate the other zero. Once again, for this to happen one should be fortunate enough to choose appropriate initial conditions nearby the corresponding zeros of (8).

Example 2

We consider here an example proposed in $\mathbf{R}^2$ by Branin [5]:

$$\begin{aligned} f_1(x_1,x_2) &= 4(x_1+x_2) \\ f_1(x_1,x_2) &= 4(x_1+x_2)+(x_1-x_2)a(x_1,x_2) \\ a(x_1,x_2) &= (x_1-2)^2+x_2^2-1 \end{aligned} \tag{10}$$

$$J(x) = \frac{-1}{det\ J(x)} \begin{bmatrix} 4,4 \\ \\ 4+a(x_1,x_2)+2(x_1-x_2)(x_1-2), 4-a(x_1,x_2)+2(x_1-x_2)x_2 \end{bmatrix},$$

$$det\ J(x) = -8\left\{a(x_1,x_2)+x_1-x_2)(x_1-x_2-2)\right\}$$

The function f possesses only one zero which is located at the origin and is a regular zero. The singular manifold in this case is an ellipse in the (x_1, x_2) plane.

We investigate the global behavior of Branin dynamical systems for this example by the adjoining cell mapping method. The region of interest $-0.5 \leq x_1 < 3.5$ and

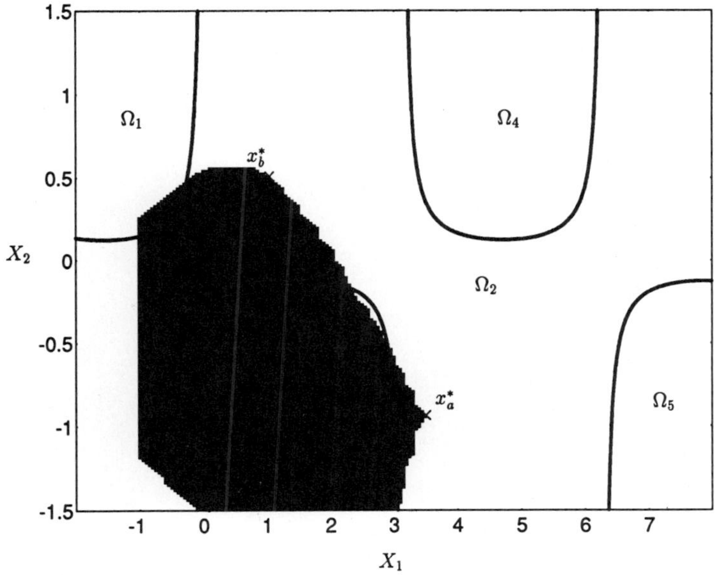

Figure 6. Domain of attraction associated with the equilibrium point x_c^* for system (6). Zeros of f are shown by "$\times$".

$-1.5 \leq x_2 < 2.0$ is divided into 200 uniform intervals with interval size of $h_1 = 0.02$ and 0.0175 along the x_1 and x_2 axes, respectively. The cell state space contains 40,000 cells. The initial integration time is chosen to be $T = 0.01$ and integration step size $\Delta t = T/5$.

Figure 8 shows the location of equilibrium cells for the system (2). Note that along with the two equilibrium cells (indicated by "$\times$") representing the asymptotically stable equilibrium point at the origin, several fake periodic cells (67 period-two and one period-three, indicated by "+") are detected near the singular manifold. These fake periodic cells are found not to lead to any other solution of (1). Figure 9 displays the domains of attraction associated with the equilibrium cells (marked by the darker region) and fake periodic cells (marked by various grey scales). Of noticeable interest is the region near (1.75, -0.5) which behaves like a center.

For the system (3), the origin is now an unstable equilibrium point. Adjoining cell mapping analysis finds no equilibrium cells to represent this point. However, many fake periodic cells exist near the singular manifolds (115 of them, all of period-two), see Figure 10. As before, none of them leads to any other solution of (1). The regions of attraction corresponding to these cells are shown in Figure 11. Notice the center-like region of attraction very similar to the one in Figure 9.

Cell mapping results for system (6) indicate again that no equilibrium cells are found to represent the unstable equilibrium at the origin. Only fake periodic cells are detected near the singular manifold, see Figure 12. There is one each of period-8, period-22, period-57, period-65, and period-81. Since each periodic group is connected, they appear as a "closed-curve" in state space, thus creating an impression of periodic solutions surrounding a center-like point near (1.75, -0.5). These fake periodic solutions do not lead to any other solution of (1). Figure 13 delineates the domains of attraction associated with the fake periodic cells and prominently displays the center-like attraction region once again.

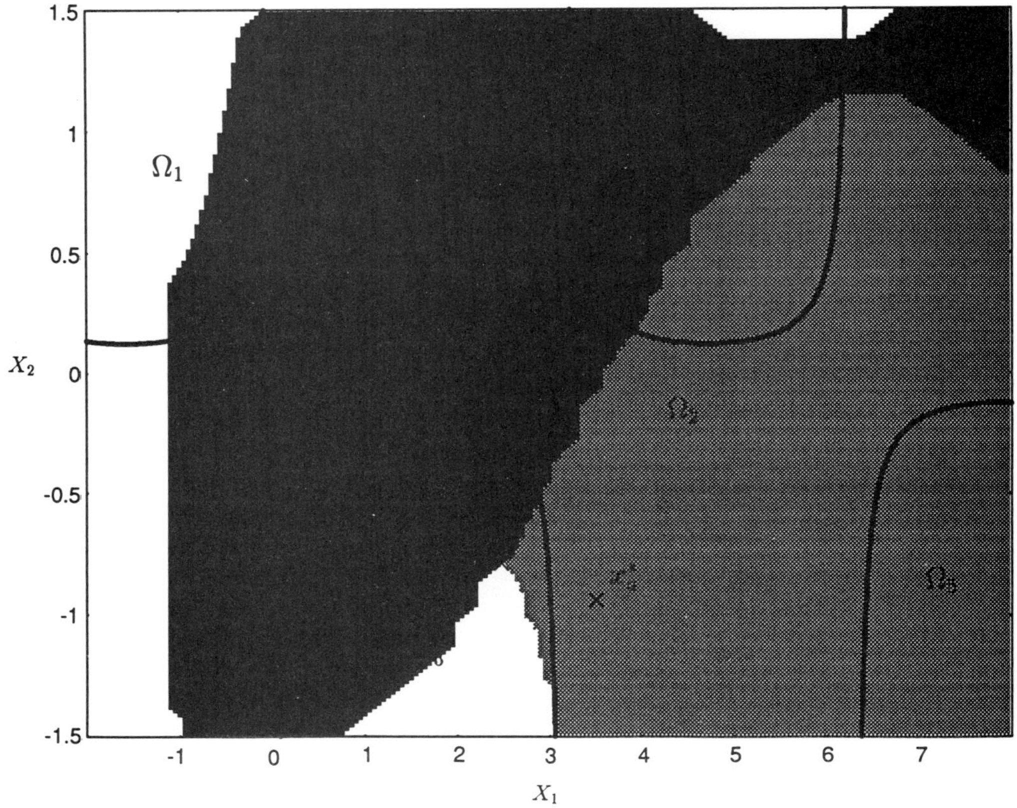

Figure 7. Domain of attraction of equilibrium points x_a^* and x_b^* for system (7). Equilibrium cells are shown by "×".

Regarding system (7), the adjoining cell mapping method discovers two equilibrium cells (both of period-one) representing the origin and fake periodic cells near the singular manifold, see Figure 14. In this case, the periodic cells have high periods; one each of period-310, period-243, period-130, period-104, period-85, period-61, and period-30. They again seem to resemble closed-orbits surrounding a center-like point near (1.75, -0.5). Figure 15 delineates the domains of attraction associated with the equilibrium cells and fake periodic cells.

We now elaborate on the behavior of Branin's method for this example. The global behavior of the resulting dynamical systems are quite complicated due to the presence of an extraneous singularity located on the ellipsoidal singular manifold. This extraneous singularity behaves like a center (as one can conclude by examining the domains of attraction of the fake periodic cells located on the singular manifold). On one hand, both Figure 9 and Figure 11 provide some useful information regarding the qualitative dynamic nature of such a type of extraneous singularity. The adjoining cell mapping method discovers a very large number of periodic cells (usually of period-2) hinting that some kind of switching behavior exists around different parts of the singular manifold. The totality of this dynamic behavior suggests the center-like behavior in the neighborhood of the extraneous singularity.

On the other hand, the results shown in Figure 13 and Figure 15 provide a better understanding of the above mentioned switching behavior. Adjoining cell mapping analysis discovers several concentric periodic solutions (of large periods) for both the

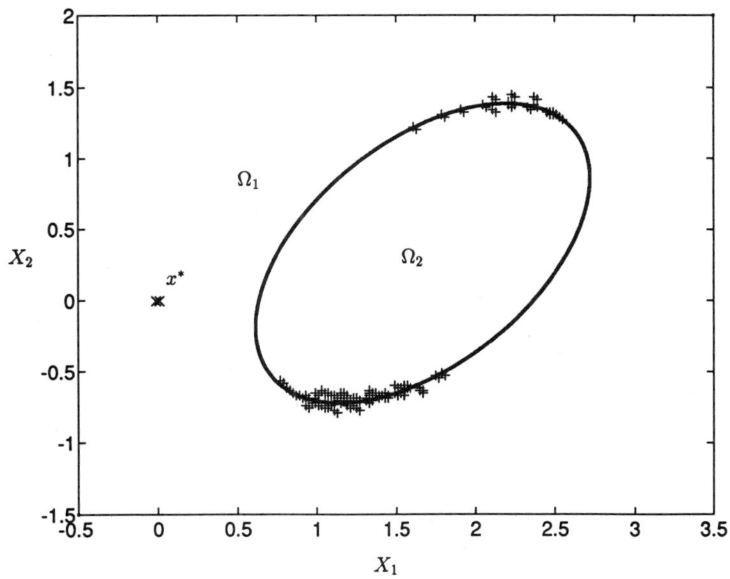

Figure 8. Location of equilibrium cells ($\times$) and fake periodic cells ($+$) for system (2) corresponding to Example 2. Singular manifold is the ellipse shown by the solid curve.

dynamical systems (6) and (7). The regions of attraction in these figures provide a concrete evidence of the existence of center-like behavior around an extraneous singularity. This behavior is clearly corroborated by the visual information contained in Figure 13 and Figure 15.

In summary, the two examples illustrated here indicate the enormous potential of considering dynamical systems (6) and (7) in addition to considering systems (2) and (3) for characterizing the global behavior of Branin's method.

CONCLUDING REMARKS

We have presented an analysis of the performance of a method due to Branin [5] for finding all the zeros of a nonlinear vector function. The analysis is based on the global study of the trajectories of a dynamical system constructed for the purpose of locating all the zeros. New numerical techniques for the purpose of locating all the zeros of a vector function can be developed where the switch of sign of $det\ J(x)$, as required by Branin [5], is unnecessary. For a detailed treatment of one such technique, the reader is referred to Zufiria and Guttalu [25].

Acknowledgments

The work reported here is partially supported by a grant from the National Science Foundation and the Faculty Research and Innovation Fund from USC. P. J. Zufiria also thanks the support of a Formación de Postgrado fellowship of the Programa Nacional de F.P.I. provided by the D.G.I.C.T. of the Ministerio de Educación y Ciencia of Spain, as well as the support by Project C91 0930 01 of the Comisión Interministerial de Ciencia y Tecnología and by Project M91 0930 03 of the Comunidad Autónoma de Madrid.

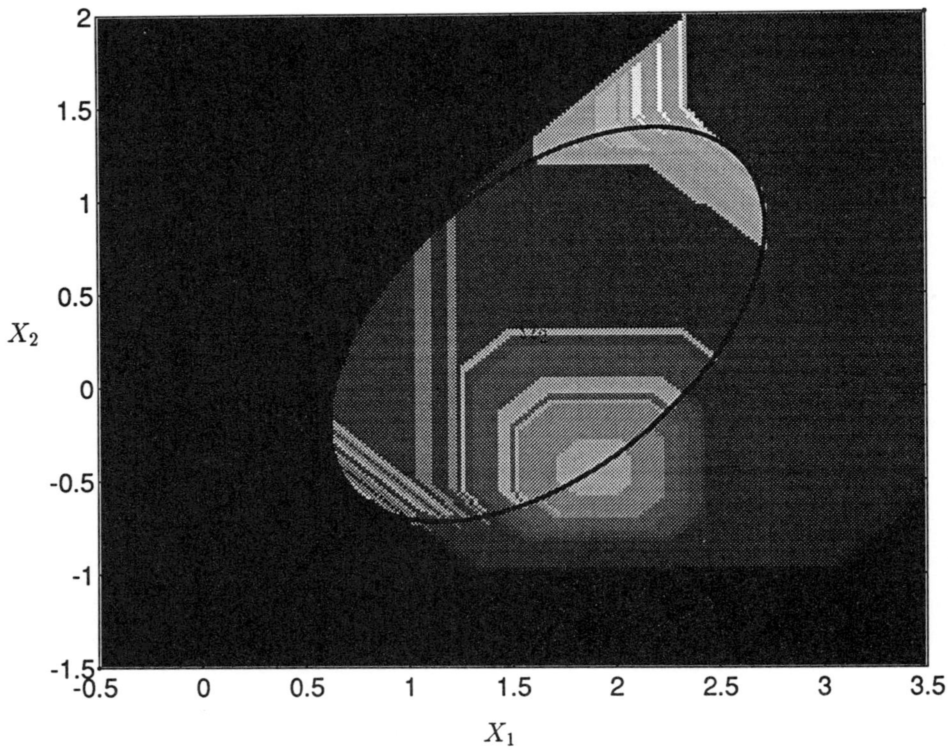

Figure 9. Domains of attraction of equilibrium point and fake periodic groups for system (2) of Example 2.

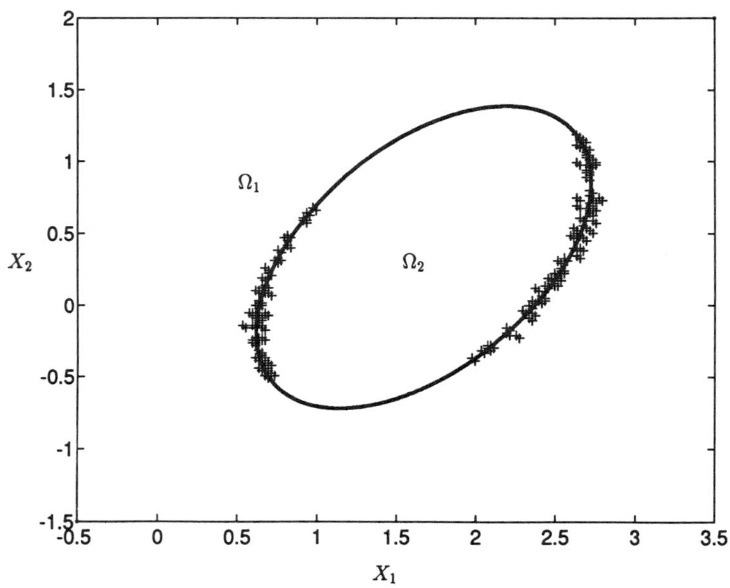

Figure 10. Location fake periodic cells (+) for system (3) corresponding to Example 2.

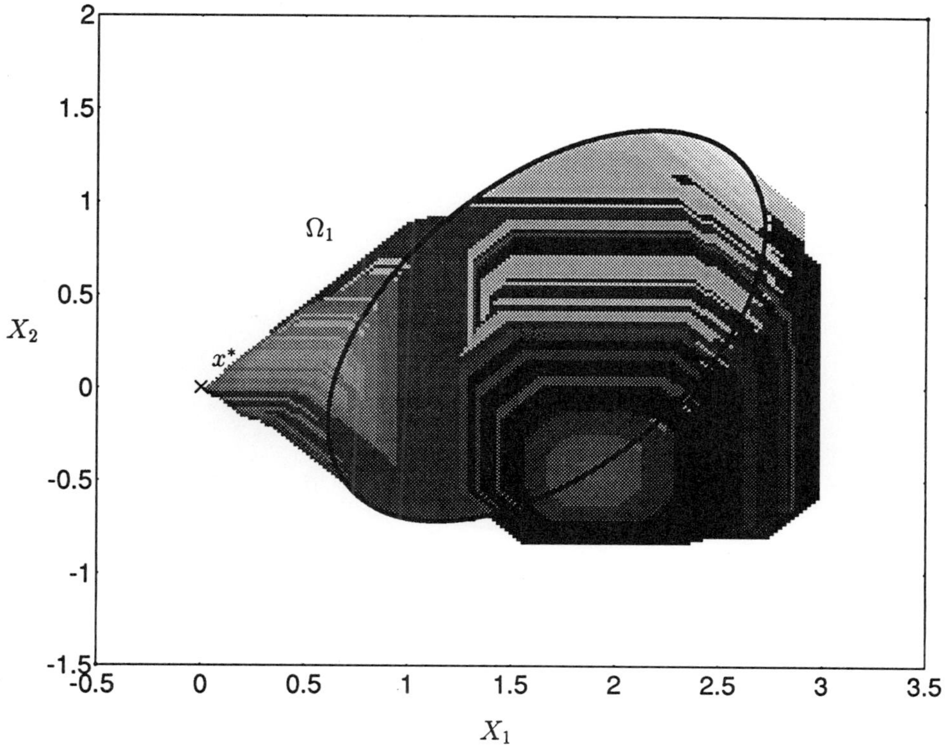

Figure 11. Domain of attraction of the fake periodic groups for system (3) of Example 2.

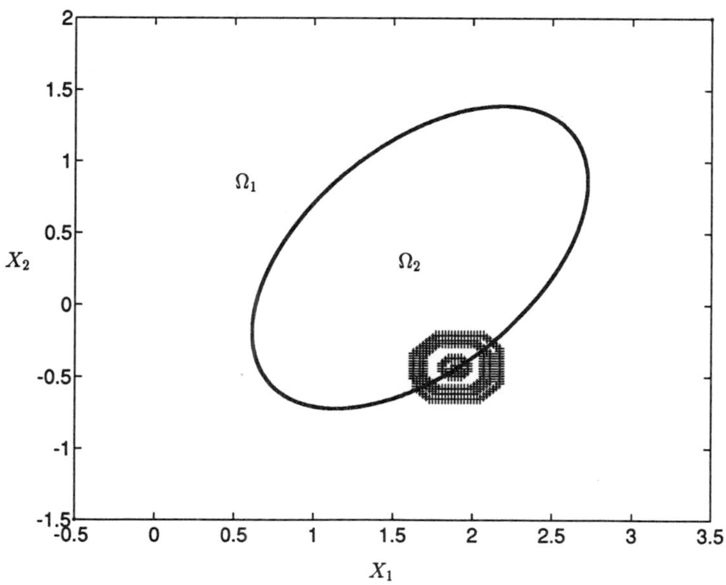

Figure 12. Location fake periodic cells (+) for system (6) corresponding to Example 2.

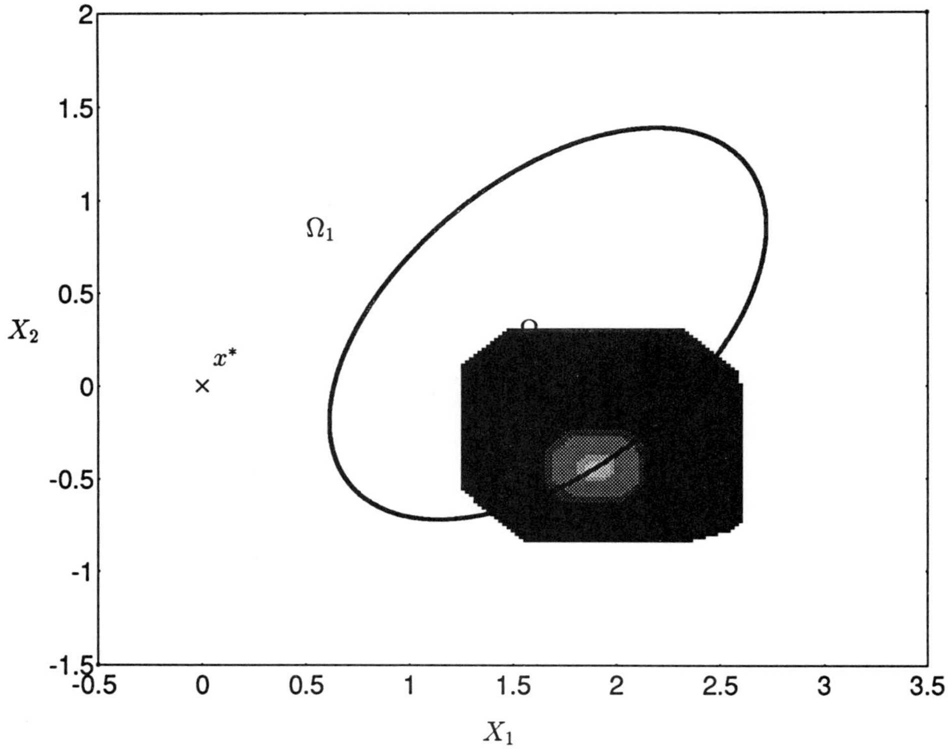

Figure 13. Domains of attraction associated with the fake periodic groups for system (6) of Example 2.

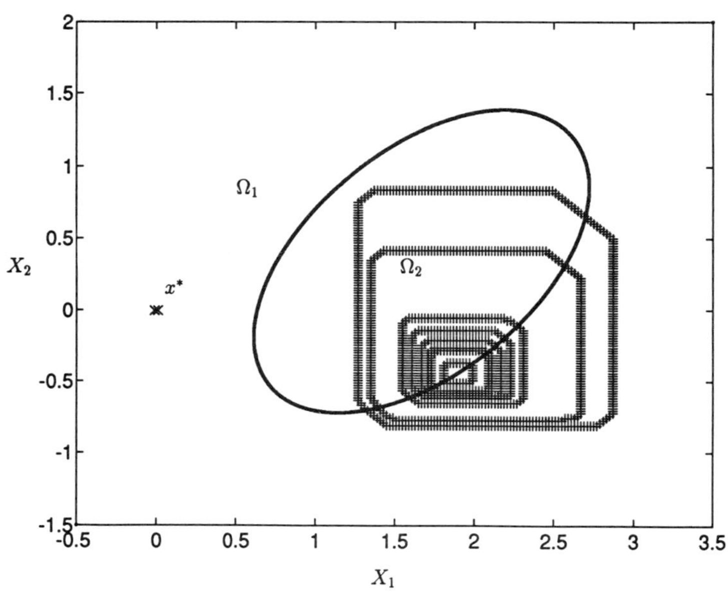

Figure 14. Location of equilibrium cells ($\times$) and fake periodic cells ($+$) for system (7) corresponding to Example 2.

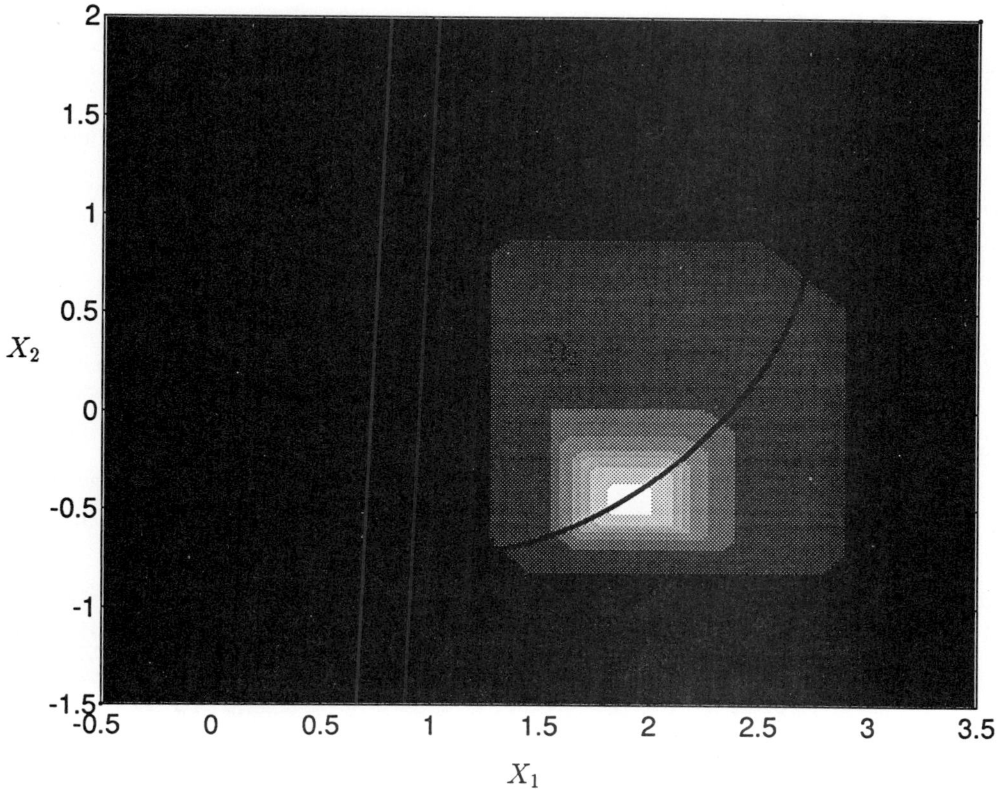

Figure 15. Domains of attraction associated with the fake periodic groups for system (7) of Example 2.

REFERENCES

1. E. L. Allgower and K. Georg, "Simplicial and continuation methods for approximating fixed points and solutions to systems of equations," *SIAM Rev.*, Vol. 22, pp. 28-85, 1980.

2. E. L. Allgower and K. Georg, "Homotopy methods for approximating several solutions to nonlinear systems of equations," in *Numerical Solutions of Highly Nonlinear Problems* (W. Forrester, ed.), pp. 253-270. North-Holland: Amsterdam, 1980.

3. E. L. Allgower and K. Georg, "Relationships between deflation and global methods in the problem of approximating additional zeros of a system of nonlinear equations," in *Homotopy methods and global convergence* (B. C. Eves et al, eds.), pp. 31-42. Plenum Press: New York, 1983.

4. E. L. Allgower and K. Georg, *Numerical Continuation Methods.* Springer-Verlag: New York, 1990.

5. F. H. Branin, "Widely convergent method for finding multiple solutions of simultaneous nonlinear equations," *IBM J. Res. Develop.*, Vol. 16, pp. 504-522, 1972.

6. R. Brent, "On the Davidenko-Branin method for solving simultaneous nonlinear equations," *IBM J. Res. Develop.*, Vol. 16, pp. 434-436, 1972.

7. D. Davidenko, "On a new method of numerical solution of systems of nonlinear equations" (in Russian), *Dokl. Akad. Nauk USSR*, Vol. 88, pp. 601-602, 1953.

8. D. Davidenko, "On approximate solution of systems of nonlinear equations" (in Russian), *Ukraine Mat. Z.*, Vol. 5, pp. 196-206, 1953.

9. D. W. Decker and H. B. Keller, "Path following near bifurcation," *Comm. Pure Appl. Math.*, Vol. 34, pp. 149-175 1981.

10. J. Gomulka, "Remarks on Branin's method for solving nonlinear equations," in *Towards Global Optimisation* (I. C. W. Dixon and G. P. Szego, eds.), pp. 96-106. North-Holland: Amsterdam, 1975.

11. J. Gomulka, "Two implementations of Branin's method: numerical experience," in *Towards Global Optimisation 2* (I. C. W. Dixon and G. P. Szego, eds.), pp. 151-164. North-Holland: Amsterdam, 1978.

12. R. S. Guttalu and P. J. Zufiria, "On a class of nonstandard dynamical systems: singularity issues," in *Adv. Cont. Dyna. Sys.* (C. T. Leondes, ed.), Vol. 34, pp. 279-324. Academic Press: New York, 1990.

13. R. S. Guttalu and P. J. Zufiria, " The adjoining cell mapping and its recursive unraveling, Part II: Application to selected problems," *Nonlinear Dynamics*, Vol. 4, pp. 309-336, 1993.

14. J. W. Hardy, "An implemented extension of Branin's method," in *Towards Global Optimisation* (I. C. W. Dixon and G. P. Szego, eds.), pp. 117-142. North-Holland: Amsterdam, 1975.

15. M. W. Hirsch and S. Smale, "On algorithms for solving $f(x) = 0$," *Comm. Pure Appl. Math.*, Vol. 32, pp. 281-312, 1979.

16. C. S. Hsu and R. S. Guttalu, "Index evaluation for dynamical systems and its application to locating all the zeros of a vector function," *ASME J. Appl. Mech.*, Vol. 50, pp. 858-862, 1982.

17. C. S. Hsu and W. H. Zhu, "A simplicial mapping method for locating the zeros of a function." *Quar. Appl. Math.*, Vol. 42, pp. 41-59, 1984.

18. C. S. Hsu, *Cell-to-Cell Mapping: A Method of Global Analysis for Nonlinear Systems.* Springer-Verlag: New York, 1987.

19. J. M. Ortega and W. C. Rheinboldt, *Iterative Solution of Nonlinear Equations in Several Variables.* Academic Press: New York, 1970.

20. S. Smale, "A convergent process of price adjustment and global Newton methods," *J. Math. Econ.*, Vol. 3, pp. 107-120, 1976.

21. G. Treccani, "On the convergence of Branin's method: a counter-example," in *Towards Global Optimisation* (I. C. W. Dixon and G. P. Szego, eds.), pp. 107-116. North-Holland: Amsterdam, 1975.

22. L. T. Watson, S. C. Billups and A. P. Morgan, "Algorithm 652. HOMPACK: A suite of codes for globally convergent homotopy algorithms," *ACM Tran. Math. Software*, Vol. 13, pp. 281-310, 1987.

23. X. Ying, and I. N. Katz, "A Reliable Principle Algorithm to Find the Number of Zeros of an Analytic Function in a Bounded Domain," *Numer. Math.*, Vol. 53, pp. 143-163, 1988.

24. P. J. Zufiria and R. S. Guttalu, "On an application of dynamical systems theory to determine all the zeros of a vector function," *J. Math. Anal. Appl.*, Vol. 152, pp. 269-295, 1990.

25. P. J. Zufiria and R. S. Guttalu, "A computational method for finding all the roots of a vector function," *Appl. Math. Comp.*, Vol. 35, pp. 13-59, 1990.

26. P. J. Zufiria and R. S. Guttalu, "The adjoining cell mapping and its recursive unraveling, Part I: Description of adaptive and recursive algorithms," *Nonlinear Dynamics*, Vol. 4, pp. 207-226, 1993.

A NUMERICAL SOLUTION FOR THE NONLINEAR EIGENVALUE SYSTEM

Dan S. Cheng and Allen R. Stubberud

Center for High-Speed Image/Signal Processing (CHIP)
Department of Electrical and Computer Engineering
University of California, Irvine
Irvine, CA 92717

ABSTRACT

In the application of the classical techniques of the calculus of variations to the nonlinear optimization problem, the nonlinear eigenvalue problem $-\lambda \cdot \Delta U = f(U)$,where Δ is an elliptic differential operator, is often encountered. A new numerical technique, using a finite-element-based algorithm, for finding the solution $U(X)$ where $X \in R^n$, the eigenvalue λ, and the maximum of the corresponding cost functional $J(U)$, where $J(U) = \int_\Omega F(U)dx$, $f(U) = \frac{\partial F(U)}{\partial U}$, is presented. This algorithm is applicable to a class of problems for which the nonlinear function $f(U)$ has specific properties. The algorithm is simple but fast and both convergence and uniqueness of the solution are demonstrated in several examples.

INTRODUCTION

This paper is concerned with a numerical technique, which uses a finite element method, and which is applicable to the nonlinear eigenvalue problem:

$$\begin{cases} -\lambda \cdot \Delta U = f(U) & \text{in } \Omega \\ U|_{\partial\Omega} = 0, \end{cases} \tag{1}$$

where Ω is a bounded domain in R^n, Δ is an elliptic differential operator, and λ is an eigenvalue. Suppose the corresponding cost functional for (1)

$$I_1(U) = \int_\Omega F(U)dx \tag{2}$$

is to be optimized, subject to the constraint functional

$$I_2(U) = \int_\Omega G(U)dx = t.$$

Mechanics and Control, Edited by R.S. Guttalu
Plenum Press, New York, 1994

For our problem we will assume that $f(U) = \frac{\partial F(U)}{\partial U}$ is continuous on Ω, and that $G(U) = |\nabla U|^2$, where ∇ is the gradient operator, that is, $I_2(U) = \|\nabla U\|_2^2 = t$, with $t > 0$. Under these assumptions $I_1(U)$ is to be optimized. For computational reasons, the case where the eigenvalue is pointwise continuous is not considered in this paper, for example, the case when

$$f(U) = \begin{cases} f_1(U) & \text{if } U \leq U^* \\ f_2(U) & \text{if } U \geq U^* \end{cases}$$

where $f_1(U)$ and $f_2(U)$ are continuous over all U, $f_1(U^*) = f_2(U^*)$, but $f_1(U) \neq f_2(U)$ for all U is not considered. Now, we introduce a finite element method by which the eigenvalue equation can be solved numerically so that the eigenvalue λ is computed, and the corresponding maximum of the functional $I_1(U)$ is found. We discuss the algorithm and some analysis in section 2 and numerical examples in both one and two dimensions are used to demonstrate this algorithm in section 3.

ALGORITHM AND MATHEMATICAL ANALYSIS

In this section, we present an algorithm and some related analysis for solving the nonlinear eigenvalue problem presented in equations (1) and (2). We assume that $U \in H_0^1(\Omega)$ where $H_0^1(\Omega) \equiv \{U : U|_{\partial\Omega} = 0, U \in L^2(\Omega), \nabla U \in L^2(\Omega)\}$, and $f(U) = \frac{\partial F(U)}{\partial U}$ is continuous on Ω. It is known that any stationary point U of (2) is a solution of the equation (1)[5]. Since both the eigenvalue λ and the maximum of $I_1(U)$ are dependent on t, we denote $\lambda = \lambda(t)$, and Max $I_1(U) = \text{M(t)}$.

Theorem 1: For the optimization problem defined by (1) and (2), then $\frac{\partial M(t)}{\partial t} = M'(t) = 0.5\lambda(t)$.

Proof: We multiply both sides of equation (1) by U, then integrate over Ω, thus obtaining $\lambda(t) \cdot t = \int_\Omega f(U)U dx$, where we have used $\|\nabla U\|_2^2 = t$. Let $U = \sqrt{t} \cdot W$; then $M(t) = Max \int_\Omega F(U)dx$, subject to $\|\nabla W\|_2^2 = 1$. Now we form

$$M'(t) = \frac{0.5}{\sqrt{t}} \cdot \int_\Omega f(\sqrt{t} \cdot W)W dx = 0.5 \cdot t^{-1} \cdot \int_\Omega f(U)U dx,$$

where $\|\nabla U\|_2^2 = t$. Hence, $\frac{\partial M(t)}{\partial t} = M'(t) = 0.5\lambda(t)$.

ALGORITHM

In the following algorithm, let $X \in R^n$, and $\varphi(X)$ and $\nabla\varphi(X)$ be orthogonal and orthonormal bases, respectively, in the Hilbert space. The advantage of such a basis is that the global solution is obtained if the solution exists.

Step 1 We let $j = 1, 2, \cdots, n$, and choose

$$U(X) = \sum_{i_j} \alpha_{i_j} \varphi_{i_j}(X)$$

thus,

$$\|\nabla U\|_2^2 = \sum_{i_j} \alpha_{i_j}^2 = t$$

Step 2 We let

$$\phi_1(U) = \int_\Omega F(U)dx - \rho \cdot (\sum_{i_j} \alpha_{i_j}^2 - t)$$

where ρ is a Lagrange multiplier. Then,

$$\frac{\partial \phi_1(U)}{\partial \alpha_{i_j}} = \int_\Omega \frac{\partial F}{\partial U} \cdot \frac{\partial U}{\partial \alpha_{i_j}} dx - 2 \cdot \rho \cdot \alpha_{i_j} = 0$$

which implies that

$$\alpha_{i_j} = \frac{\sqrt{t} \cdot \int_\Omega f(U) \cdot \frac{\partial U}{\partial \alpha_{i_j}} dx}{\sqrt{\sum_{i_j} \{\int_\Omega f(U) \cdot \frac{\partial U}{\partial \alpha_{i_j}} dx\}^2}}$$

where

$$\frac{\partial U}{\partial \alpha_{i_j}} = \varphi_{i_j}$$

as described in Step 1.

Step 3 We assume the sum square error for the unknown parameter α_{i_j} is

$$E_{sum} = 0.5 \cdot \sum_{i_j} [\alpha_{i_j} - \frac{\sqrt{t} \cdot \int_\Omega f(U) \cdot \frac{\partial U}{\partial \alpha_{i_j}} dx}{\sqrt{\sum_{i_j} [\int_\Omega f(u) \cdot \frac{\partial U}{\partial \alpha_{i_j}} dx]^2}}]^2$$

Now let

$$g_{i_j} = \int_\Omega f(u) \cdot \frac{\partial U}{\partial \alpha_{i_j}} dx$$

then

$$E_{sum} = 0.5 \cdot \sum_{i_j} [\alpha_{i_j} - \frac{\sqrt{t} \cdot g_{i_j}}{\sqrt{\sum_{k_l} g_{k_l}^2}}]^2$$

Step 4 We calculate the gradient matrix for the sum square error

$$\frac{\partial E_{sum}}{\partial \alpha_{m_k}} = \sum_{i_j} (\alpha_{i_j} - \frac{\sqrt{t} \cdot g_{i_j}}{\sqrt{\sum_{k_j} g_{k_j}^2}}) * (\chi - A_{i_j, m_k})$$

where $k = 1, 2, \cdots, n$,

$$\chi = \begin{cases} 1 & \text{if } m_k = i_j \\ 0 & \text{otherwise} \end{cases}$$

and

$$A_{i_j, m_k} = \frac{\sqrt{t}\{\frac{\partial g_{i_j}}{\partial U} \cdot \frac{\partial U}{\partial \alpha_{m_k}} \cdot \sqrt{\sum_{k_j} g_{k_j}^2} - B\}}{\sum_{k_j} g_{k_j}^2}$$

where

$$B = g_{i_j} \cdot [\sum_{i_j} g_{i_j}^2]^{-0.5} \cdot \sum_{i_j} (g_{i_j}) \cdot (\frac{\partial g_{i_j}}{\partial U}) \cdot (\frac{\partial U}{\partial \alpha_{m_k}})$$

Finally then

$$A_{i_j, m_k} = \frac{\sqrt{t}\{\frac{\partial g_{i_j}}{\partial U} \cdot \frac{\partial U}{\partial \alpha_{m_k}} \cdot \sum_{i_j} g_{i_j}^2(U) - g_{i_j} \cdot \sum_{n_j} (g_{n_j}) \cdot (\frac{\partial g_{n_j}}{\partial U}) \cdot (\frac{\partial U}{\partial \alpha_{m_k}})\}}{\{\sum_{i_j} g_{i_j}^2\}^{1.5}}$$

Step 5 Now, we employ the steepest descent method by letting

$$\alpha_{k_j}^+ = \alpha_{k_j} - \eta \cdot \frac{\partial E_{sum}}{\partial \alpha_{k_j}}$$

Step 6 We form the new sum square error by letting $Suw(i_j) = \frac{\partial E_{sum}}{\partial \alpha_{i_j}}$ and

$$E_{sum} = 0.5 \cdot \sum_{i_j} [\alpha_{i_j} - \eta \cdot Suw(i_j) - \frac{\sqrt{t} \cdot g_{i_j}}{\sqrt{\sum_{k_j} g_{k_j}^2}}]^2$$

Step 7 We calculate the derivative of E_{sum} with respect to η and obtain

$$\frac{dE_{sum}}{d\eta} = \sum_{i_j} [\alpha_{i_j} - \eta \cdot Suw(i_j) - \frac{\sqrt{t} \cdot g_{i_j}}{\sqrt{\sum_{k_j} g_{k_j}^2}}] *$$

$$[-Suw(i_j) - \frac{\sqrt{t} \{\frac{\partial g_{i_j}}{\partial \eta} \cdot \sum_{k_j} g_{k_j}^2 - g_{i_j} \cdot \sum_{k_j} (g_{k_j}) \cdot (\frac{\partial g_{k_j}}{\partial \eta})\}}{\{\sum_{k_j} g_{k_j}^2\}^{1.5}}]$$

where

$$\frac{\partial g_{i_j}}{\partial \eta} = \int_\Omega f'(U) \cdot \frac{\partial U}{\partial \eta} \cdot \frac{\partial U}{\partial \alpha_{i_j}} dx$$

and

$$\frac{\partial u}{\partial \eta} = -\sum_{i_j} Suw(i_j) \varphi_{i_j}$$

Step 8 We now use Newton's method to determine η from the algorithm

$$\eta^+ = \eta - \frac{E_{sum}}{\frac{dE_{sum}}{d\eta}}$$

Step 9 We continue the process sequentially until we find the "Minimum" for E_{sum}.

In the next section, we illustrate these techniques in a set of examples.

NUMERICAL EXAMPLES

Suppose $F(U) = F(-U)$ is a continuous nonnegative nonlinear function defined on a unit cube. Then, the trial functions [8] for the n-dimensional case are:

$$\begin{cases} L_{k_j}(X) = \frac{2^{n/2} \cdot \phi_{k_j}(X)}{\|KNORM\|_2 \cdot \pi} \\ U_N(t, X) = \sum_{k_j=1}^{N} \alpha_{k_j} L_{k_j}(X) \end{cases} \tag{3}$$

$\forall j = 1, 2, \cdots, n$, and where

$$\|KNORM\|_2 = \sqrt{k_1^2 + k_2^2 + \cdots + k_n^2}$$

and

$$\phi_{k_j}(X) = \prod_{i=1}^{n} sin(k_i \cdot \pi \cdot x_i)$$

then

$$\|\nabla U\|_2^2 = \int_\Omega |\nabla U(x)|^2 dx = \sum_{k_j=1}^{N} \alpha_{k_j}^2 = t$$

In the next few examples, we consider $X \in R^1$ and $X = (x, y) \in R^2$ for simple illustrations of our algorithm. Again, notice that $\lambda(t)$ is the eigenvalue of (1) and $M(t)$

is the maximum of $I_1(u)$. The result of Theorem 1 that $M'(t) = 0.5 \cdot \lambda(t)$ can help to check the numerical computations for both the eigenvalue and the maximum for $t \in (0, \infty)$. Consider the following two cases:

Case 1: $X \in R^1$. Let $\Omega = (0,1)$,

$$\begin{cases} L_k(x) = \frac{\sqrt{2} sin(k\pi x)}{k \cdot \pi} \\ U_N(t,x) = \sum_{k=1}^{N} \alpha_k \mathrm{L_k(x)} \end{cases} \tag{4}$$

then

$$\|\dot{U}\|_2^2 = \int_0^1 |U'(x)|^2 dx = \sum_{k=1}^{N} \alpha_k^2 = t$$

Case 2: $X \in R^2$. Let $\Omega = (x,y) \in (0,1) \times (0,1)$,

$$\begin{cases} L_{j,k}(x,y) = \frac{2 \cdot Sin(k\pi x) \cdot Sin(j\pi y)}{\pi \cdot \sqrt{i^2 + j^2}} \\ U_N(t,x,y) = \sum_{j,k=1}^{N} \alpha_{j,k} L_{j,k}(x,y) \end{cases} \tag{5}$$

then

$$\|\nabla U\|_2^2 = \int_\Omega |\nabla U(x)|^2 dx = \sum_{j,k=1}^{N} \alpha_{j,k}^2 = t$$

Notice the trivial solution of (1) is for the situation where all the coefficients are the same or where all the coefficients have same magnitude, that is, the solution is assumed to be zero. Since these solutions are trivial, we assume that all the coefficients are not the same and thus we can obtain a global maximum $M(t)$ for the functional $I_1(u)$. By choosing the trial functions [8] as (3), the even coefficients are zero, that is, when k_j is an even integer, $\alpha_{k_j} = 0$. Furthermore, the coefficients are symmetric in higher dimensions. For instance, $\alpha_{k,j} = \alpha_{j,k}$ in two dimensions. In particular for the examples of cases 1 and 2, the even coefficients in the approximated solutions are zero, that is, the coefficients $\alpha_{2k} = 0$, and $\alpha_{j,2k} = \alpha_{2j,k} = 0$ with $\alpha_{n,m} = \alpha_{m,n} \quad \forall n,m = 1,3,5,\cdots,N + \frac{(-1)^N+1}{2}$. In the algorithm, we applied the Newton's method followed by the steepest descent method. We know that the finite element method will enable us to get the solution as close as necessary to the actual solution. Indeed, our algorithm can be programmed dynamically so that the eigenvalue $\lambda(t)$ and the maximum $M(t)$ can be obtained as close as we desire. However, if we consider $U(X)$ where X is of high dimension, the computational time will take much longer than for lower dimensional cases. The case of $F(-U) \neq F(U)$ is not considered in the example. This problem is still under investigation, since the trial functions [8] chosen will not work in this case.

REMARKS: The value of the term N in the approximated solutions in the finite element method depends on the accuracy that we desire. One possible problem is that "underflow" from computing might be produced. Also, the number S_M of terms selected for Simpson's integration is dependent on the nonlinear function $F(U)$. The solution of the combination of the steepest descent method and Newton's method depends on the initial guess. For a fixed error in the numerical maximum $M(t)$ and the eigenvalue $\lambda(t)$, the number N of terms for the approximate solution is an increasing function of t for the nonhomogeneous function $F(U)$. For the homogeneous case, e.g. $f(U) = U^{2p-1}$, where p is a natural number, we only have to solve the case where $t = 1$. We then denote the solution in this case $U(1,X)$, then $U(t,X) = \sqrt{t} \cdot U(1,X)$, and thus can be used to solve for $\lambda(t)$ and $M(t)$. We can easily see that this saves substantially in the number of iterations. Thus different functions $F(U)$ will affect the number of terms N needed for convergence. Under our assumptions, this finite-element-based algorithm gives convergent results, in fact, it converges rapidly with a very small number of odd terms N for the following simple one- and two-dimensional numerical examples.

In the following tables, we present some numerical results of these examples. In the graphs below the eigenvalue $\lambda(t)$ is the dotted line and the maximum $M(t)$ is the solid line versus the t axis.

Case 1: $X \in R^1$:

(3.1) $f(U) = 2 \cdot U \cdot exp(U^2)$			(3.2) $f(U) = sin(2 \cdot U)$		
t	$\lambda(t)$	M(t)	t	$\lambda(t)$	M(t)
0.00010	0.2031	1.00287	0.00121	0.2862	0.00001
0.09608	0.2054	1.01267	0.06706	0.2845	0.00682
0.37210	0.2144	1.04166	0.26010	0.2793	0.02520
0.65602	0.2254	1.07278	0.57759	0.2704	0.05669
0.82808	0.2302	1.09232	1.01798	0.2582	0.09805
0.99981	0.2363	1.11245	1.58757	0.2433	0.14842
1.46417	0.2543	1.16930	2.63247	0.2170	0.23342
1.71605	0.2647	1.20198	3.24279	0.2035	0.27862
1.98802	0.2765	1.23878	5.28939	0.1614	0.41035
2.24966	0.2885	1.27573	7.83820	0.1189	0.53585
3.99957	0.3852	1.56822	10.88340	0.0811	0.64244
6.24959	0.5687	2.09758	11.70921	0.0728	0.66478
7.56220	0.7204	2.51842	11.96976	0.0707	0.67140
8.99907	0.9387	3.11130	15.44982	0.0449	0.74197
12.24947	1.7468	5.22010	17.34349	0.0361	0.76969
15.99977	3.6881	10.07733	19.49517	0.0273	0.79475

Case 2: $X \in R^2$:

(3.3) $f(U) = 4 \cdot U^3$			(3.4) $f(U) = Sin(2 \cdot U) \cdot log(1 + U^2) + \frac{2 \cdot U \cdot Sin^2(U)}{1.0+U^2}$		
t	$\lambda(t)$	M(t)	t	$\lambda(t)$	M(t)
1	0.026666	0.006617	0.52	0.012225	0.001646
2	0.053333	0.026469	1.00	0.021497	0.005717
3	0.079999	0.059556	1.50	0.028924	0.012033
4	0.106666	0.105877	2.00	0.035213	0.020073
5	0.133332	0.165433	2.50	0.040210	0.029547
6	0.159999	0.238223	3.00	0.044393	0.040124
7	0.186665	0.324249	3.38	0.047061	0.048924
8	0.213331	0.423508	4.00	0.050435	0.063894
9	0.239998	0.536003	4.58	0.052655	0.079029
10	0.266664	0.661732	5.00	0.054306	0.090136
11	0.293331	0.800696	5.50	0.055553	0.103859
12	0.319997	0.952894	6.00	0.056483	0.117951
13	0.346664	1.118327	6.66	0.057230	0.136692
14	0.373330	1.296995	7.00	0.057474	0.146440
15	0.399996	1.488897	8.00	0.057681	0.175167

This graph is for Example 3.1

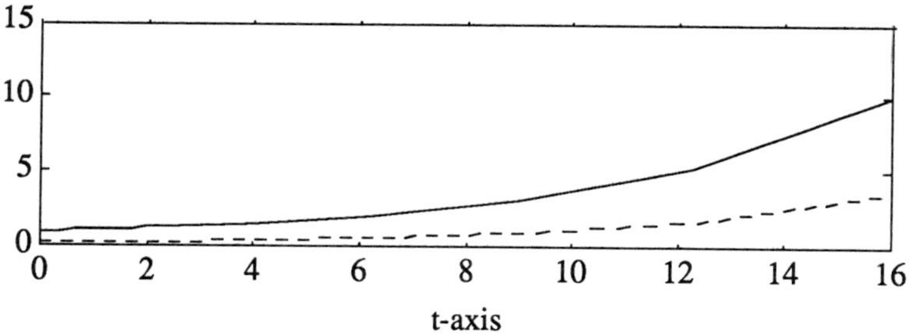

This graph is for Example 3.2

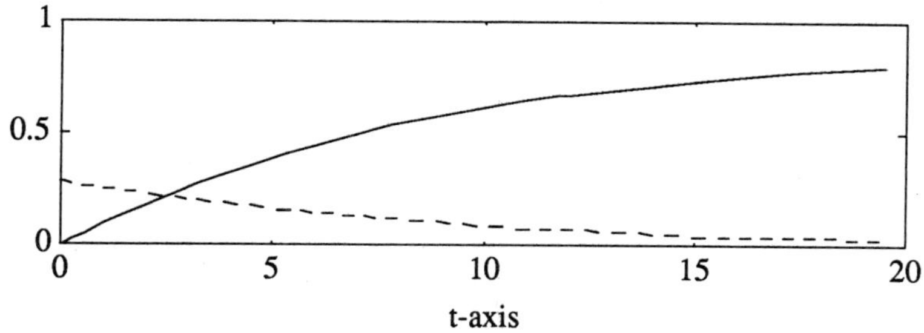

This graph is for Example 3.3

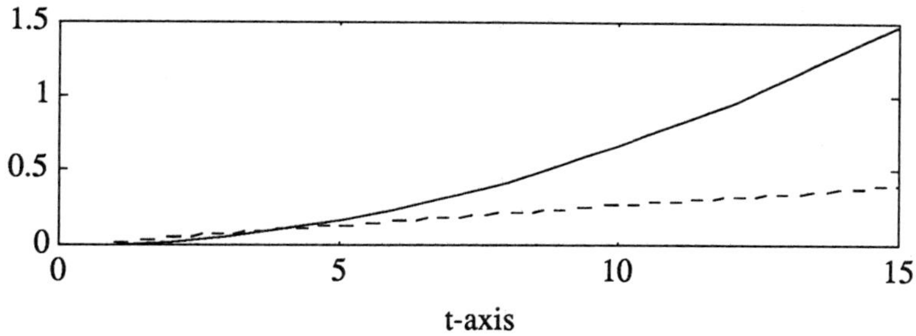

This graph is for Example 3.4

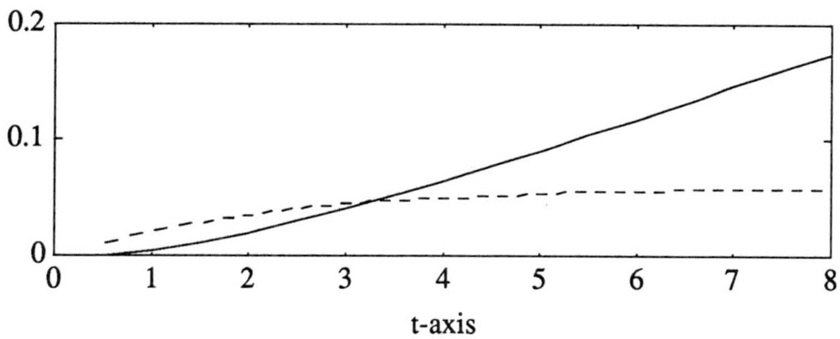

All of the above calculations were done on a DEC VAX 6310 and a Sun workstation at the University of California, Irvine using double precision with a FORTRAN Compiler. The data are rounded off, and as shown in the above. The graphics were done on a Macintosh.

REFERENCES

1. I. Babuska and J.E. Osborn, "Estimates for the errors in eigenvalue and eigenvector approximation by Galerkin methods, with particular attention to the case of multiple eigenvalues," *SIAM J. Num. Anal.*, Vol. 24, pp.1249-1276, 1987.

2. C. Bardos, J. M. Lasry and M. Schatzman, *Bifurcation and Nonlinear Eigenvalue Problems.* Springer- Verlag: New York, 1980.

3. R. Courant and D. Hilbert, *Methods of Mathematical Physics.* Interscience Publishers: New York, Vol. 1, 1953.

4. I. Fried, "Accuracy of finite element eigenproblem," *Journal of Sound and Vibration.* Vol. 18, pp. 289-295, 1971.

5. R. Glowinski, *Numerical Methods for Nonlinear Variational Problems.* Springer-Verlag: New York, 1984.

6. K. M. Liu and E. L. Ortiz, "Numerical solution of ordinary and partial functional-differential eigenvalue problems with Tau method," *Computing*, Vol. 41, pp. 205-217, 1989.

7. K. Rektorys, *The method of discretization in time and partial differential equations.* D. Reidel Publishing Company: Boston, 1982.

8. G. Strang and G. J. Fix, *An Analysis of the Finite Element Method.* Prentice-Hall: Englewood Cliffs, 1973.

9. O. C. Zienkiewicz and K. Morgan, *Finite Elements and Approximation.* John Wiley and Sons: New York, 1983.

CONTRIBUTORS

SHAHEEN AHMAD, School of Electrical Engineering, Purdue University, West Lafayette, Indiana 47907-1285

NASIR U. AHMED, Department of Electrical Engineering, and Mathematics University of Ottawa Ottawa, Ontario K1N 6N5, Canada

DAVID M. AUSLANDER, Department of Mechanical Engineering, University of California, Berkeley, CA 94720

DAVID J. BELL, Department of Mathematics, UMIST, P. O. Box 88, Sackville Street, Manchester, M60 1QD, United Kingdom

AUSTIN BLAQUIÈRE, Université de Paris, 22 rue de la Prairie, 92410 Ville d'Avray, France

DAN S. CHENG, Department of Electrical and Computer Engineering, University of California at Irvine, Irvine, California 92717

CHINGYEI CHUNG, Department of Automatic Control Engineering, Feng–Chia University, Tai–Chung City, Taiwan

MARTIN CORLESS, School of Aeronautics and Astronautics, Purdue University, West Lafayette, Indiana 47907

RUI J. P. DE FIGUEIREDO, Department of Electrical and Computer Engineering, and Mathematics, University of California at Irvine, Irvine, California 92717

CHRISTOPHER N. D'SOUZA, Wright Labs, Eglin AFB, Florida 32542

WODEK GAWRONSKI, Jet Propulsion Laboratory, California Institute of Technology, Pasadena, CA 91109

NIGEL J. C. GREENWOOD, Department of Mathematics, University of Queensland, St. Lucia, Queensland 4072, Australia

RAMESH S. GUTTALU, Department of Mechanical Engineering University of Southern California Los Angeles, CA 90089-1453

DAVID G. HULL, Department of Aerospace Engineering and Engineering Mechanics, The University of Texas, Austin, Texas 78712

MARC R. ILGEN, The Aerospace Corporation, El Segundo, CA 90245-4691

ROBERT E. KALABA, Department of Biomedical Engineering, Electrical Engineering, and Economics, University of Southern California, Los Angeles, California 90089

GEORGE LEITMANN, School of Engineering, University of California, Berkeley, California 94720

ORESTES LLANES-SANTIAGO, Departamento Sistemas de Control, Universidad de Los Andes, Mérida, Venezuela

SHENGWU LUO, School of Electrical Engineering, Purdue University, West Lafayette, Indiana 47907-1285

KENNETH D. MEASE, Department of Mechanical and Aerospace Engineering, Princeton University, Princeton, New Jersey 08544

SUDHAKAR MEDEPALLI, Mechanical Dynamics Inc., 2301 Commonwealth Blvd., Ann Arbor, Michigan 48105

RAINER MEHLHORN, Institute of Flight Mechanics and Flight Control, Technische Universität München Arcisstr, 21, 8000 München 2, Germany

LEONARD M. MEIROVITCH, Department of Engineering Science and Mechanics, Virginia Polytechnic Institute and State University, Blacksburg, Virginia 24061

DAN S. NECSULESCU, Department of Mechanical Engineering, University of Ottawa, Ottawa, Ontario K1N 6N5, Canada

ANDREW OLAS, Olas & Associates, 1135 N.W. Circle, Corvallis, Oregon 97330

ELA OLAS, Olas & Associates, 1135 N.W. Circle, Corvallis, Oregon 97330

ANIL V. RAO, Department of Mechanical and Aerospace Engineering, Princeton University, Princeton, New Jersey 08544

DAVID J. W. RUXTON, Department of Mathematics and Computing, University of Central Queensland, Rockhampton, Queensland 4702, Australia

GOTTFRIED SACHS, Institute of Flight Mechanics and Flight Control, Technische Universität München Arcisstr, 21, 8000 München 2, Germany

THOMAS J. SCHMITT, Optical Storage Development, IBM AdStar, Tucson, Arizona 85744

HAMID SHABAN-ZANJANI, Department of Mechanical Engineering, University of Ottawa, Ottawa, Ontario K1N 6N5, Canada

M. SIGAL-PAUCHARD, Université du Maine, Faculté des Sciences, Département de Mathématique et d'Informatique, Avenue Olivier Messiaen BP 535, 72017 Le Mans Cedex, France

HERBERTT SIRA-RAMÍREZ, Departamento Sistemas de Control, Universidad de Los Andes, Mérida, Venezuela

JANISLAW M. SKOWRONSKI, Department of Mechanical Engineering, University of Southern California, Los Angeles, CA 90089-1453

ARGIRIS G. SOLDATOS, Department of Mechanical Engineering, University of California, Berkeley, California 94720

JASON L. SPEYER, School of Engineering and Applied Science, University of California, Los Angeles, CA 90024-1592

WOLF STADLER, Engineering Division, San Francisco State University, 1600 Holloway Avenue, San Francisco, California 94132

ALLEN R. STUBBERUD, Department of Electrical and Computer Engineering, University of California at Irvine, Irvine, California 92717

FIRDAUS E. UDWADIA, Department of Mechanical Engineering, Civil Engineering, and Decision Systems, University of Southern California, Los Angeles, California 90089-1453

THOMAS L. VINCENT, Department of Aerospace and Mechanical Engineering, University of Arizona, Tucson, Arizona 85745

TYRONE L. VINCENT, Department of Electrical and Computer Engineering, University of Arizona, Tucson, Arizona 85745

N. X. VINH, Department of Aerospace Engineering, The University of Michigan, Ann Arbor, Michigan 48109-2140

MOHAMED ZRIBI, School of Electrical Engineering, Purdue University, West Lafayette, Indiana 47907-1285

PEDRO J. ZUFIRIA, Departamento de Matemática Aplicada a las Tecnologías de la Información, and GTI, ETSI de Telecomunicación, UPM, 28040 Madrid, Spain

INDEX